A First Course in
Control System Design

RIVER PUBLISHERS SERIES IN AUTOMATION, CONTROL AND ROBOTICS

Series Editors

SRIKANTA PATNAIK
SOA University, Bhubaneswar
India

ISHWAR K. SETHI
Oakland University
USA

QUAN MIN ZHU
University of the West of England
UK

Advisor

Tarek Sobh
University of Bridgeport, USA

Indexing: All books published in this series are submitted to Thomson Reuters Book Citation Index (BkCI), CrossRef and to Google Scholar

The "River Publishers Series in Automation, Control and Robotics" is a series of comprehensive academic and professional books which focus on the theory and applications of automation, control and robotics. The series focuses on topics ranging from the theory and use of control systems, automation engineering, robotics and intelligent machines.

Books published in the series include research monographs, edited volumes, handbooks and textbooks. The books provide professionals, researchers, educators, and advanced students in the field with an invaluable insight into the latest research and developments.

Topics covered in the series include, but are by no means restricted to the following:

- Robots and Intelligent Machines
- Robotics
- Control Systems
- Control Theory
- Automation Engineering

For a list of other books in this series, visit www.riverpublishers.com

A First Course in
Control System Design

Kamran Iqbal

University of Arkansas at Little Rock
Little Rock, AR
USA

LONDON AND NEW YORK

Published 2017 by River Publishers

River Publishers

Alsbjergvej 10, 9260 Gistrup, Denmark

www.riverpublishers.com

Distributed exclusively by Routledge

4 Park Square, Milton Park, Abingdon, Oxon OX14 4RN

605 Third Avenue, New York, NY 10017, USA

First issued in paperback 2023

A First Course in Control System Design / by Kamran Iqbal.

Routledge is an imprint of the Taylor & Francis Group, an informa business

Publisher's Note

The publisher has gone to great lengths to ensure the quality of this reprint but points out that some imperfections in the original copies may be apparent.

While every effort is made to provide dependable information, the publisher, authors, and editors cannot be held responsible for any errors or omissions.

ISBN 13: 978-87-7022-981-4 (pbk)

ISBN 13: 978-87-93609-05-1 (hbk)

ISBN 13: 978-1-003-33689-1 (ebk)

Contents

Preface

Control systems have become pervasive in our lives. Our homes have environmental controls. The appliances we use, such as the washing machine, microwave, etc., carry embedded controllers. We fly in airplanes and drive automobiles, which both make extensive use of control systems technology. Our body regulates essential functions like blood pressure, heartbeat, breathing, and insulin levels in blood, through biological control systems. The cells in the body regulate our metabolism and energy production using nutrient levels and electrolytes. The postural stability of the body depends on regulating body's center of gravity over the base of support. Fine motor control underlies the manipulation and locomotion tasks we undertake as part of daily living. We essentially perform the control function as we walk or drive a car; the control objective in both cases being to follow a desired course at a preferred speed.

The industrial revolution in the eighteenth century ushered in the age of machines that needed automatic controllers. Ingenious solutions to the control problems were developed as a result. An early example involved the use of centrifugal fly ball governor for throttle adjustment to regulate the speed of the steam engine. Though control technology quickly developed to solve practical problems, the theoretical understanding of the control systems and its design process took longer, and later appeared in the early and mid-twentieth century. The post WWII era launched the space age that focused its attention on the optimal design of control systems, particularly their implementation via the computing machines. The quest for boosting the industrial output through factory automation has enabled advancement in industrial process control and in assembly line robots that make extensive use of control systems. The growing automation in the past few decades has increased our reliance on automatic control systems.

A control system aims at realizing a desired behavior at the output of a device or system (the plant) by manipulating its input through a controller. Feedback based on observation of the process output via sensing elements plays an important role in all control systems (Figure 1). In the feedback

xi

Figure 1 A generic control system block diagram that includes the controller, process to be controlled, the actuator and the sensor. The output is fed back and compared with the reference signal in the comparator.

control systems, the controller monitors the difference between the desired and actual output variables, and accordingly adjusts the system inputs by employing various control schemes. Often, the control objective is to reduce the error to zero at sufficiently fast rate and maintain it there. The desired output may be expressed as a set point, a constant value that the controller will maintain at the output. Alternatively, in tracking systems, the objective is to track a time-varying reference input. An example of the latter is the control system used to make a drone-mounted camera follow a moving point of interest.

The control system design is invariably undertaken to achieve multiple objectives. First and foremost among them is the stability of the closed-loop system, i.e., the system that has outputs affecting the inputs in real-time. The next objective is attaining dynamic stability, which refers to the ability of the controller to damp out the oscillations at the output of the plant, and is characterized by the damping ratio of the dominant response modes. Furthermore, the controller aims to improve the speed of response, as also reflected by the system bandwidth. The steady-state behavior of the closed-loop system, ideally, has a transfer function of unity, i.e., the closed-loop system operates with zero steady-state error. A final objective in the control design is to impart robustness to the system, which implies an ability to maintain performance levels in the presence of disturbance inputs, as well as its ability to withstand parameter variations and certain modeled dynamics.

The controller itself may be of static or dynamic type. In simple cases, a static gain controller may be adequate to achieve a desired level of performance. An example is the automobile in cruise control, where the gas intake is adjusted to suit the selected speed. In other cases, a simple gain control is inadequate, and a dynamic controller is warranted. A dynamic controller is a dynamic system in its own right, which generates a time-varying controller output that translates into the plant input. For example, the variation in the gas pedal while driving an automobile in cruise control in response to the climb or descent condition represents a time-varying controller output. An alternate understanding of the dynamic controller is a frequency-selective filter that enhances certain frequencies while curbing others.

The dynamic controllers are further distinguished as phase-lead or phase-lag type. These respectively provide improvements to the transient or the steady-state response of the system. The two designs can be combined when needed. In the contemporary design methodology, the controller combines one or more of the three basic control modes: a proportional (P), an integral (I), and a derivative (D) mode. The resulting proportional-integral-derivative (PID) controller is a general-purpose controller that has the ability to meet many of the control objectives defined above. The PID controller is robust against variations in plant parameters, and is therefore popular in industrial control systems. In the traditional design, the controller is commonly implemented using analog circuits build with operational amplifiers and resistive-capacitive (RC) networks. In contemporary control systems, the controller is implemented as a software routine on a computer, microcontroller, or a programmable logic controller (PLC).

A digital controller suitable for computer implementation might be obtained through the emulation of an existing analog controller design. At high enough sampling rates, digital approximation of the analog controller provides comparable performance to the original controller. Alternatively, the design of the digital controller can be based on the pulse transfer function of the plant, i.e., a transfer function obtained through z-transform that is valid at the sampling intervals. Computer implementation of a controller invariably adds a phase lag to the feedback loop that compromises the available stability margins. Hence, a more conservative controller design with additional stability margins may be necessary if digital implementation of the controller is intended. The controller design and implementation for the analog as well as the sampled-data systems (Figure 2) are addressed in this book.

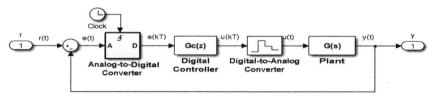

Figure 2 Block diagram of a digital control system that additionally includes an analog-to-digital converter (ADC) and a digital-to-analog converter (DAC).

This book covers the control system design as applicable to single-input single-output (SISO) systems. It places emphasis on understanding and applying effective control system design techniques. The controller design is performed on the mathematical model of the plant (the device or process to be controlled). System models are described in the frequency-domain using the transfer functions, or in the time domain using ordinary differential equations (ODEs). The state variable models describe the system in terms of time derivatives of a set of state variables. Control system design can be performed in either time or frequency-domain; essential design techniques for both are covered in this book. A limited number of skill assessment exercises are provided at the end of each chapter. Additional exercises can be found in standard control systems textbooks (listed as References).

The control systems concepts are applicable to the various engineering disciplines. These concepts are typically covered at a junior or senior level in the engineering curriculum. Students in the scientific disciplines can also benefit from the control systems design concepts. A typical audience of this book includes inquisitive readers with interest in science, technology, engineering, and mathematics (STEM) fields. The mathematical background required for understanding the material, and hence benefitting from this book, includes knowledge of linear algebra, complex numbers, and elementary differential equations. Additionally, some familiarity with the Laplace transform is desired; a brief overview of the same is provided in the Appendix.

The organization of this monograph is as follows: Chapter 1 discusses the modeling of physical systems. The dynamic character of such systems is captured through ODE models. Application of the Laplace transform converts the ODEs to algebraic equations in the Laplace transform variable 's', that are then manipulated to obtain the input-output system description in the form of a transfer function (TF). The examples in this chapter include electrical, mechanical, and electromechanical systems, e.g., the DC motor model. The chapter culminates with a discussion about linearizing a nonlinear system model.

Chapter 2 addresses the methods used to analyze the transfer function models. These models have been characterized in terms of their poles and zeros. The poles effectively determine the modes of system response. The stability characterization requires the poles to be located in the open left half (complex) plane (OLHP). The system response comprises natural and forced response components. Alternatively, it includes the transient and steady-state response, where the first one dies out with time, and the latter persists.

System frequency response characterizes its response to sinusoidal inputs, and represents a sinusoid at the same frequency.

Chapter 3 addresses the methods of analysis for the state variable models. State variables are the natural variables, like inductor current and capacitor voltage in the electrical systems, or position and velocity of the inertial mass in the mechanical systems. The state variable system description is characterized by a set of first order ODEs, or equivalently by the system, the input, and the output matrices. A solution to the first order state equations is obtained through the Laplace transform, and includes a convolution integral involving the state-transition matrix of the system. The choice of the state-variables for a given system is not unique, giving rise to several equivalent system descriptions, some of which may be preferred over others. The popular descriptions are the controller form, the observer form, and the modal form descriptions.

Chapter 4 discusses the objectives of control system design. These include closed-loop stability, transient response improvement, steady-state response improvement, and improvement to the sensitivity robustness. The chapter discusses various ways to characterize these objectives.

Chapter 5 introduces the cascade controller models. The static controller includes a simple gain that multiplies the error signal. The dynamic controllers include phase-lead, phase-lag, and lead-lag types. An alternate description of the dynamic controllers includes the PD, PI, and PID controllers.

Chapter 6 discusses the root locus technique for designing cascade controllers for the transfer function models. The root locus (RL) is the loci of the roots of the closed-loop characteristic polynomial as a function of variation in the controller gain. The RL technique primarily addresses the design of static controllers, but is easily extended to the design of dynamic controllers. In addition, it covers design for improvements to the transient as well as the steady-state response. The chapter includes a discussion of the rate feedback design that is often superior to the cascade controller design. The chapter ends with a discussion of controller realization using analog circuits using operational amplifiers and resister-capacitor networks.

Chapter 7 discusses the techniques to analyze the sampled-data systems, i.e., systems that include a clock-driven device, such as a microprocessor in the loop. The sampled-data systems are characterized by their pulse transfer function, that is obtained via application of the z-transform technique, and is applicable at the sampling intervals. The stability of the sampled-data control system requires the poles of the closed-loop characteristic polynomial to be located inside the unit circle in the complex z-plane.

Chapter 8 discusses the digital controller design for sampled-data systems described by their pulse transfer function models. For high enough sampling rates, an equivalent digital controller may be obtained by emulating the analog controller designed for the continuous-time system. Alternatively, the pulse transfer function can be used to design a cascade digital controller via the root locus technique.

Chapter 9 discusses the controller design for state variable models using full state feedback that allows arbitrary placement of roots of the closed-loop characteristic polynomial. The pole placement design is facilitated by first transforming the state variable model to the controller form. The pole placement method extends to the tracking system design that often requires placing an integrator in the loop to achieve zero steady-state error.

Chapter 10 discusses the controller design for state variable models of the sampled-data systems. The continuous-time state-space model is converted to discrete-time by assuming a zero-order-hold at the input. The resulting discrete state equations can be solved through iteration. A digital controller for the discrete state variable model can be designed through pole placement similar to the continuous-time case. Placing all poles of the closed-loop pulse transfer function at the origin results in a deadbeat design that forces the system response to settle in the n time period.

Chapter 11 discusses compensator design via frequency response modification. Frequency-domain methods, which predate the time-domain methods, utilize gain and phase margins to characterize the relative stability of the closed-loop system. A phase-lag compensator improves the DC gain and/or the phase margin of the system, resulting in the steady state or the transient response improvement. A phase-lead compensator improves the system bandwidth, and hence its transient response. The two can be combined if both transient and steady state response improvements are desired.

Throughout this book, symbols appearing in equations in regular font represent the scalar variables; symbols appearing in boldface letters represent arrays of variables; lower case letters represent vectors, and upper case letters represent matrices. Control systems simulations presented in the book was performed in MATLAB and Simulink (Mathworks, Inc.). For MATLAB simulations, the code is provided with the examples. The figures representing control systems configurations were drawn in the Simulink GUI. The circuit models were drawn in TINA–TI V.9 (Texas Instruments).

Acknowledgement

I would like to extend my gratitude to my teachers for instilling the love of Control Systems in me.

I would like to thank my employer, College of Engineering and Information Technology, University of Arkansas at Little Rock for facilitating my work.

I would like to thank the editor and the publisher for helping me with this project.

I would like to thank my family and friends for there love and support.

I would like to thank my daughter Eeman for helping me with proof reading.

List of Figures

List of Table

List of Abbreviations

DC	Direct current
AC	Alternating current
RLC	Resistor, inductor, and capacitor
OL	Open loop
CL	Closed-loop
CLCP	Closed-loop characteristic polynomial
OLHP	Open left half-plane
RL	Root locus
PD	Proportional-derivative
PI	Proportional-integral
PID	Proportional-integral-derivative
SISO	Single-input single-output
MIMO	Multi-input multi-output
BIBO	Bounded-input bounded-output

Common Symbols used in the book

$G(s)$	Plant transfer function
$K(s)$	Controller/compensator transfer function
$G_c(s)$	Compensator transfer function
$H(s)$	Sensor transfer function
$T(s)$	Closed-loop transfer function
$u(t)$	Plant input
$y(t)$	Plant output
$e(t)$	Error signal
A	System matrix in state variable representation
b	Input distributions vector
c	Output contributions vector

1

Physical System Models

Learning Objectives

1. Obtain a physical system model from the component descriptions.
2. Obtain the system transfer function from its differential equation model.
3. Obtain a physical system model in state variable form.
4. Linearize a nonlinear system model.

Physical systems of interest to engineers include electrical, mechanical, electromechanical, thermal, and fluid systems, among others. The behavior of these systems is mathematically described by the dynamic equations, i.e., ordinary linear differential equations (ODEs), if lumped parameter assumption is made.

To model a system with interconnected components, individual component models can be assembled to formulate the system model. For electrical systems, these elements include resistors, capacitors, and inductors. For mechanical systems, these include inertias (masses), springs, and dampers (or friction elements). For thermal systems, these include thermal capacitance and thermal resistance. For fluid systems, these include the reservoir capacity and the flow resistance. All of these elements either store or dissipate energy, which gives rise to the time-varying or dynamic behavior of the systems.

Modeling of the physical system behavior involves two kinds of variables: flow variables that 'flow' through the components, and across variables that are measured across the components. For the electrical circuits, voltage or potential is measured across the circuit nodes, whereas current or electrical charge flows through the circuit branches. In mechanical linkage systems, displacement and velocity are measured across the connecting nodes, whereas force or effort 'flows' through the linkages. For thermal and fluid systems,

1

heat and mass serve as the flow variables, and temperature and pressure constitute the across variables.

The relationship between flow and across variables defines the type of physical component being modeled. The elementary types are the resistive, the inductive, and the capacitive components. This terminology, taken from electrical circuits, also extends to other types of physical systems.

1.1 Physical Component Models

Let $x(t)$ denote an across variable and $q(t)$ denote a flow variable; then, the elementary component types are defined by their respective relationships, given as follows:

1. **The resistive element:** $x(t) = k\, q(t)$.

For example, the voltage and current relationship through a resister is described by the Ohm's law: $V(t) = R\, i(t)$. Or, the force–velocity relationship though a linear damper is given as: $v(t) = \frac{1}{b} f(t)$.

2. **The capacitive element:** $x(t) = k \int q(t)\mathrm{d}t + x_0$.

For example, the voltage and current relationship through a capacitor in electric circuit is given as: $V(t) = \frac{1}{C} \int i(t)\mathrm{d}t + V_0$. Similarly, the force–velocity relationship through an inertial mass element is given as: $v(t) = \frac{1}{m} \int f(t)\mathrm{d}t + v_0$.

3. **The inductive element:** $x(t) = k \frac{\mathrm{d}q(t)}{\mathrm{d}t}$.

For example, the voltage–current relationship through an inductive coil in an electric circuit is given as: $V(t) = L\frac{\mathrm{d}i(t)}{\mathrm{d}t}$. Or, the force–velocity relationship though a linear spring is given as: $v(t) = \frac{1}{K} \frac{\mathrm{d}f(t)}{\mathrm{d}t}$.

1.1.1 First-Order Models

Electrical, mechanical, thermal, and fluid systems that contain a single energy storage element are described by first-order models. The order here refers to the order of the ODE used to describe the system. The mathematical model of each component is described in terms of the output variable, i.e., the variable that represents the output of the energy storage element.

Example 1.1: Series RC circuit

Consider a series RC circuit connected across a constant voltage source, V_s (Figure 1.1). Using Kirchhoff's voltage law (KVL): $v_R + v_C = V_s$, where capital letters represent constant values and small letters represent time-varying quantities.

By substituting $v_C = v_0$ and $v_R = iR = RC\frac{dv_0}{dt}$, we obtain the first-order ODE that describes the circuit model as:

$$RC\frac{dv_0(t)}{dt} + v_0(t) = V_s.$$

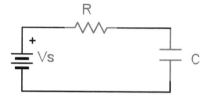

Figure 1.1 An RC circuit.

Example 1.2: Parallel RL circuit

A parallel RL circuit fed by a constant current source, I_s (Figure 1.2), is similarly modeled, via a first-order ODE, where the variable of interest is the inductor current i_L, and Kirchhoff's current law (KCL) is used to sum the currents at the node, i.e., $i_R + i_L = I_s$. Then, by substituting $i_R = \frac{v}{R} = \frac{L}{R}\frac{di_L}{dt}$ we obtain the differential equation for the circuit as:

$$\frac{L}{R}\frac{di_L(t)}{dt} + i_L(t) = I_s$$

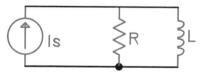

Figure 1.2 An RL circuit.

We may note that the constant multiplier appearing with the derivative term in the above RC and RL circuits defines the time constant of the circuit, i.e., the time when the system response rises to 63.2% of the final value in response to a step input.

Example 1.3: Inertial mass acted on by a force

The motion of an inertial mass acted on by a force $f(t)$ in the presence of kinetic friction is governed by Newton's second law of motion. The resultant force on the mass is $f - bv$, where v is the velocity variable and b is the friction constant. The resulting model is given as:

$$m\frac{dv(t)}{dt} + bv(t) = f(t).$$

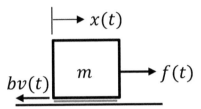

Figure 1.3   Motion of an inertial mass with surface friction.

We may note that the first-order ODE models in the above examples can be generalized by using a generic input $u(t)$, output $y(t)$, and a time constant τ to describe a general first-order model as:

$$\tau\frac{dy(t)}{dt} + y(t) = u(t).$$

Other examples of first-order models include the thermal and fluid systems.

Example 1.4: Room heating

Assume that the heat flow into the room is denoted as q_i, the thermal capacity of the room is C_r, the temperature of the room is θ_r, the ambient temperature is θ_a, and the wall insulation represents a thermal resistance R_w to the heat flow. Then, from the energy balance, we have:

$$C_r\frac{d\theta_r}{dt} + \frac{\theta_r - \theta_a}{R_w} = q_i.$$

In terms of the temperature differential, $\Delta\theta = \theta_r - \theta_a$,

$$R_w C_r\frac{d\Delta\theta}{dt} + \Delta\theta = q_i\Delta\theta.$$

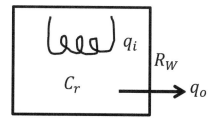

Figure 1.4 Room heating with heatflow through walls.

Note the similarity with the general first-order model, $\tau\frac{dy}{dt} + y = u$, where the thermal heating time constant is given as: $\tau = R_w C_r$.

Example 1.5: Hydraulic reservoir

Consider a cylindrical reservoir filled with an incompressible fluid with a control valve at the bottom exit. Let P denote the base pressure, A denote the area, h denote the height, V denote the volume, ρ denote the density, R_L denote the valve resistance; Q_{in}, Q_{out} denote the volumetric flow rates, and g denote the gravitational constant.

Then, the base pressure is given as: $P = P_{atm} + \rho g h = P_{atm} + \frac{\rho g}{A}V$.

The capacitance of the fluid reservoir is defined as: $C = \frac{dV}{dP} = \frac{\rho g}{A}$

The governing equation of the hydraulic flow is obtained as:
$$C\frac{dP}{dt} = Q_{in} - Q_{out} = Q_{in} - \frac{P - P_{atm}}{R_L}$$

Or $R_L C\frac{d\Delta P}{dt} + \Delta P = Q_{in}R_L$

The above equation is in the standard first-order model form with $\tau = R_L C$.

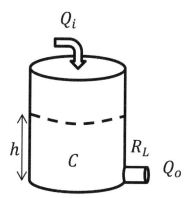

Figure 1.5 Fluid reservoir with constricted outflow.

1.1.2 Second-Order Models

A physical system with two energy storage elements is described by a second-order system model, as illustrated by the following examples.

Example 1.6: Series RLC circuit

A series RLC circuit with voltage input and current output (Figure 1.6) has a governing relationship:

$$L\frac{di(t)}{dt} + Ri(t) + \frac{1}{C}\int i(t)dt = V_s(t)$$

This equation can be written in terms of electric charge $q(t)$ as:

$$L\frac{d^2q(t)}{dt^2} + R\frac{dq(t)}{dt} + \frac{1}{C}q(t) = V_s(t)$$

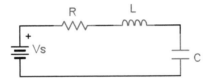

Figure 1.6   A series RLC circuit.

Example 1.7: Inertial mass with position output

An inertial mass moving in a constant gravitational field has both kinetic and potential energies and is modeled as a second-order system. The dynamic equation of a mass with weight mg, pulled upward by a force $f(t)$, is described in terms of the position output $y(t)$ as:

$$m\frac{d^2y(t)}{dt^2} + mg = f(t).$$

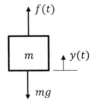

Figure 1.7   Motion of an inertial mass under gravity.

Example 1.8: The mass–spring–damper system

Amass–spring–damper system (Figure 1.8) with applied force $f(t)$ and displacement output $x(t)$ has a familiar second-order model given as:

$$m\frac{\mathrm{d}^2 x(t)}{\mathrm{d}t^2} + b\frac{\mathrm{d}x(t)}{\mathrm{d}t} + kx(t) = f(t).$$

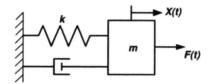

Figure 1.8   A mass–spring–damper system.

1.2 Transfer Function Models

The Laplace transform is extensively used as an analysis tool in the study of control systems due to its ability to transform a linear differential equation into an algebraic equation. The resulting equation can be manipulated to obtain the relationship between the input and output variables as a transfer function.

From the Laplace transform properties, the Laplace transform of successive time derivatives of a variable $y(t)$, ignoring any initial conditions, is defined as:

$$\mathcal{L}\left[\frac{\mathrm{d}^n y(t)}{\mathrm{d}t^n}\right] = s^n y(s).$$

The general first-order model with input $u(t)$ and output $y(t)$, is described as: $\tau\frac{\mathrm{d}y(t)}{\mathrm{d}t} + y(t) = u(t)$. We may use the Laplace transform to write it as:

$$(\tau s + 1)y(s) = u(s)$$

The resulting input–output transfer function of the first-order system is given as:

$$\frac{y(s)}{u(s)} = \frac{1}{\tau s + 1}$$

Example 1.9: The mass–spring–damper system

The mass–spring–damper model is described by the second order ODE, $m\ddot{x} + b\dot{x} + kx = f$, where the dot above a variable represents its time derivative. The model has a Laplace transform description:

$$ms^2 x(s) + bsx(s) + kx(s) = f(s)$$

The input–output relationship (or transfer function) for the spring-mass-damper system with force input and displacement output is given as:

$$\frac{x(s)}{f(s)} = \frac{1}{ms^2 + bs + k}.$$

Higher order models described by linear ODE's can be similarly transformed, and transfer functions between designated input-output pairs can be defined.

1.2.1 DC Motor Model

A DC motor (Figure 1.9) is an example of an electro-mechanical system that inputs electrical energy and outputs mechanical energy. In an armature-controlled DC motor, the motor input is the armature voltage $V_a(t)$ and its output is motor speed $\omega(t)$, or the armature angular position $\theta(t)$.

In order to develop a model of the DC motor, let $i_a(t)$ denote the armature current, and L and R denote the electrical side inductance and resistance of the coil. The mechanical side inertia and friction are denoted as J and b, respectively. Let k_t denote the torque constant and k_b the motor constant; then, the dynamic equations of the DC motor for the electrical and mechanical sides are given as:

$$L\frac{di_a(t)}{dt} + Ri_a(t) + k_b\omega(t) = V_a(t)$$

$$J\frac{d\omega(t)}{dt} + b\omega(t) - k_t i_a(t) = 0.$$

Using Laplace transform, these equations are transformed as:

$$(Ls + R)i_a(s) + k_b\omega(s) = V_a(s)$$

$$(Js + b)\omega(s) - k_t i_a(s) = 0.$$

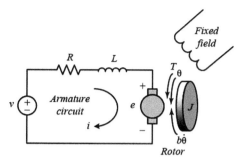

Figure 1.9   An armature-controlled DC motor.

We multiply the first equation by k_t, the second equation by $(Ls + R)$, and add them together to get

$$(Ls + R)(Js + b)\omega(s) + k_t k_b \omega(s) = k_t V_a(s).$$

From the above equation, the transfer function of the DC motor with angular velocity output is derived as:

$$\frac{\omega(s)}{V_a(s)} = \frac{k_t}{(Ls + R)(Js + b) + k_t k_b}.$$

Alternatively, the transfer function from $V_a(t)$ to the angular displacement $\theta(t)$ is given as:

$$\frac{\theta(s)}{V_a(s)} = \frac{k_t}{s\left[(Ls + R)(Js + b) + k_t k_b\right]}.$$

The denominator polynomial in the motor model typically has real and distinct roots that represent the reciprocal of time constants of the electrical and the mechanical sides (τ_e, τ_m). The DC motor model is alternatively described in terms of the two time constants as:

$$\frac{\omega(s)}{V_a(s)} = \frac{k_t/JL}{(s + 1/\tau_e)(s + 1/\tau_m)}$$

Of these time constants, the electrical one is usually much smaller, so the slower mechanical time constant dominates the motor response.

Example 1.10: For a small DC motor let: $R = 1\ \Omega$, $L = 1\ \text{mH}$, $J = 0.01\ kgm^2$, $b = 0.1\ \frac{Ns}{rad}$, and $k_t = k_b = 0.05$; then the transfer function of the motor is obtained as:

$$\frac{\omega(s)}{V_a(s)} = \frac{5000}{(s + 1000)(s + 10) + 250} = \frac{5000}{(s + 10.25)(s + 999.75)}$$

The two time constants of the motor are given as: $\tau_e \cong 1$ ms, $\tau_m \cong 98$ ms.

1.2.2 Simplified Model of a DC Motor

A simplified model of the DC motor can be developed by ignoring the coil inductance ($L \to 0$) that in turn defines the electrical time constant. The resulting first-order model is given as:

$$\frac{\omega(s)}{V_a(s)} = \frac{k_t/R}{Js + b + k_t k_b/R}$$

The reduced-order model includes a single time constant $\left(\tau_m = \frac{JR}{bR + k_t k_b}\right)$, i.e.,

$$\frac{\omega(s)}{V_a(s)} = \frac{k_t/JR}{s + 1/\tau_m}$$

Example 1.11: A small DC motor

Using the above parameter values for a small DC motor, its simplified model is obtained as:

$$\frac{\omega(s)}{V_a(s)} = \frac{5}{s + 10.25}$$

The motor time constant is given as: $\tau_m \cong 97.6$ ms, which coincides with the slower time constant in the second-order model.

1.2.3 Industrial Process Models

Industrial process dynamics, in most cases, can be represented by a first-order lag. Additionally, an industrial processes often displays a dead-time, i.e., there is a finite time delay between the application of input and the appearance of the process output.

Let τ represent the time constant associated with an industrial process, τ_d represent the dead time (or the transport lag), and K represent the process dc gain; then, industrial process dynamics are represented by the following first-order plus dead time (FOPDT) model:

$$G(s) = \frac{Ke^{-\tau_d s}}{\tau s + 1}$$

The process parameters $\{K, \tau, \tau_d\}$ can be identified from the process response to a unit step input.

We note that for analysis and controller design purposes, the delay term in the process model may be represented by its rational approximation. Two such approximations are: $e^{-\tau_d s} \approx 1 - \tau_d s$ and $e^{-\tau_{dt} s} = \frac{1}{1+\tau_d s}$.

1.3 State Variable Models

State variable models of the system are time-domain models that express the system behavior as time derivatives of a set of state variables. The state variables are often the natural variables associated with the energy storage elements. The system order equals the number of such elements in the system.

In the case of electrical circuits, capacitor voltage and inductor currents serve as natural state variables. In the case of mechanical systems with inertial elements, position and velocity of the inertial mass serve as natural state variables. In thermal systems, the heat flow is a natural state variable. In hydraulic systems, the head (height of the liquid in the reservoir) is a natural state variable.

The Examples 1.12–1.14 illustrate the state-variable models in the case of the electrical, mechanical, and electromechanical systems.

Example 1.12: Series RLC circuit

The governing equation of a series RLC circuit driven by a voltage source, V_s, is given as:

$$L\frac{di(t)}{dt} + Ri(t) + \frac{1}{C}\int i(t)dt = V_s$$

Let the inductor current $i(t)$, and the capacitor voltage $v_c(t)$ serve as the state variables for the circuit; then, the state equations (i.e., the time derivatives of state variables) are written as:

$$C\frac{dv_c}{dt} = i$$

$$L\frac{di}{dt} = V_s - v_c - Ri.$$

The state equations are commonly expressed in a matrix form. For the series RLC circuit these are:

$$\frac{d}{dt}\begin{bmatrix} v_c \\ i \end{bmatrix} = \begin{bmatrix} 0 & 1/C \\ -1/L & -R/L \end{bmatrix}\begin{bmatrix} v_c \\ i \end{bmatrix} + \begin{bmatrix} 0 \\ 1/L \end{bmatrix}V_s.$$

Let v_c denote the system output; then, the same is expressed through the output equation:

$$v_c = \begin{bmatrix} 1 & 0 \end{bmatrix} \begin{bmatrix} v_c \\ i \end{bmatrix}.$$

Example 1.13: The mass–spring–damper system

The dynamic equation of the mass–spring–damper system is given as:

$$m\frac{d^2x(t)}{dt^2} + b\frac{dx(t)}{dt} + kx(t) = f(t).$$

Let the state variables be selected as the mass position, $x(t)$, and the mass velocity, $v(t) = \dot{x}(t)$; and, let the output variable be the position $x(t)$. The resulting state variable model of the system is given in terms of the state and output equations as:

$$\frac{d}{dt}\begin{bmatrix} x \\ v \end{bmatrix} = \begin{bmatrix} 0 & 1 \\ -k/m & -b/m \end{bmatrix}\begin{bmatrix} x \\ v \end{bmatrix} + \begin{bmatrix} 0 \\ 1/m \end{bmatrix} f$$

$$x = \begin{bmatrix} 1 & 0 \end{bmatrix}\begin{bmatrix} x \\ v \end{bmatrix}.$$

Example 1.14: The DC motor model

The dynamic equations for the DC motor are given as:

$$L\frac{di_a(t)}{dt} + Ri_a(t) + k_b\omega(t) = V_a(t).$$

$$J\frac{d\omega(t)}{dt} + b\omega(t) - k_t i_a(t) = 0.$$

Let $i_a(t)$, $\omega(t)$ serve as the state variables for DC motor; then, the state variable model of the DC motor is given as:

$$\frac{d}{dt}\begin{bmatrix} i_a \\ \omega \end{bmatrix} = \begin{bmatrix} -R/L & -k_b/L \\ k_t/J & -b/J \end{bmatrix}\begin{bmatrix} i_a \\ \omega \end{bmatrix} + \begin{bmatrix} 1/L \\ 0 \end{bmatrix} V_a.$$

$$\omega = \begin{bmatrix} 0 & 1 \end{bmatrix}\begin{bmatrix} i_a \\ \omega \end{bmatrix}.$$

Using the following parameter values for a small DC motor: $R = 1\,\Omega$, $L = 1$ mH, $J = 0.01$ kgm^2, $b = 0.1\,\frac{\text{N·s}}{\text{rad}}$, $k_t = k_b = 0.05$, the state variable model is given as:

$$\frac{d}{dt}\begin{bmatrix} i_a \\ \omega \end{bmatrix} = \begin{bmatrix} -1000 & -50 \\ 5 & -10 \end{bmatrix}\begin{bmatrix} i_a \\ \omega \end{bmatrix} + \begin{bmatrix} 1000 \\ 0 \end{bmatrix} V_a$$

$$\omega = \begin{bmatrix} 0 & 1 \end{bmatrix} \begin{bmatrix} i_a \\ \omega \end{bmatrix}.$$

1.4 Linearization of Nonlinear Models

Some physical system models are inherently nonlinear; an example is a simple pendulum model, as described by the following equation:

$$ml^2\ddot{\theta}(t) + mgl \sin \theta(t) = T(t),$$

where $\theta(t)$ is the pendulum angle, $T(t)$ is the applied torque; m, l represent the mass and the length of the pendulum, and g is the gravitational constant.

Using (θ, ω) as state variables, the nonlinear pendulum model is expressed as:

$$\frac{d}{dt} \begin{pmatrix} \theta \\ \omega \end{pmatrix} = \begin{pmatrix} \omega \\ -\frac{g}{l}\sin \theta \end{pmatrix} + \begin{pmatrix} 0 \\ T(t) \end{pmatrix}.$$

The two components on the right side of the equation, respectively, define the contribution to the derivative vector from the state and input variables.

As the Laplace transform is only applicable to linear models, in order to obtain a transfer function description, the nonlinear model first needs to be linearized.

Linearization of a nonlinear model is carried out through its first-order Taylor series expansion about an equilibrium point, i.e., the point where the derivative terms take on zero values. Two such equilibrium points are identified in the case of a simple pendulum: $\theta_e = 0°, 180°$.

The linearized model is expressed in terms of state variable deviations that are measured relative to the equilibrium point. Alternatively, the linearize model may be obtained via the Jacobian (partial derivatives of the vector function with respect to the state variable vector).

In the case of the pendulum model the Jacobian is given as:
$$\begin{pmatrix} 0 & 1 \\ -\frac{g}{l}\cos \theta & 0 \end{pmatrix}.$$

These linearized models defined at equilibrium points: $\theta_e = 0°, 180°$ are, respectively, given as:

$$\frac{d}{dt} \begin{bmatrix} \theta \\ \omega \end{bmatrix} = \begin{bmatrix} 0 & 1 \\ -\frac{g}{l} & 0 \end{bmatrix} \begin{bmatrix} \theta \\ \omega \end{bmatrix} + \begin{bmatrix} 0 \\ 1 \end{bmatrix} T(t)$$

$$\frac{d}{dt} \begin{bmatrix} \theta \\ \omega \end{bmatrix} = \begin{bmatrix} 0 & 1 \\ \frac{g}{l} & 0 \end{bmatrix} \begin{bmatrix} \theta \\ \omega \end{bmatrix} + \begin{bmatrix} 0 \\ 1 \end{bmatrix} T(t)$$

1.4.1 The General Nonlinear Case

To generalize the linearization procedure, we may consider a nonlinear state variable model described by the following state and output equations:

$$\dot{x}(t) = f(x, u)$$

$$y(t) = g(x, u),$$

where x is a vector of state variables, u is the input variable, y is the output variable, f is a vector function of the state and input variables, and g is a scalar function of the state and input variables.

The linear state-space model is developed as follows: Assume that a stationary/equilibrium point exists and is defined by: $f(x_e, u_e) = 0$.

Next, let $x(t) = x_e(t) + \delta x(t)$; $u(t) = u_e(t) + \delta u(t)$. Then, using $\delta x, \delta u$ as the new state and input variables, the linearized model is expressed as:

$$\dot{\delta x}(t) = [\partial f_i / \partial x_j]\,|_{(x_e, u_e)}\delta x(t) + [\partial f_i / \partial u]\,|_{(x_e, u_e)}u(t)$$

$$y(t) = [\partial g / \partial x_j]\,|_{(x_e, u_e)}\delta x(t) + [\partial g / \partial u]\,|_{(x_e, u_e)}\delta u(t),$$

where $[\partial f_i / \partial x_j]$ is a Jacobian matrix, $[\partial f_i / \partial u], [\partial g / \partial x_j]$ are the gradient vectors, and $[\partial g / \partial u]$ is a scalar involving partial derivatives, all computed at the equilibrium point.

The linearized model is expressed in its more familiar matrix form as:

$$\dot{\delta x}(t) = A\delta x(t) + bu(t)$$

$$y(t) = c^T \delta x(t) + d\delta u(t).$$

In the above description, A is an $n \times n$ system matrix, b is a column vector for input distributions, c^T is a row vector of output contributions, and d is a scalar feed through term.

The above model represents a single-input single-output (SISO) system. It can be further generalized to represent a multi-input multi-output (MIMO) system (see references).

Skill Assessment Questions

Link to the answers:
http://www.riverpublishers.com/book_details.php?book_id=449

1.1 Consider a series RLC circuit with a voltage source as input and capacitor voltage as output.

    a. Obtain a differential equation model for the circuit.

    b. Obtain the input–output transfer function from the circuit model.

    c. Define inductor current and capacitor voltage as state variables, and obtain a state variable description of the circuit model.

1.2 Consider a parallel RLC circuit with current source as input and capacitor voltage as output:

    a. Obtain a differential equation model for the circuit.

    b. Obtain the input–output transfer function from the circuit model.

    c. Define inductor current and capacitor voltage as state variables, and obtain a state variable description of the circuit model.

1.3 Consider the model of a small DC motor, where the following parameter values are assumed: $R = 1\,\Omega$, $L = 10$ mH, $J = 0.01$ kgm^2, $b = 0.1\,\frac{\text{Ns}}{\text{rad}}$, $k_t = k_{\text{b}} = 0.02$.

    a. Obtain the differential equations for the DC motor.

    b. Obtain the motor transfer functions with angular velocity and displacement as outputs.

    c. Obtain a state variable model of the motor with $\{i_{\text{a}}, \omega, \theta\}$ as state variables.

1.4 Human postural dynamics are modeled as a rigid-body inverted pendulum, described by:

$$(I + ml^2)\ddot{\theta} - mgl\sin(\theta) = T$$

where m is the mass of the body, l is the vertical position of center of mass, g is the gravity, θ is the angular displacement from the vertical, and T is applied torque at the ankles. Assume the following parameter values: $m = 80$ kg, $l = 1\,m$, $I = 4$ kgm^2, $g = 9.8$ m/s^2.

   a. Linearize the model about the vertical $(\theta = \dot{\theta} = 0)$.

   b. Obtain the transfer function from ankle torque input to angular displacement output.

   c. Obtain a state variable description of the model with $\{\theta, \dot{\theta}\}$ as state variables.

1.5 The nonlinear model of a pendulum over cart is given as (Friedland, 1986):

$$(M + m)\ddot{y} + ml\ddot{\theta} \cos \theta - ml\dot{\theta}^2 \sin \theta = f$$

$$ml \cos \theta \, \ddot{y} + ml^2\ddot{\theta} - mgl \sin \theta = 0$$

   a. Define the state variables and express the nonlinear model in the standard form: $\dot{x}(t) = f(x, u)$, $y = g(x)$, where $x = [y, \dot{y}, \theta, \dot{\theta}]$; $u = f$.

   b. Linearize the model about the equilibrium point: $\theta_e = 0$, $\dot{\theta} = 0$, $y = 0$, $\dot{y} = 0$.

2

Analysis of Transfer Function Models

Learning Objectives

1. Characterize a system model by its transfer function poles and zeros.
2. Determine the system response to step, impulse, and sinusoidal inputs.
3. Determine the transient and steady-state response of the system.
4. Characterize the bounded-input bounded-output (BIBO) stability of the system.

This chapter introduces the mathematical methods for analyzing the transfer function models. We begin with a description of system poles and zeros. The poles (roots of the denominator polynomial) determine its natural response that is elicited whenever the system is excited through input, or through the initial conditions. The complex poles, in particular, result in an oscillatory natural response that eventually dies out in the case of stable systems. A complementary input response typically follows the forcing function.

System step response, i.e., its response to a unit step input, comprises transient and steady-state components, and is often used to characterize the system. System impulse response similarly describes its response to a unit impulse input. System response to sinusoidal inputs in the steady-state is a sinusoid at the input frequency, scaled by the system gain.

Bounded-input bounded-output stability is an important characteristic of the systems that refers to the system response to every finite input staying finite. A necessary and sufficient condition for BIBO stability is that the system poles are located in the open left-half s-plane.

2.1 System Poles and Zeros

The transfer function $G(s)$, of a system is a rational function in the Laplace transform variable s, that describes its input–output relation, i.e., the system response $y(s)$ to an input, $u(s)$. The transfer function is expressed as a ratio of the numerator and denominator polynomials, i.e., $G(s) = \frac{N(s)}{D(s)}$.

Given a transfer function model, the roots of the numerator polynomial define the system zeros, i.e., those frequencies at which the system response goes to zero. The roots of the denominator polynomial define the system poles, i.e., those frequencies at which the system response becomes infinite.

For example, the general first-order system described by the transfer function: $G(s) = \frac{1}{\tau s+1}$ has no finite zeros and one real pole at $s = \frac{1}{\tau}$.

The second-order system model of a DC motor is given as: $G(s) = \frac{K}{(s+1/\tau_e)(s+1/\tau_m)}$; this system has no finite zeros and two real poles at $s_1 = \frac{1}{\tau_m}$ and $s_2 = \frac{1}{\tau_e}$.

The second-order system model of a spring–mass–damper system has the transfer function: $G(s) = \frac{1}{ms^2+bs+k}$. In this case, the system poles are characterized by the discriminant, $\Delta = b^2 - 4mk$. Specifically, for $\Delta > 0$, the system has real poles, given as: $s_{1,2} = -\frac{b}{2m} \pm \sqrt{\left(\frac{b}{2m}\right)^2 - \frac{k}{m}}$; whereas, for $\Delta < 0$, the system poles are complex, and are given as: $s_{1,2} = -\frac{b}{2m} \pm j\sqrt{\frac{k}{m} - \left(\frac{b}{2m}\right)^2} = -\sigma \pm j\omega_d$, where ω_d refers to the damped natural frequency of the system.

2.2 System Step Response

The system step response is defined as its response to a unit step input $u(t)$. For systems with complex dominant poles, the step response is used to characterize the dynamic stability of the system.

Using the Laplace transform, the unit step response of a system described by its transfer function $G(s)$, is given as: $y(s) = G(s)\frac{1}{s}$. The inverse Laplace transform can then be used to obtain the system response in the time-domain.

For a first-order system, let $G(s) = \frac{1}{\tau s+1}$; then, the response to a step input is given as: $y(s) = \frac{1}{s(\tau s+1)}$. We use the partial fraction expansion (PFE) to write: $y(s) = \frac{A}{s} + \frac{B}{\tau s+1}$, where $A = 1$, $B = \tau$; then, using the inverse Laplace transform,

$$y(t) = (A + Be^{-t/\tau})u(t).$$

For a second-order system with real poles, e.g., the DC motor, let $G(s) = \frac{K}{(s+1/\tau_e)(s+1/\tau_m)}$; then, the step response is: $Y(s) = \frac{K}{s(s+1/\tau_e)(s+1/\tau_m)} = \frac{A}{s} + \frac{B}{\tau_m s+1} + \frac{C}{\tau_e s+1}$, where $A = K\tau_m\tau_e$. Using the inverse Laplace transform,

$$y(t) = (A + Be^{-t/\tau_{\mathrm{m}}} + Ce^{-t/\tau_{\mathrm{e}}})u(t).$$

For a second-order system with complex poles, we may express the system transfer function as: $G(s) = \frac{1}{ms^2+bs+k} = \frac{1/m}{(s+\sigma)^2+\omega_{\mathrm{d}}^2}$. Then, the step response is: $Y(s) = \frac{1/m}{s[(s+\sigma)^2+\omega_{\mathrm{d}}^2]} = \frac{A}{s} + \frac{B(s+\sigma)+C\omega_{\mathrm{d}}}{(s+\sigma)^2+\omega_{\mathrm{d}}^2}$, where $A = 1/k$.
Using the inverse Laplace transform,

$$y(t) = (A + Be^{-\sigma t} \cos \omega_{\mathrm{d}} t + Ce^{-\sigma t} \sin \omega_{\mathrm{d}} t)u(t)$$
$$= (A + De^{-\sigma t} \cos(\omega_{\mathrm{d}} t - \phi))u(t).$$

2.2.1 Transient and Steady-State Components

The system response to a step input comprises two components, a transient component that dies out with time, and a steady-state component that persists, i.e., $y(t) = y_{\mathrm{tr}}(t) + y_{\mathrm{ss}}(t)$.

The transient response component arises from the excitation of inherent system modes via input, or via the initial conditions; the steady-state response component arises solely because of the input, i.e., the forcing function.

From the above first and second-order system examples, the transient and steady-state components of system response are given as:

For a first-order system, $y_{\mathrm{tr}}(t) = Be^{-t/\tau}u(t)$, and $y_{\mathrm{ss}}(t) = A$.

For a second-order system with real poles, $y_{\mathrm{tr}}(t) = (Be^{-t/\tau_{\mathrm{m}}} + Ce^{-t/\tau_{\mathrm{e}}})u(t)$, and $y_{\mathrm{ss}}(t) = A$.

For a second-order system with complex poles, $y_{\mathrm{tr}}(t) = (B \cos \omega_{\mathrm{d}} t + C \sin \omega_{\mathrm{d}} t)e^{-\sigma t}u(t)$, and $y_{\mathrm{ss}}(t) = A$.

From the above examples, we note that for a unit step input, the steady-state component of the system response is a constant, given as: $y_{\mathrm{ss}}(t) = A$. Further, its value can be computed as $A = G(s)|_{s=0}$.

Since the transient response of a stable system dies out with time, we may only be interested in its qualitative aspects; these are described in terms of the rise time (i.e., time when the step response first reaches unity, t_{r}), the maximum overshoot (%OS), the peak time (t_{p}), and the settling time (t_{s}).

2.3 System Impulse Response

The system impulse response is defined as its response to a unit impulse input $\delta(t)$, and may be, equivalently, used to characterize the system.

In terms of Laplace transform, the system response to a unit impulse input is given as: $y(s) = G(s)$. Then, the impulse response in the time domain is given as: $y(t) = \mathcal{L}^{-1}[G(s)]$.

For a first-order system, with $y(s) = G(s) = \frac{1}{\tau s + 1}$, we have $y(t) = Ae^{-t/\tau}u(t)$, where $A = \frac{1}{\tau}$.

For a second-order system with real poles, e.g., the DC motor, let: $y(s) = G(s) = \frac{K}{(s+1/\tau_e)(s+1/\tau_m)}$; then, using inverse Laplace transform, $y(t) = (Ae^{-t/\tau_m} + Be^{-t/\tau_e})u(t)$.

For a second order system with complex poles, let: $y(s) = G(s) = \frac{1}{ms^2+bs+k} = \frac{1/m}{(s+\sigma)^2+\omega_d^2}$; then, $y(t) = Ae^{-\sigma t}\sin(\omega_d t)u(t)$, where $A = \frac{1}{m\omega_d}$.

From the above examples, we note that the impulse response comprises only the transient part of the system response. Further, the impulse response is a characteristic of the physical system that arises from the inherent system characteristics. Hence, the impulse response is also termed as the natural response of the system, and is used to identify the modes of system response.

2.4 BIBO Stability

Stability is a desired characteristic of a system; it refers to the system output staying finite when the system is excited via an input, or via the initial conditions on the energy storage elements that characterize the residual energy of the system. Stability can be deduced if the impulse response of the system is observed to stay bounded.

Stability is guaranteed (and can be mathematically proven) in the case of linear systems built with passive components, which either store or dissipate energy. For systems involving feedback, the stability requires that the feedback is negative.

The BIBO stability implies that for every bounded input, $u(t): |u(t)| < M_1 < \infty$, the output is bounded, i.e., $y(t): |y(t)| < M_2 < \infty$.

A bounded-input will generate a bounded-output if the system transient response dies out with time. Thus, a necessary and sufficient condition for BIBO stability is that: $\lim_{t\to\infty} y_{tr}(t) = 0$.

As the transient response depends on the pole locations; in particular, $y_{tr}(t) = \sum_{i=1}^{n} A_i e^{p_i t}$, the requirement for BIBO stability translates as, $\text{Re}[p_i] < 0$, where p_i, is a pole of the system.

For physical systems built with passive components, there is a dissipation of energy due to friction, electrical resistance, thermal resistance, etc. In such cases, the resulting system model has damping terms in it, and the condition $\text{Re}[p_i] < 0$ is duly satisfied.

In the ideal case of a simple harmonic oscillator, if no damping is assumed, the system model is described by the ODE: $\ddot{y} + w_n^2 y = 0$, where w_n represents the natural frequency of the oscillator. The resulting system transfer function is given as: $G(s) = \frac{1}{s^2 + w_n^2}$.

The system poles in this case lie on the stability boundary, i.e., the $j\omega$-axis. The natural response, i.e., the response to an impulse input, in the case of such systems persists in the form of oscillations at the natural frequency. Such systems are termed as marginally stable. The linear methods of stability analysis extend to the nonlinear systems under certain conditions. In the more general terms, nonlinear system stability is analyzed through the Lyapunov methods.

2.5 Sinusoidal Response of the System

The sinusoidal response of a system is its response to a sinusoidal input, described as: $u(t) = \cos w_0 t$ or $u(t) = \sin w_0 t$.

Let the system transfer function be given as $G(s)$; then, using the Laplace transform, the sinusoidal response of the system is obtained as: $y(s) = G(s)\frac{s}{s^2 + w_0^2}$, or $y(s) = G(s)\frac{w_0}{s^2 + w_0^2}$.

For a first-order system, let $G(s) = \frac{1}{\tau s + 1}$, and let $u(t) = \sin w_0 t$; then

$$y(s) = \frac{w_0}{(\tau s + 1)(s^2 + w_0^2)} = \frac{A}{\tau s + 1} + \frac{Bs + Cw_0}{(s^2 + w_0^2)}.$$

Using the inverse Laplace transform, $y(t) = (Ae^{-t/\tau} + B \cos w_0 t + C \sin w_0 t)u(t)$. We may note that: $y_{\text{ss}(t)} = B \cos w_0 t + C \sin w_0 t$.

For a second-order system with real poles, let $G(s) = \frac{K}{(s + 1/\tau_e)(s + 1/\tau_m)}$, $u(t) = \sin w_0 t$.

Then, $y(s) = \frac{Kw_0}{(s + 1/\tau_e)(s + 1/\tau_m)(s^2 + w_0^2)} = \frac{A}{\tau_m s + 1} + \frac{B}{\tau_e s + 1} + \frac{Cs + Dw_0}{(s^2 + w_0^2)}.$

Using the inverse Laplace transform, $y(t) = (Ae^{-t/\tau_m} + Be^{-t/\tau_e} + C \cos w_0 t + D \sin w_0 t)u(t)$.

The steady-state component of the response is given as: $y_{\text{ss}(t)} = C \cos w_0 t + D \sin w_0 t$.

For a second order system with complex poles, let $G(s) = \frac{1}{ms^2 + bs + k} = \frac{1/m}{(s + \sigma)^2 + w_d^2}$, $u(t) = \sin w_0 t$.

Then, $y(s) = \frac{w_0/m}{[(s + \sigma)^2 + w_d^2](s^2 + w_0^2)} = \frac{A(s + \sigma) + Bw_d}{(s + \sigma)^2 + w_d^2} + \frac{Cs + Dw_0}{(s^2 + w_0^2)}.$

Using the inverse Laplace transform,

$$y(t) = (Ae^{-\sigma t} \cos \omega_d t + Be^{-\sigma t} \sin \omega_d t + C \cos \omega_0 t + D \sin \omega_0 t)u(t)$$

Once again, the steady-state component of the response is given as: $y_{ss(t)} = C \cos \omega_0 t + D \sin \omega_0 t$.

From the above examples we learn that in the steady-state, the sinusoidal response of the system contains sinusoidal terms at the input frequency.

2.5.1 The Frequency Response Function

Assume that a given system has transfer function $G(s)$; then, its frequency response function is defined as: $G(j\omega) = G(s)|_{s=j\omega}$.

Note that the frequency response function is a complex function of a real frequency variable ω, i.e., $G(j\omega) = |G(j\omega)| e^{\angle G(j\omega)}$.

Next, assume a complex exponential input of the form: $u(t) = e^{j\omega_0 t}$, $u(s) = \frac{1}{s - j\omega_0}$. Then, the system output is given as: $y(s) = \frac{G(s)}{s - j\omega_0}$. Since the transient response of a stable system dies out, its steady-state response is computed as: $y_{ss}(s) = \frac{A}{s - j\omega_0}$, where $A = G(j\omega_0)$.

Thus, the steady-state response to a complex sinusoid $u(t) = e^{j\omega_0 t}$ is given as:

$$y_{ss}(t) = G(j\omega_0)e^{j\omega_0 t} = |G(j\omega_0)| e^{j\omega_0 t + \angle G(j\omega)}$$

Since $\cos \omega_0 t = \frac{1}{2}(e^{j\omega_0 t} + e^{-j\omega_0 t})$, the steady-state response to an input: $u(t) = \cos \omega_0 t$ is given as:

$$y_{ss}(t) = |G(j\omega_0)| \cos(\omega_0 t + \angle G(j\omega_0))$$

Thus, in the case of stable linear systems, the steady-state response to a sinusoid of a certain frequency is a sinusoid at the same frequency scaled by the magnitude of the frequency response function computed at that frequency.

Skill Assessment Questions

Link to the answers:
http://www.riverpublishers.com/book_details.php?book_id=449

2.1 Consider a series RLC circuit with a constant voltage input and capacitor voltage as output (assume the following component values $R = 2\,\Omega, C = 0.1$ F, $L = 1$ H):

    a. Obtain the input-output transfer function for the system.
    b. Compute the impulse response of the system.
    c. Compute the step response of the system.
    d. Compute steady-state system response to a sinusoidal input, $\sin \omega_0 t$.

2.2 Consider a parallel RLC circuit with constant current input and voltage output (assume the following component values: $R = 2\,\Omega, C = 0.1$ F, $L = 1$ H):

    a. Obtain the input-output transfer function for the system.
    b. Compute the impulse response of the system.
    c. Compute the step response of the system.
    d. Compute steady-state system response to a sinusoidal input, $\sin \omega_0 t$.

2.3 Consider the model of a small dc motor, where the following parameter values are assumed: $R = 1\,\Omega$, $L = 10$ mH, $J = 0.01$ kgm^2, $b = 0.1\,\frac{\text{Ns}}{\text{rad}}$, $k_t = k_b = 0.02$.

    a. Obtain the input-output transfer function for the motor with voltage input and angular velocity output.
    b. Compute the impulse response of the system.
    c. Compute the step response of the system.
    d. Compute steady-state system response to a sinusoidal input, $\sin \omega_0 t$.

2.4 Consider the model of the small dc motor in 2.3. Assume that the coil inductance, $L \to 0$.

    a. Obtain the input-output transfer function for the motor model.
    b. Compute the impulse response of the system.
    c. Compute the step response of the system.
    d. Compute steady-state system response to a sinusoidal input $\sin \omega_0 t$.

2.5 Consider the model of a spring-mass-damper system, where the following values are assumed: $m = 1$ kg, $b = 2 \frac{\text{Ns}}{\text{m}}$, $k = 2 \frac{\text{N}}{\text{m}}$.

    a. Obtain the input-output transfer function for the system.
    b. Compute the impulse response of the system.
    c. Compute the step response of the system.
    d. Compute steady-state system response to a sinusoidal input $\sin \omega_0 t$.

3

Analysis of State Variable Models

Learning Objectives

1. Obtain the system transfer function from its state variable description.
2. Obtain the state transition matrix of the system and solve for the state variables.
3. Obtain state-space realization of a transfer function in alternate forms.

This chapter discusses the algebraic methods to analyze the state variable models. These models are described in terms of a set of first-order ODEs involving time derivatives of the system state variables (Chapter 1).

The state equations can be collectively integrated using linear algebraic methods. The solution is given as a convolutional integral that involves the state-transition matrix of the system.

The choice of the state variables for a given system is not unique. Hence, a given system transfer function can be realized into multiple equivalent state variable models. The state variable vectors for these models are related through linear transformations. Certain realization structures may be preferred for ease of computing the system response, or for determining its stability.

3.1 System Transfer Function

Let $x(t)$ be the vector of state variables, $u(t)$ the input, and $y(t)$ the output of a single-input single-output (SISO) system, then the state-space model of the systems is written in generic form as:

$$\dot{x}(t) = Ax(t) + bu(t)$$
$$y(t) = c^T x(t) + du(t).$$

In the above model, A is the system matrix, b is the input matrix (a column vector), c^T is the output matrix (a row vector), and d represents a scalar feedforward term from system input to its output. We will assume that there is no direct contribution from input to output (hence $d = 0$).

The state–space model relates to the system transfer function via the Laplace transform. Indeed, the Laplace transform of the state equations, ignoring any initial conditions, is given as:

$$sx(s) = Ax(s) + bu(s).$$

Let I denote $n \times n$ identity matrix, then the solution to the state vector is obtained as:

$$x(s) = (sI - A)^{-1}bu(s).$$

The system output is computed as:

$$y(s) = c^T(sI - A)^{-1}bu(s).$$

Since $y(s) = G(s)u(s)$, where $G(s)$ is the system transfer function, it is obtained from its state variable model as:

$$G(s) = c^T(sI - A)^{-1}b.$$

Using inverse Laplace transform, the impulse response of the system is obtained as:

$$g(t) = c^T e^{At}b.$$

3.2 Solution to the State Equations

In order to develop the time-domain solution to the state equations, we begin with a scalar equation:

$$\dot{x}(t) = ax(t) + bu(t).$$

Using the integrating factor e^{-at}, the equation is written as:

$$\frac{d}{dt}(e^{-at}x(t)) = e^{-at}bu(t)$$

After integrating it from 0 to τ, we obtain

$$e^{-at}x(t) - x_0 = \int_0^\tau e^{-a\tau}bu(\tau)d\tau.$$

Thus, the solution to the state variable $x(t)$ is given as:

$$x(t) = e^{at}x_0 + \int_0^\tau e^{a(t-\tau)}bu(\tau)d\tau.$$

Next, we generalize this solution to the matrix case. Toward this aim, we first define the matrix exponential as:

$$e^{At} = \sum_{i=0}^\infty \frac{A^i t^i}{i!} = I + At + \dots$$

Then, assuming that the matrix A and its exponential commute, i.e., $Ae^{At} = e^{At}A$, the solution to the matrix equation, $\dot{x}(t) = Ax(t) + bu(t)$, is written as:

$$x(t) = e^{At}x_0 + \int_0^\tau e^{A(t-\tau)}bu(\tau)d\tau.$$

We note that the corresponding Laplace transform solution with non-zero initial conditions is given as:

$$x(s) = (sI - A)^{-1}x_0 + (sI - A)^{-1}bu(s).$$

Comparing the two solutions, we observe that:

$$\mathcal{L}[e^{At}] = (sI - A)^{-1}$$

$$\mathcal{L}\left[\int_0^\tau e^{A(t-\tau)}bu(\tau)d\tau\right] = (sI - A)^{-1}bu(s).$$

In the absence of any input, i.e., for $u(t) = 0$, we obtain the following solution to the state equations: $x(t) = e^{At}x_0$, i.e., the matrix exponential e^{At} relates $x(t)$ to x_0, where x_0 represents the initial condition vector that embodies the residual energy in the system.

3.3 The State-Transition Matrix

The matrix exponential e^{At} is commonly referred as the state-transition matrix of the system; its Laplace transform $(sI - A)^{-1}$ is called the characteristic matrix of A (or the resolvent of A). The computation of the state-transition matrix, e^{At}, constitutes the first step toward writing the time-domain solution to the state equations.

The state-transition matrix can be computed in the following ways:

1. $e^{At} = \mathcal{L}^{-1}[(s\boldsymbol{I} - \boldsymbol{A})^{-1}]$
2. $e^{At} = \sum_0^\infty \frac{\boldsymbol{A}^i t^i}{i!}$ (assuming series convergence)
3. Via the Cayley–Hamilton theorem (see References)

The resolvent of $\boldsymbol{A}$ can be computed in the following ways:

1. $(s\boldsymbol{I} - \boldsymbol{A})^{-1} = s^{-1}(\boldsymbol{I} + s^{-1}\boldsymbol{A} + \ldots)$
2. $(s\boldsymbol{I} - \boldsymbol{A})^{-1} = \frac{\mathrm{adj}(s\boldsymbol{I} - \boldsymbol{A})}{\det(s\boldsymbol{I} - \boldsymbol{A})}$

The characteristic polynomial of $\boldsymbol{A}$ is a nth order polynomial obtained as the determinant of $(s\boldsymbol{I} - \boldsymbol{A})$, i.e.,

$$\Delta(s) = |s\boldsymbol{I} - \boldsymbol{A}|$$

Let s_i, $i = 1, \ldots, n$ denote the roots of the characteristic polynomial (the eigenvalues of the system matrix, $\boldsymbol{A}$); then, the general form of the system response using arbitrary constants is written as:

$$y(t) = \sum_{i=1}^n C_i e^{s_i t}$$

In the above, the terms $e^{s_i t}$, $i = 1, \ldots, n$, represent the modes of the system response. These modes are present in all system outputs.

Example 3.1: The mass–spring–damper system

For the mass–spring–damper system, the resolvent of $\boldsymbol{A}$ is computed as:

$$s\boldsymbol{I} - \boldsymbol{A} = \begin{bmatrix} s & -1 \\ \frac{k}{m} & s + \frac{b}{m} \end{bmatrix}.$$

$$(s\boldsymbol{I} - \boldsymbol{A})^{-1} = \frac{1}{\Delta(s)} \begin{bmatrix} s + b/m & 1 \\ -k/m & s \end{bmatrix}; \quad \Delta(s) = s^2 + \frac{b}{m}s + \frac{k}{m}.$$

Let $m = 1$, $k = 2$, $b = 3$, then

$$e^{At} = \begin{bmatrix} 2e^{-t} - e^{-2t} & e^{-t} - e^{-2t} \\ 2e^{-2t} - 2e^{-t} & 2e^{-2t} - e^{-t} \end{bmatrix} = \begin{bmatrix} 2 & 1 \\ -2 & -1 \end{bmatrix} e^{-t} + \begin{bmatrix} -1 & -1 \\ 2 & 2 \end{bmatrix} e^{-2t}.$$

Note that the terms in the state transition matrix are of the form: $c_1 e^{s_1 t} + c_2 e^{s_2 t}$. These terms comprise the system response modes: $\{e^{s_1 t}, e^{s_2 t}\}$, where s_1, s_2 are the roots of the characteristic polynomial, which is given as: $\Delta(s) = |s\boldsymbol{I} - \boldsymbol{A}|$.

In particular, for the parameter values taken above, s_1, $s_2 = -1, -2$, and the modes of system response are: $\{e^{-t}, e^{-2t}\}$.

Example 3.2: The DC motor model

Consider the DC motor model, where the following component values are assumed: $R = 1\ \Omega$, $L = 1\ \text{mH}$, $J = 0.01\ \text{kgm}^2$, $b = 0.1\ \frac{\text{Ns}}{\text{rad}}$, $k_t = k_b = 0.05$.

The state variables for the motor model are selected as the armature current and the motor angular velocity. Then, the state and output equations for the DC motor are given as:

$$\frac{d}{dt}\begin{bmatrix} i_a \\ \omega \end{bmatrix} = \begin{bmatrix} -1000 & -50 \\ 5 & -10 \end{bmatrix}\begin{bmatrix} i_a \\ \omega \end{bmatrix} + \begin{bmatrix} 1000 \\ 0 \end{bmatrix} V_a$$

$$\omega = \begin{bmatrix} 0 & 1 \end{bmatrix}\begin{bmatrix} i_a \\ \omega \end{bmatrix}.$$

Then,

$$(s\boldsymbol{I} - \boldsymbol{A})^{-1} = \frac{1}{\Delta(s)}\begin{bmatrix} s+10 & -50 \\ 5 & s+1000 \end{bmatrix}; \quad \Delta(s) = s^2 + 1010s + 10250$$

$$G(s) = \boldsymbol{c}^T(s\boldsymbol{I} - \boldsymbol{A})^{-1}\boldsymbol{b} = \frac{5000}{s^2 + 1010s + 10250}$$

$$e^{\boldsymbol{A}t} = \begin{bmatrix} 1.0003 & 0.0505 \\ -0.0051 & -0.0003 \end{bmatrix}e^{-999.75t} + \begin{bmatrix} -0.0003 & -0.0505 \\ 0.0051 & 1.0003 \end{bmatrix}e^{-10.25t},$$

where $\{e^{-999.75t}, e^{-10.25t}\}$ represent the system response modes that correspond to the electrical and mechanical time constants of the dc motor: $\tau_e \cong 0.001s$, $\tau_m \cong 0.1s$

3.4 Linear Transformation of the State Variables

A linear transformation of the state variable vector x is defined by multiplication with a constant invertible matrix $\boldsymbol{P}$, resulting in a new set of state variables defined by vector z. Then,

$$z = \boldsymbol{P}x, \quad x = \boldsymbol{P}^{-1}z$$

Using the above linear transformation, the state and output equations are transformed as:

$$P^{-1}\dot{z} = AP^{-1}z + bu$$

$$y = c^T P^{-1}z$$

Or,

$$\dot{z} = PAP^{-1}z + Pbu$$

$$y = c^T P^{-1}z.$$

The above equations describe an equivalent state variable model of the system in terms of the (new) state vector $z(t)$. We may note that the two models share the same transfer function, i.e.

$$G(s) = c^T P^{-1}(sI - PAP^{-1})^{-1}Pb = c^T (sI - A)^{-1}b.$$

We may also note that the new system matrix PAP^{-1} has a different structure than then the original matrix A. This fact may be exploited to impart a desired structure to the system matrix through linear transformation of the state variables.

One such structure is the controller form structure, whereby the state variables for a single-input single-output (SISO) system are selected as output and its derivatives: $y(t)$, $\dot{y}(t), \ldots, y^{(n-1)}(t)$. The resulting system and input matrices have the following special form:

$$A = \begin{bmatrix} 0 & 1 & 0 & \cdots \\ 0 & 0 & 1 & \cdots \\ \vdots & \vdots & & \ddots & 1 \\ -a_n & -a_{n-1} & \cdots & -a_1 \end{bmatrix}, \quad b = \begin{bmatrix} 0 \\ 0 \\ \vdots \\ 1 \end{bmatrix}$$

The controller form structure is so named because it facilitates the controller design in the state-space. In the controller form structure, the last row of the system matrix contains the coefficients of the characteristic polynomial, which can be written by inspection and is given as:

$$\Delta(s) = s^n + a_1 s^{n-1} + \cdots + a_{n-1}s + a_n.$$

Finally, the multiplicity of the state variable models that corresponds to the same transfer function shows that the choice of state variables for a given system is not unique.

3.5 State-Space Realization of Transfer Function Models

The realization problem refers to obtaining a state variable description of the system from a given transfer function model. This can be done in several ways resulting in equivalent but different model structures. Two ways of doing so, that generate the controller form and the modal form structures, are discussed below.

Obtaining state variable description of the system is facilitated by first drawing its simulation diagram. A simulation diagram is a visualization tool that helps visualize the structure of the dynamic system.

Before the development of digital computer, analog computers were used to simulate the dynamic system behavior. These computers were built with electronic circuits designed to work as integrators and adders, with simple gains realized via potentiometers. A simulation diagram, analogously, represents the system structure in terms of these three components.

3.5.1 Controller Form Realization

To illustrate a controller form realization via the simulation diagram, we consider a third order transfer function: $G(s) = \frac{b_1 s^2 + b_2 s + b_3}{s^3 + a_1 s^2 + a_2 s + a_3}$, and write it in cascade form using auxiliary variable $v(s)$ as:

$$G(s) = \left(\frac{1}{s^3 + a_1 s^2 + a_2 s + a_3} \right) \left(\frac{b_1 s^2 + b_2 s + b_3}{1} \right) = \frac{v(s)}{u(s)} \frac{y(s)}{v(s)}.$$

The transfer function $v(s)/u(s)$ corresponds to a differential equation:

$$\dddot{v}(t) + a_1 \ddot{v}(t) + a_2 \dot{v}(t) + a_3 v(t) = u(t).$$

In order to obtain the simulation diagram for this system, the highest derivative term in the ODE is realized using a summer as:

$$\dddot{v}(t) = -a_1 \ddot{v}(t) - a_2 \dot{v}(t) - a_3 v(t) + u(t).$$

The remaining derivative terms are realized using integrators and gain elements (Figure 3.1).

The $\frac{y(s)}{v(s)}$ term corresponds to:

$$y(t) = b_1 \ddot{v}(t) + b_2 \dot{v}(t) + b_3 v(t)$$

where $v(t)$, $\dot{v}(t)$, $\ddot{v}(t)$ are the outputs of the integrators that are included in the simulation. Hence, the composite transfer function $G(s)$ is realized as shown in the Figure 3.1.

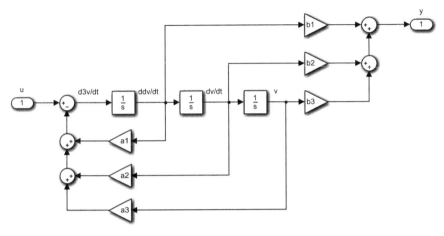

Figure 3.1   A simulation diagram for the controller form realization of a transfer function model.

Next, we designate the outputs of the integrators as state variables, i.e., let $x_1(t) = v(t)$, $x_2(t) = \dot{v}(t)$, $x_3(t) = \ddot{v}(t)$; then, the following state variable model is obtained:

$$\frac{d}{dt}\begin{bmatrix} x_1 \\ x_2 \\ x_3 \end{bmatrix} = \begin{bmatrix} 0 & 1 & 0 \\ 0 & 0 & 1 \\ a_3 & a_2 & a_1 \end{bmatrix}\begin{bmatrix} x_1 \\ x_2 \\ x_3 \end{bmatrix} + \begin{bmatrix} 0 \\ 0 \\ 1 \end{bmatrix} u$$

$$y = \begin{bmatrix} b_1 & b_2 & b_3 \end{bmatrix}\begin{bmatrix} x_1 \\ x_2 \\ x_3 \end{bmatrix}.$$

The above state-space model is in the controller form. Further, the coefficients of the characteristic polynomial appear in the last row of the system matrix. Hence, given a transfer function model, the equivalent controller form realization may be written by inspection.

We may also note that the integrators outputs can be numbered in different ways (i.e., from right to left, left to right, etc.). Such arrangements result in equivalent, but different state variable models.

In the MATLAB, the controller form realization is called the companion form. The MATLAB command 'canon' (for canonical) is used to transform a given model into its companion form, as given below:

```
>> G=tf(num, den);
>> canon(G,'companion')
```

A minor difference being in the MATLAB companion form the coefficients of the characteristic polynomial appear in the last column.

3.5.2 Modal Form Realization

A modal form realization represents a parallel structure obtained via partial fraction expansion (PFE) of the system transfer function into a summation of first and second order factors, that respectively contain the real and complex poles of the system. Each of these factors may be realized through a controller form realization, and the results concatenated to obtain the modal form realization for the system.

The system matrix for the modal form realization is block diagonal, i.e., it consists of 1×1 and 2×2 blocks that correspond to the first and second order factors in the PFE. The eigenvalues in these blocks represent the system modes of response, that are represented in the system output $y(t)$ through the output matrix.

Example 3.3: To illustrate the modal form realization, we consider the following partial fraction expansion of a third order system model:

$$G(s) = \frac{s^2 + s + 1}{(s + 2)(s^2 + 2s + 2)} = \frac{3/2}{s + 2} - \frac{1/2s + 1}{(s^2 + 2s + 2)}$$

Both factors are realized through the controller form realizations, and the results are concatenated to obtain:

$$\frac{d}{dt} \begin{bmatrix} x_1 \\ x_2 \\ x_3 \end{bmatrix} = \begin{bmatrix} -2 & 0 & 0 \\ 0 & 0 & 1 \\ 0 & -2 & -2 \end{bmatrix} \begin{bmatrix} x_1 \\ x_2 \\ x_3 \end{bmatrix} + \begin{bmatrix} 1 \\ 0 \\ 1 \end{bmatrix} u$$

$$y = \begin{bmatrix} 3/2 & -1 & -1/2 \end{bmatrix} \begin{bmatrix} x_1 \\ x_2 \\ x_3 \end{bmatrix}$$

An equivalent modal form realization (somewhat different from the above) can be obtained in the MATLAB via the 'canon' command, as shown below:

```
>> G=tf([1 1 1], conv([1 2],[1 2 2]));
>> canon(G,'modal')
```

In the MATLAB modal form, the 2×2 system blocks have the structure $A_i = \begin{bmatrix} \sigma & \omega \\ -\omega & \sigma \end{bmatrix}$, so that $|sI - A_i|$ represents a second-order polynomial of the form: $(s + \sigma)^2 + \omega^2$.

3.5.3 Diagonal Form Realization

When the characteristic equation of the system has real and distinct roots, the modal form matrix reduces to a diagonal matrix. The resulting system structure is completely decoupled, described by a set of decoupled first-order ODEs.

The diagonal form realization from an existing state-space model can be obtained through a linear transformation matrix P, when the eigenvectors of A matrix are selected as the columns of P.

Example 3.4: Consider the model of the small dc motor, given as:

$$\frac{d}{dt}\begin{bmatrix} i_a \\ \omega \end{bmatrix} = \begin{bmatrix} -1000 & -50 \\ 5 & -10 \end{bmatrix}\begin{bmatrix} i_a \\ \omega \end{bmatrix} + \begin{bmatrix} 1000 \\ 0 \end{bmatrix}V_a$$

$$\omega = \begin{bmatrix} 0 & 1 \end{bmatrix}\begin{bmatrix} i_a \\ \omega \end{bmatrix}.$$

We may use the MATLAB 'eig' command to diagonalize the system:

```
>> A=[-1000 -50; 5 -10];
>> [P, D]=eig(A)
```

The matrix P obtained from MATLAB is used to transform the input b and the output c^T vectors in the original state-space model. The result is a decoupled system model, given as:

$$\begin{bmatrix} \dot{x}_1 \\ \dot{x}_2 \end{bmatrix} = \begin{bmatrix} -999.75 & 0 \\ 0 & -10.25 \end{bmatrix}\begin{bmatrix} x_1 \\ x_2 \end{bmatrix} + \begin{bmatrix} -1000.3 \\ -5.1 \end{bmatrix}V_a$$

$$\omega = \begin{bmatrix} 5.1 & -0.999 \end{bmatrix}\begin{bmatrix} x_1 \\ x_2 \end{bmatrix}.$$

The decoupled system of equations includes two scalar ODEs that can be easily integrated. For example, assuming a unit step input, the solution to the decoupled state variables is given as:

$$x_1(t) = 1.003 + (x_{10} - 1.003)e^{-t/999.75}$$
$$x_2(t) = 0.498 + (x_{20} - 0.498)e^{-t/10.25}$$

Where x_{10}, x_{20} represent the initial conditions on the state variables. The system output (motor speed) represents a combination of the state variables, and is computed as:

$$\omega(t) = 5.1x_1(t) - 0.999x_2(t).$$

Skill Assessment Questions

Link to the answers:
http://www.riverpublishers.com/book_details.php?book_id=449

3.1 Obtain the state-transition matrix for the simple pendulum model given below. Find the modes of the system response.

$$\frac{d}{dt}\begin{bmatrix} \theta \\ \omega \end{bmatrix} = \begin{bmatrix} 0 & 1 \\ -\frac{g}{l} & 0 \end{bmatrix}\begin{bmatrix} \theta \\ \omega \end{bmatrix} + \begin{bmatrix} 0 \\ 1 \end{bmatrix} T$$

3.2 Obtain the state-transition matrix for the simplified dc motor model given below. Find the modes of the system response.

$$\frac{d}{dt}\begin{bmatrix} \theta \\ \omega \end{bmatrix} = \begin{bmatrix} 0 & 1 \\ 0 & -\frac{1}{\tau_m} \end{bmatrix}\begin{bmatrix} \theta \\ \omega \end{bmatrix} + \begin{bmatrix} 0 \\ \frac{k_t}{JR} \end{bmatrix} V_a$$

3.3 Obtain the controller form and diagonal form realizations for the following transfer function model: $G(s) = \frac{s+1}{s(s+2)(s+5)}$. Find the modes of the system response.

3.4 Obtain the modal form realization for the following transfer function model: $G(s) = \frac{s+1}{s(s^2+2s+2)}$. Find the modes of the system response.

3.5 Obtain the controller form and modal form realizations for the following transfer function model: $G(s) = \frac{28s+120}{s^2+7s+14}$. Find the modes of the system response.

3.6 Obtain the state transition matrix and the modes of system response for the model of human postural dynamics described as an inverted pendulum: $G(s) = 1/(s^2 - \Omega^2)$.

4

Control System Design Objectives

Learning Objectives

1. Characterize the stability of the closed-loop system.
2. Characterize the transient and steady-state response of the closed-loop system.
3. Determine the sensitivity of the closed-loop system to parameter variations in the plant model.

The goal of the control system design is to obtain a desired behavior at the output of a physical system, which may be an automobile, an airplane, a dc motor, etc. In addition, it is desirable that the control system works automatically without human intervention, like the cruise control of an automobile. An automatic control system involves feedback, i.e., it uses the observation of the output to affect the input to the plant (the system or the process under control), usually through a comparator that generates an error signal.

The feedback in beneficial in several ways: it permits precise control of the overall system gain and its time/frequency response; it makes the system robust against parameter variations; and, it can compensate for signal distortions and non-linearities, etc.

The feedback control system model considered in this chapter (Figure 4.1) includes a plant $G(s)$, an observation model $H(s)$, and a controller gain K that multiplies the error signal, obtained by comparing the feedback signal against a reference. The controller gain acts as the design parameter as it affects the closed-loop performance of the system.

The static gain controller K has its limitations. It it falls short in providing the desired system improvements, it may be replaced by, e.g., a PID controller (more on this in the next chapter).

The feedback controller design for a given system model is aimed to realize a set of objectives (the design specifications). Common design objectives include closed-loop system stability, transient response shaping, steady-state

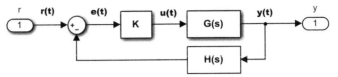

Figure 4.1 Feedback control system with plant $G(s)$, sensor $H(s)$, and static gain controller represented by K.

error reduction, and sensitivity reduction to model parameter variations. These objectives are discussed below.

4.1 Stability of the Closed-Loop System

Ensuring closed-loop stability is the first and foremost control system design objective. Even though the plant $G(s)$ is stable, the presence of feedback may cause the closed-loop system to become unstable.

To explore the stability of the feedback control system (Figure 4.1), we first determine the overall system transfer function from r to y, by considering the error signal:

$$e = r - Hy = r - KGHe.$$

Then, $y = KGe = \frac{KGr}{1+KGH}$.

The resulting closed-loop transfer function is given as:

$$T(s) = \frac{KG(s)}{1 + KGH(s)}.$$

The determination of stability is based on the closed-loop characteristic polynomial, given as:

$$\Delta(s) = 1 + KGH(s).$$

The $\Delta(s)$ is a polynomial of degree n, and is written as:

$$\Delta(s) = s^n + a_1 s^{n-1} + \cdots + a_{n-1}s + a_n.$$

Next, the stability of the above polynomial is determined by algebraic methods. A necessary condition for stability is that all coefficients of the polynomial are nonzero and have the same sign. The sufficient conditions for polynomial stability are obtained through the following criteria.

4.1.1 The Hurwitz Criterion

According to the Hurwitz's criterion, a polynomial is stable, i.e., it has its roots in the open left-half plane (OLHP), if and only if the following series of determinants have positive values:

$$|a_1|, \quad \begin{vmatrix} a_1 & a_3 \\ 1 & a_2 \end{vmatrix}, \quad \begin{vmatrix} a_1 & a_3 & a_5 \\ 1 & a_2 & a_4 \\ 0 & a_1 & a_3 \end{vmatrix}, \dots$$

4.1.2 The Routh's Criterion

An equivalent Routh's criterion requires construction of the following Routh's array:

$$
\begin{array}{c|ccc}
s^n & 1 & a_2 & \dots \\
s^{n-1} & a_1 & a_3 & \dots \\
\vdots & b_1 & b_2 & \dots \\
 & c_1 & c_2 & \dots \\
s^1 & \dots & & \\
s^0 & \dots & &
\end{array}
$$

The first two rows in the array are filled by alternating the coefficients of the characteristic polynomial. The elements appearing in the subsequent rows are computed as: $b_1 = -\dfrac{1}{a_1}\begin{vmatrix} 1 & a_3 \\ a_1 & a_2 \end{vmatrix}$, $b_2 = -\dfrac{1}{a_3}\begin{vmatrix} a_2 & a_4 \\ a_3 & a_5 \end{vmatrix}$, $c_1 = -\dfrac{1}{b_1}\begin{vmatrix} a_1 & a_3 \\ b_1 & b_2 \end{vmatrix}$, etc.

The Routh's criterion states that the number of unstable roots of the polynomial equals the number of sign changes in the first column of the array.

Low-order polynomials ($n = 2, 3$) are often encountered in model-based control system design. Using either criterion, the stability conditions for low-order polynomials are given as:

$$n = 2 : a_1 > 0, \ a_2 > 0$$

$$n = 3 : a_1 > 0, \quad a_2 > 0, \quad a_3 > 0, \quad a_1 a_2 - a_3 > 0.$$

These stability conditions are then used to determine the range of the controller gain K that ensures stability of the closed-loop system, i.e., to select those values of K that cause the roots of the closed-loop characteristic polynomial to lie in the OLHP.

The stability conditions on K are illustrated through the following examples.

Example 4.1: Let $\Delta(s) = s^2 + 2s + K$. Then, $\Delta(s)$ is stable for $K > 0$.

Example 4.2: Let $\Delta(s) = s^3 + 3s^2 + 2s + K$. Then, $\Delta(s)$ is stable for $K > 0$ and $6 - K > 0$ or $0 < K < 6$.

Example 4.3: PID controller for a DC motor.

Let $G(s) = \frac{\theta(s)}{V_a(s)} = \frac{10}{s(s+6)}$, $K(s) = k_\mathrm{p} + k_\mathrm{d}s + \frac{k_\mathrm{p}}{s}$. The motor is connected in a unity gain feedback configuration, i.e., $H(s) = 1$.

The closed-loop motor transfer function is given as: $\frac{y}{r} = \frac{K(s)G(s)}{1+K(s)G(s)}$, and the closed-loop characteristic polynomial is given as:

$$\Delta(s) = s^2(s+6) + 10(k_\mathrm{d}s^2 + k_\mathrm{p}s + k_i) = s^3 + (6+10k_\mathrm{d})s^2 + 10k_\mathrm{p}s + 10k_i.$$

The resulting constraints on the controller gains for stability of the third-order polynomial are given as:

$$k_\mathrm{p}, \; k_i, \; 6 + 10k_\mathrm{d} > 0$$

$$k_\mathrm{p}(6 + 10k_\mathrm{d}) - k_i > 0.$$

4.2 System Transient Response

The transient response of the closed-loop system is required to be dynamically stable, i.e., any output oscillations should die out with time. Additionally, the closed-loop system response should have adequate damping, so that these oscillations die out in a reasonably short time.

The transient response of a dynamic system includes the modes of system response that, in turn, depend on the roots of the characteristic equation; these roots can be real or complex. The system response modes are characterized as follows.

4.2.1 Modes of System Response

First, assume that the characteristic polynomial has real and distinct roots and can be factored as:

$$\Delta(s) = (s - s_1)(s - s_2)\dots(s - s_n).$$

Then, the system response modes are given as: $\left\{e^{s_1 t}, \; e^{s_1 t}, \; \dots\right\}$.

Next, assume that the characteristic polynomial has real and repeated roots, i.e., it contains factors of the form:

$$\Delta_m(s) = (s - s_1)^m$$

The contribution of the above factor to system modes is given as: $\left\{e^{s_1 t}, \; te^{s_1 t}, \; \dots, \; t^{m-1}e^{s_1 t}\right\}$.

In the case of complex roots, each quadratic factor with complex roots can be written as:

$$\Delta_c(s) = (s + \sigma)^2 + \omega^2.$$

The corresponding system modes are given as: $\{e^{-\sigma t}\cos \omega t, \ e^{-\sigma t}\sin \omega t\}$.

To generalize, assume that the system response modes are given as: $\phi_k(t), \ k = 1, \ldots, n$. Then, using arbitrary constants, c_k, the transient response of the closed-loop system is given as:

$$y(t) = \sum_{k=1}^{n} c_k \phi_k(t).$$

4.2.2 System Design Specifications

The design specifications for transient response are usually given in terms of one or more of the following parameters: rise time (t_r), settling time (t_s), damping ratio (ζ), percentage overshoot to a step input (%OS), etc (Figure 4.2).

The settling time is normally defined as the time when the system response settles to within 2% of its final value. The rise time can be defined in multiple ways, one being when the response first reaches 100% of its final value (applicable in the case when system step response has an overshoot). In order to relate the design specifications to the system response, we may consider a prototype second-order transfer function of the form:

$$T(s) = \frac{\omega_n^2}{s^2 + 2\zeta\omega_n s + \omega_n^2}.$$

The denominator polynomial in the above transfer function has complex roots at: $s = -\zeta\omega_n \pm j\omega_d$, where $\omega_d = \omega_n\sqrt{1 - \zeta^2}$.

For the prototype second-order system, we have:

$$t_s \cong \frac{4}{\zeta\omega_n} = \frac{4}{\text{Re}\,[s]}$$

$$\%OS = 100\,e^{-\pi\zeta\omega_n/\omega_d}.$$

A table of ζ vs. %OS is given as:

ζ	0.9	0.8	0.7	0.6	0.5
%OS	0.2	1.5	4.6	9.5	16.3

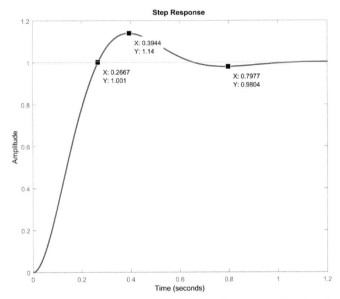

Figure 4.2   Step response of the prototype second-order system showing rise time (t_r), percentage overshoot (%OS) and settling time (t_s).

The rise time is related to the closed-loop system bandwidth ω_B via $\omega_B t_r \approx 1$. For the prototype second-order system, the rise time can also be approximated from: $\omega_n t_r \cong 3\zeta$ in the range $0.6 < \zeta < 0.8$.

Example 4.4: Let $G(s) = 89/(s^2 + 10s + 89)$. The system transfer function has poles at $-5 \pm j8$. The step response of the system (Figure 4.2) shows that for this system: $t_r \cong 0.267$s, $t_s \cong 0.8$s, and % OS $= 14\%$.

Example 4.5: Let $\Delta(s) = s^2 + 2s + K$; the objective is to choose K for $\zeta = 0.7$. Then, since $\zeta \omega_n = \zeta \sqrt{K} = 1$, $K = 2$. Note that in this case, $t_r \cong 1.5$ s and $t_s \cong 4$s.

Example 4.6: Let $\Delta(s) = s^2 + K_1 s + K$; choose K, K_1 for 5% OS ($\zeta = 0.7$), $t_s = 2$s.
   The desired characteristic polynomial for $\zeta = 0.7$, $\zeta \omega_n = 2$ is given as: $\Delta_{des}(s) = (s + 2)^2 + 2^2$. Accordingly, $K_1 = 4$, $K = 8$.

4.2.3 Performance Indices

An alternate way to characterize the transient response of the system is to define a performance index and choose the controller gain value that minimizes that index. Toward this end, three popular performance indices are defined as:

$$\text{IAE} : \int_0^{t_s} |e(t)|\, dt$$

$$\text{ISE} : \int_0^{t_s} |e(t)|^2 dt$$

$$\text{ITAE} : \int_0^{t_s} t\,|e(t)|\, dt.$$

Of these, the ITAE index is most popular. For a second-order system with a step input $(r(s) = \frac{1}{s})$, the ITAE index results in an optimal design with $\zeta = 0.7$ (%OS = 4.6%).

The optimum coefficients of the desired denominator polynomial of varying degree based on ITAE index can be computed; the first few coefficients are given below:

$$s + \omega_n$$
$$s^2 + 1.4\omega_n s + \omega_n^2$$
$$s^3 + 1.75\omega_n s^2 + 2.15\omega_n^2 s + \omega_n^3$$
$$s^4 + 2.1\omega_n s^3 + 3.4\omega_n^2 s^2 + 2.7\omega_n^3 s + \omega_n^3.$$

4.3 System Steady-State Response

The steady-state response of the feedback control system is required to have a low tracking error with respect to a constant (step) or linearly varying (ramp) reference input. In addition, we may require that the system response to a step disturbance stays small.

To characterize the tracking error, we consider a unity gain feedback system (i.e., $H(s) = 1$). In terms of the Laplace transform variables, the error $e(s)$ to a reference signal $r(s)$ is given as:

$$e(s) = r(s) - y(s) = r(s) - KGe(s).$$

Then, the expression for the output error is obtained as:

$$e(s) = \frac{1}{1 + KG(s)} r(s).$$

The steady-state error is computed using the final-value theorem (FVT), and is given as:

$$e(\infty) = \lim_{s \to 0} se(s).$$

4.3.1 Error Constants

Traditionally, the steady-state error is characterized in terms of system position and velocity error constants. These are defined as:

$$K_{\mathrm{p}} = \lim_{s \to 0} KG(s)$$

$$K = \lim_{s \to 0} sKG(s).$$

Using the above error constants, the steady-state error to a step $\left(r(s) = \frac{1}{s}\right)$ or a ramp input $\left(r(s) = \frac{1}{s^2}\right)$ is given as:

$$e(\infty)|_{\mathrm{step}} = \frac{1}{1 + K_{\mathrm{p}}}$$

$$e(\infty)|_{\mathrm{ramp}} = \frac{1}{K}.$$

We may note that the presence of an integrator in the loop forces the steady-state error to a step input to go to zero, as shown in the following example.

Example 4.7: Let $KG(s) = \frac{K}{s(s+2)}$.
Then, $K_{\mathrm{p}} = \infty$, $K = \frac{K}{2}$. Accordingly, $e(\infty)|_{\mathrm{step}} = 0$, $e(\infty)|_{\mathrm{ramp}} = \frac{2}{K}$.

4.3.2 Steady-State Error to Ramp Input

Assuming that the steady-state error to a step input is zero, the expression for steady-state error to a ramp input can be developed in an alternate way as follows. Let the error be expressed in terms of the closed-loop transfer function as:

$$e(s) = [1 - T(s)]\, r(s); \quad r(s) = \frac{1}{s^2}$$

Then, $e(\infty)|_{ramp} = \lim_{s \to 0} \frac{1 - T(s)}{s}$.

Using the L' Hospital's rule: $e(\infty)|_{ramp} = \lim_{s \to 0} -\frac{dT(s)}{ds}$. To compute the RHS, we use natural logarithm to write: $\frac{d}{ds} \ln T(s) = \frac{1}{T(s)} \frac{dT(s)}{ds}$.
So that, $\lim_{s \to 0} \frac{d}{ds} \ln T(s) = \lim_{s \to 0} \frac{1}{T(s)} \frac{dT(s)}{ds} = \lim_{s \to 0} \frac{dT(s)}{ds}$.
Next, assume that $T(s)$ is expressed as:

$$T(s) = \frac{K(s - z_1) \ldots (s - z_m)}{(s - p_1) \ldots (s - p_n)}.$$

Then,

$$\ln T(s) = \ln K + \sum_{\ln}(s - z_i) - \sum_{\ln}(s - p_i)$$

$$= \ln K + \sum \frac{1}{s - z_i} - \sum \frac{1}{s - p_i}.$$

It follows that

$$e(\infty)|_{\text{ramp}} = \sum \frac{1}{p_i} - \sum \frac{1}{z_i}$$

where z_i are the CL zeros (same as OL zeros), and p_i are the CL poles of the system. Thus, the steady-state error to a ramp input is independent of controller gain K.

The above expression is valid for the non-unity gain configurations as well.

4.4 Disturbance Rejection

Disturbance inputs are unavoidable in physical systems. Common examples of disturbance inputs are: road bumps while driving, turbulence in airplanes, and machinery vibrations in mechanical plants, etc.

In order to characterize the effects of disturbance inputs, let $r(t)$ denote the reference input, and $d(t)$ is a disturbance input (Figure 4.3); then, the output of the closed-loop system is given as:

$$y(s) = \frac{KG(s)}{1 + KGH(s)}r(s) + \frac{G(s)}{1 + KGH(s)}d(s)$$

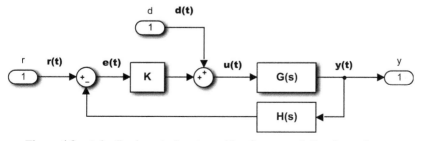

Figure 4.3   A feedback control system with reference and disturbance inputs.

In unity feedback configuration $(H(s) = 1)$, the output error is given as

$$e(s) = \frac{1}{1 + KG(s)}r(s) - \frac{G(s)}{1 + KG(s)}d(s).$$

The steady-state error to step input $(r(s) = 1/s)$ and/or step disturbance $(d(s) = 1/s)$ is computed as:

$$e(\infty) = \frac{1}{1 + K_p}r(\infty) - \frac{G(0)}{1 + K_p}d(\infty).$$

Thus, in order to reduce the effects of the disturbance input on the system output, the loop gain should be large over the frequency band of that input. The control system designer can affect it via choosing a large K.

However, a large loop gain may undermine closed-loop stability in the case of higher order plants. Hence, when using a static gain controller, a trade-off exists between ensuring the closed-loop stability, with adequate stability margins, and achieving adequate disturbance rejection. More complex controller models may be employed to ensure that competing design objectives are suitably met.

4.5 Robustness

The robustness of the closed-loop control system refers to its ability to withstand parameter variations and certain unmodeled dynamics in the plant transfer function, and still maintain stability and performance goals.

Robustness to parameter variations may be characterized in terms of the sensitivity of the closed-loop system gain $T(s)$ to variation in one or more of the plant parameters.

The system sensitivity function is defined as the ratio of the percentage change in the closed-loop system transfer function to the percentage change in the process transfer function.

$$S_G^T = \frac{\partial T/T}{\partial G/G} = \frac{\partial T}{\partial G}\frac{G}{T} = \frac{1}{1 + KG(s)}.$$

The sensitivity of the system transfer function to percentage change in a parameter value α is given as:

$$S_\alpha^T = S_G^T S_\alpha^G.$$

The sensitivity calculations are illustrated via the following example.

Example 4.8: Let $KG(s) = \frac{K}{s(s+a)}$, $H(s) = 1$.

Then, $T(s) = \frac{K}{s(s+a)+K}$; $S_G^T = \frac{s(s+a)}{s(s+a)+K}$; $S_a^G = -\frac{a}{s+a}$; $S_a^T = -\frac{as}{s(s+a)+K}$.

Note that a high loop gain (i.e., large K) reduces the sensitivity of $T(s)$ to a.

Skill Assessment Questions

Link to the answers:
http://www.riverpublishers.com/book_details.php?book_id=449

4.1 The characteristic polynomial of a closed-loop system is given as: $s(s+1)(s+2)+K=0$. Find the range of K for stability.

4.2 Use the proto-type second-order system to determine the desired characteristic polynomial for the following design objectives:

    a. $\omega_n = 1\frac{rad}{s}$; $\zeta = 0.8$
    b. $\omega_n = 1\frac{rad}{s}$; $\%OS = 5\%$
    c. $t_r = 0.5s$; $\zeta = 0.7$
    d. $t_s = 2s$; $\%OS = 5\%$

4.3 Find the steady-state error to a unit step input for the following cases. For (b) and (d), also find the steady-state error to a ramp input.

    a. $G(s) = \frac{1}{s+1}$
    b. $G(s) = \frac{1}{s(s+1)}$
    c. $G(s) = \frac{1}{(s+1)(s+2)}$
    d. $G(s) = \frac{s+1}{s(s+2)}$

4.4 Assume that a plant with $KG(s) = \frac{25K}{s(s+2)(s+25)}$ is connected in unity gain feedback configuration.

    a. Write the closed-loop characteristic equation and find the range of K for stability.

    b. Find the steady-state error to: (i) step input, (ii) step disturbance, (iii) ramp input.

4.5 Let $KG(s) = \frac{K(s+a)}{s(s+1)}$; find the sensitivity S_a^T of the closed-loop system to variations in a.

4.6 Consider the model of human postural dynamics described as an inverted pendulum, given as: $G(s) = 1/(s^2 - \Omega^2)$, where $\Omega^2 = ml/(I + ml^2)$. Find the sensitivity of the closed-loop transfer function to variations in m, I, and l.

5

Cascade Controller Models

Learning Objectives

1. Characterize the various controller models.
2. Characterize the three basic controller modes.

The traditional design of a single-input single-output (SISO) control system involves placing a controller $K(s)$ in cascade with the plant, $G(s)$, in the feedback loop (Figure 5.1). The controller suitably modifies the error signal, $e = r - Hy$, obtained from a comparator that monitors the output, $y(t)$, and compares it with a reference input, $r(t)$.

The elementary controller for the SISO case is a static gain K that affects the closed-loop characteristic polynomial and hence, the output behavior of the system. The static gain controller is simple and is effective in many a control problems. If, however, a simple gain controller does not meet the control objectives, then a dynamic controller $K(s)$ may be considered.

A dynamic controller is a dynamic system in its own right, and is described by either a transfer function or a state variable model. The input to the dynamic controller is the error signal, $e(t)$, and its output is $u(t)$, the input to the plant. These two controller types are described below.

5.1 The Static Controller

The static controller consists of a static gain K in the feedback loop that can affect the loop gain, and, hence the closed-loop response of the system. The gain K can be selected keeping in view the desired roots of the closed-loop characteristic polynomial that, in turn, determine the closed-loop system response.

The advantage of the static controller is its simplicity and ease of use. Its disadvantage is that it is limited in scope, i.e., it offers a limited choice

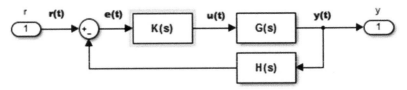

Figure 5.1   Feedback control system with dynamic controller $K(s)$, plant $G(s)$, and sensor $H(s)$.

of possible closed-loop root locations. Nevertheless, static controller is the first choice of the control systems designer in an effort to modify the plant behavior to meet the design specifications.

5.2 The Dynamic Controller

A dynamic controller is a dynamic system in its own right, represented by a transfer function, $K(s)$, that suitably modifies the error signal, thus affecting the overall loop transfer function, which is given as:

$$L(s) = KGH(s) = K(s)G(s)H(s).$$

Traditionally, the dynamic controllers are categorized as phase-lead, phase-lag, or lead–lag type. The lead and the lag terms here refer to the phase contribution from the controller to the Bode phase plot of the loop transfer function. The phase contribution is positive for phase-lead and negative for phase-lag.

Both phase-lead and phase-lag controllers are first-order controllers, i.e., they add a single pole to the loop transfer function, and can be described by their common transfer function description, given as:

$$K(s) = \frac{K(s + z_c)}{s + p_c}.$$

The phase-lead is characterized by $z_c < p_c$, and the phase-lag by $z_c > p_c$.

A lead–lag controller a second order controller that combines a phase-lead section with a phase-lag section and has the form:

$$K(s) = \frac{K(s + z_1)(s + z_2)}{(s + p_1)(s + p_2)}.$$

For example, a lead–lag controller may be given as:

$$K(s) = \frac{K(s + 0.1)(s + 10)}{(s + 1)^2}.$$

5.3 The PID Controller

The PID (Proportional-integral-derivate) controller is a general-purpose controller that combines the three basic modes of control, i.e., the proportional (P), the derivative (D), and the integral (I) modes. The PID controller with input e and output u is described by the equations:

$$u(t) = k_{\mathrm{p}} + k_{\mathrm{d}}\frac{d}{dt}e(t) + k_{\mathrm{i}}\int e(t)\mathrm{d}t$$

or, $u(s) = k_{\mathrm{p}} + K_{\mathrm{d}}s + k_{\mathrm{i}}/s$.

The gains, k_{p}, k_{d}, and k_{i}, in the above equations represent the controller gains for the three basic control modes (Figure 5.2). Of these, the proportional mode affects the roots of the closed-loop characteristic polynomial. The derivative and the integral mode, respectively, help improve the transient and the steady-state response of the system.

The control system design objectives may require using only a subset of the controller modes. These choices include the proportional control, proportional and derivative (PD) control, and proportional and integral (PI) control. A proportional controller is the same as the static gain controller. The remaining cases are described below.

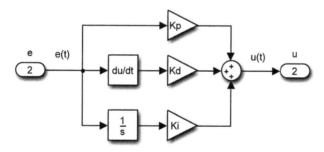

Figure 5.2   Three basic control modes represented in the PID controller.

5.3.1 Proportional–Derivative (PD)

A PD controller is described by the transfer function:

$$K(s) = k_{\mathrm{p}} + k_{\mathrm{d}}s = k_{\mathrm{d}}\left(s + \frac{k_{\mathrm{p}}}{k_{\mathrm{d}}}\right).$$

The addition of the PD controller thus adds a single zero to the loop transfer function. A pole farther (4 or 5 times) to the left of the controller zero may be added to suppress the high-frequency noise, whereby the controller is described as:

$$K(s) = \frac{(k_p + k_d s)}{(s + p_c)}, \quad p_c \gg k_p / k_d.$$

The resulting controller is equivalent to the phase-lead controller, given as:

$$K(s) = \frac{K(s + z_c)}{(s + p_c)}, \quad p_c > z_c.$$

The PD controller, similar to the phase-lead controller, improves the transient response of the system.

5.3.2 Proportional–Integral (PI)

A PI controller is described by the transfer function:

$$K(s) = k_p + \frac{k_i}{s} = \frac{k_p(s + k_i / k_p)}{s}.$$

The PI controller adds a pole at the origin (an integrator) together with a finite zero, placed close to the origin in the s-plane, to the loop transfer function. The PI controller is commonly used in designing tracking systems and servomechanisms, as the presence of the integrator in the loop forces the steady-state error to a step input to go to zero.

5.3.3 Proportional–Integral–Derivative (PID)

The PID controller is a general-purpose controller that imparts both transient and steady-state response improvements to the system output. Further, it provides stability as well as robustness to the system.

The PID controller is described by the transfer function:

$$K(s) = k_p + k_d s + \frac{k_i}{s} = \frac{k_d s^2 + k_p s + k_i}{s}.$$

The PID controller includes both PD and PI parts; it thus adds two zeros and an integrator pole to the loop transfer function.

The zero due to the PI part is normally placed close to the origin; the zero due to the PD part is placed at a suitable location for transient response improvement.

For noise suppression, a second pole may be added to the derivative term resulting in the modified controller transfer function, given as:

$$K(s) = k_{\mathrm{p}} + \frac{k_{\mathrm{i}}}{s} + \frac{k_{\mathrm{d}}s}{T_f s + 1}$$

In the MATLAB Control System Toolbox, a PID controller object is created via the following command:

```
>> C=pid(kp,ki,kd,Tf)
```

As an example, let $k_p = k_i = k_d = 1$, $T_f = 0.1$; then, the resulting PID controller object is created as:

```
>> C=pid(1,1,1,.1);
>> zpk(C)
```

5.3.4 PID Controller Tuning

PID controllers are commonly employed in the industrial settings for process control applications. Industrial process controllers can be manually tuned using control knobs provided for this purpose. Empirical tuning rules (e.g., Ziegler–Nicholas rules, see references) may be used for controller tuning.

PID controller tuning typically has the following objectives:

1. Stability of the closed-loop system.
2. Adequate performance in terms of both reference tracking and distur-bance suppression.
3. Adequate robustness, as typically measured by the stability margins (gain and phase margins).

The MATLAB provides the following command for PID controller tuning:

```
>> C=pidtune(sys,C0);
```

In the above, `sys` is a SISO model represented by the dynamic system object, created in the MATLAB Control System Toolbox, and `C0` is a base controller. The MATLAB controller tuning aims for a $60°$ phase margin that results in a moderate 8–10% overshoot to a step command.

Example 5.1: PID controller for the DC motor model.

For a small DC motor let: $R = 1\Omega$, $L = 1mH$, $J = 0.01kg \cdot m^2$, $b = 0.1 \frac{N \cdot s}{rad}$, $k_t = k_b = 0.05$; then the transfer function of the motor is given as:

$$\frac{w(s)}{V_{\mathrm{a}}(s)} = \frac{5000}{(s+1000)(s+10)+250}.$$

We will use MATLAB Control System Toolbox to design a PID controller for the DC motor model, and plot the step response of the closed-loop system. The MATLAB commands are given as follows:

```
>> G=tf(5000,[1 1010 10250]);
>> C0=pid(1,1,1,.1);
>> C=pidtune(G,C0)
>> T=feedback(C*G,1)
>> step(T)
```

The resulting step response of the PID-controlled DC motor is shown below (Figure 5.3). We may make the following observations from the step response plot:

1. The step response has about 8% overshoot.
2. The response settles in about 0.4 s.
3. The steady-state error of the step response is zero.

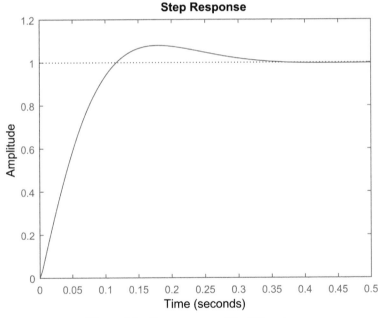

Figure 5.3   Step response of the DC motor.

Skill Assessment Questions

Link to the answers:
http://www.riverpublishers.com/book_details.php?book_id=449

5.1 The simplified model of a DC motor with voltage input and shaft angle output is described by the following equations:

$$e_a = R_a i_a + K_b \omega; \quad K_t i_a = J\dot{\omega} + D\omega; \quad \omega = \dot{\theta}. \text{ Let } R_a = 1,$$
$$J = 0.01, \ D = 0.05, \ K_t = K_b = 0.1.$$

The motor is connected in cascade with a controller $K(s)$ in a unity gain feedback configuration. Obtain the closed-loop characteristic polynomial for the following controller configurations:

a. $K(s) = K$

b. $K(s) = K(s + 10)$

c. $K(s) = K \left(5 + \dfrac{1}{s} \right)$

d. $K(s) = K \left(5 + s + \dfrac{1}{s} \right)$

e. $K(s) = K \left(\dfrac{s + 10}{s + 1} \right)$

f. $K(s) = K \left(\dfrac{s + 1}{s + 10} \right)$

5.2 Consider the motor model in Question 5.1 above. Use MATLAB to design and tune a PID controller for the motor. Plot the step response of the closed-loop system.

6

Control System Design with Root Locus

Learning Objectives

1. Sketch the root locus for a given transfer function model of the plant.
2. Design a static gain controller using the root locus technique.
3. Design a dynamic controller for transient and/or steady-state response improvement.
4. Design a rate feedback compensator through root locus technique.
5. Realize the controller design using operational amplifier circuits.

Beginning with this chapter, we discuss the techniques for designing cascade controllers for feedback control systems.

The control system is designed for close-loop stability and the desired transient and/or steady-state response improvements. The various controller design techniques include the root locus (RL) technique, frequency-domain design, and the state-space design. This chapter focuses on the design of control systems using the RL technique.

The organization of the feedback control system for the RL design (Figure 6.1) includes the process to be controlled (the plant), a sensor for observation of the output, and a controller. The plant and the sensor are described in terms of their transfer function models, $G(s)$ and $H(s)$, respectively. The controller is represented by a static gain K that multiplies the error signal.

We initially focus on the static controller design using root locus. The design technique is later extended to dynamic controllers of phase-lead, phase-lag, lead–lag, PD, PI, and PID types. With the availability of a rate sensors (rate gyroscope and tachometer), rate feedback design, which includes the design of a minor (inner) feedback loop, may be considered.

The controllers designed for process improvement may be implemented through electronic circuits built with operational amplifiers. The controller realization is discussed toward the end of the chapter.

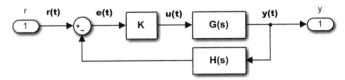

Figure 6.1   Feedback control system with plant $G(s)$, sensor $H(s)$, and a static gain controller represented by K.

6.1 The Root Locus

The root locus (RL) is a graph of the resulting root locations of the closed-loop characteristic polynomial with variation in the controller gain K. The RL is used as a graphical technique for selecting the controller gain in the case of single-input single-output (SISO) type plants.

In order to develop the RL plot, we consider a feedback control system with a static gain controller (Figure 6.1), where $G(s)$ is the plant transfer function, and $H(s)$ is the sensor transfer function. We will assume $H(s) = 1$ unless otherwise specified.

The loop transfer function including the controller, the plant, and the sensor is given as $KGH(s)$; then, the characteristic equation of the closed-loop system is given as:

$$\Delta(s) = 1 + KGH(s) = 0.$$

The roots of $\Delta(s)$ can be plotted in the complex plane for the various values of gain K. The locus of all such root locations, as K varies from $0 \to \infty$, constitutes the RL for the system.

A generalized RL constitutes the loci of all root locations for $K \in (-\infty, \infty)$, and is obtained by extending the RL to the negative values of K.

Example 6.1: Let $G(s) = \frac{1}{s}$, $H(s) = 1$. The characteristic equation of the closed-loop system is given as:

$$\Delta(s) = s + K.$$

This equation has a single root at $s_1 = -K$. Consequently, the RL, as K varies from 0 to ∞, traces a line that proceeds from the origin along the negative real-axis to infinity.

Example 6.2: Let $G(s) = \frac{1}{s(s+2)}$, $H(s) = 1$. Then, the characteristic equation of the closed-loop system is given as:

$$\Delta(s) = s^2 + 2s + K.$$

The roots of the characteristic equation are real for $K \leq 1$, and complex for $K > 1$. The complex roots are given as: $s_{1,2} = -1 \pm j\sqrt{K-1}$. These roots are tabulated for some values of K below.

K	0	0.5	1.0	1.5	2.0	2.5	3.0
s_1, s_2	$0, -2$	$-0.29, -1.71$	$-1, -1$	$-1 \pm j0.71$	-1 ± 1	-1 ± 1.22	-1 ± 1.41

The loci of these roots, as K varies from $0 \to \infty$, comprises two branches that commence at the open-loop (OL) poles $\{0, -1\}$, meet in the middle at $\sigma = -0.5$, then split up and extend to ∞ along the $\sigma = -1$ line in both upward and downward directions. These directions are called the RL asymptotes.

6.1.1 Root Locus Rules

All RL plots share some common properties, also referred as the RL rules. The prominent ones are described below:

1. The RL has n branches, where n equals the order of the denominator in the loop transfer function $KGH(s)$. These branches commence at the OL poles of the system (for $K = 0$). Of these, m branches terminate at OL zeros (as $K \to \infty$). The remaining $n - m$ branches follow the RL asymptotes (as $K \to \infty$).
2. In the case of generalized RL, the RL branches for $K < 0$ proceed from infinity and/or OL zeros toward the OL poles.
3. Root Locus is symmetric with respect to the real-axis (this is a consequence of the fact that for physical systems complex roots occur in conjugate pairs).
4. The real-axis locus lies to the left of an odd number of poles and zeros (for $K > 0$), and to the right of an odd number of poles and zeros (for $K < 0$).
5. The RL asymptotes are given at angles: $\phi_a = \frac{2k+1}{n-m}(180°)$, $k = 0, 1, \ldots, n-m-1$, (for $K > 0$), and at angles: $\phi_a = \frac{2k+1}{n-m}(360°)$, $k = 0, 1, \ldots, n-m-1$ (for $K < 0$). These asymptotes intersect at a common point on the real-axis, given as: $\sigma_a = \frac{\sum p_i - \sum z_i}{n-m}$.
6. Real-axis locus in between a pair of poles contains a break-away point where the two RL branches split. Real-axis locus in between a pair of zeros likewise contains a break-in point where the two branches approaching from opposite directions in the complex plain join. Both break-away and break-in points for $K > 0$ as well as $K < 0$ are found among the solutions to the equation: $\sum \frac{1}{\sigma - p_i} - \sum \frac{1}{\sigma - z_i} = 0$.

The above rules suffice to sketch the RL by hand for low-order plants. Additional rules to further refine the RL plot, including the angles of departure/arrival at complex poles/zeros, can be defined (see references).

The MATLAB Control Systems Toolbox provides the 'rlocus' command to plot the root locus, which takes a Dynamic System Object as an input. The latter is defined via the 'tf', 'zpk', or 'ss' commands.

The following examples illustrate the sketching of RL for the given loop transfer functions.

Example 6.3: Let $KGH(s) = \frac{K(s+3)}{s(s+2)}$; then, the RL plot is sketched using the following steps:

1. The RL has $n = 2$ branches and $n - m = 1$ asymptote. The RL branches start at the OL poles (at $0, -2$ for $K = 0$). One branch terminates at the zero at $s = -3$; the other follows the asymptote as $K \to \infty$.
2. The asymptote angle is given as: $\pm 180°$ (for $K > 0$); with an inter-section at: $\sigma_a = 1$.
3. The real-axis locus lies in the intervals: $\sigma \in [0, 2] \cup [3, \infty)$ (for $K > 0$).
4. The break-away/break-in points are given as the solution to: $\frac{1}{\sigma} + \frac{1}{\sigma+2} = \frac{1}{\sigma+3}$; the equation has two solutions at: $\sigma = -1.27$ (break-away) and -4.73 (break-in).

The resulting RL plot (for $K > 0$) can be sketched by hand. A plot, generated with the following MATLAB commands, is shown in Figure 6.2.

```
>> rlocus(zpk([-3],[0 -2],1))
>> axis('equal')
```

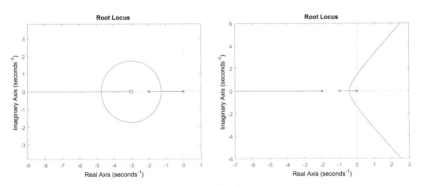

Figure 6.2   The root locus plot for $G(s) = \frac{(s+3)}{s(s+2)}$ (*left*); the root locus plot for $G(s) = \frac{1}{s(s+1)(s+2)}$ (*right*).

Example 6.4: Let $KGH(s) = \frac{K}{s(s+1)(s+2)}$; then, the RL plot is sketched using the following steps:

1. The RL has three branches and three asymptotes. The RL branches start at the OL poles (for $K = 0$) and follow the asymptotes as $K \to \infty$.
2. The asymptote angles are given as: $\pm 60°$, $180°$ (for $K > 0$) and their common intersection point is $\sigma_a = -\frac{3}{3} = -1$.
3. The real-axis locus lies along $\sigma \in [0, 1]$ (for $K > 0$).
4. The break-away/break-in point is given as the solution to: $\frac{1}{\sigma} + \frac{1}{\sigma+1} + \frac{1}{\sigma+2} = 0$, which reduces to: $3\sigma^2 + 6\sigma + 2 = 0$, and has solutions at $\sigma = -0.38, -2.62$. The first one defines the break-away point for $K > 0$; the second one defines the break-in point for $K < 0$.

The resulting RL plot (for $K > 0$) can be sketched by hand. A plot, generated with the following MATLAB command is shown in Figure 6.2.

```
>> rlocus(zpk([],[0 -1 -2],1))
```

In the event, the RL branches cross the $j\omega$-axis, the controller gain K at the stability boundary determines the maximum permissible value of K for closed-loop stability and can be determined from the polynomial stability conditions or via the Routh's array.

For the above example, the stability condition for the third-order polynomial: $\Delta(s) = s^3 + 3s^2 + 2s + K$ reduces to: $6 - K = 0$. Thus, the range of K for stability is: $0 < K < 6$. The roots of $\Delta(s)$ for $K = 6$ are given as: $s = -3, \pm j\sqrt{2}$, i.e., the $j\omega$-axis crossing occurs at $\pm j\sqrt{2}$, and the closed-loop system for $K = 6$ will oscillate at the frequency $\sqrt{2}$ rad/sec.

6.2 Static Controller Design

The RL plot for a given loop transfer function $KGH(s)$ describes the roots of the closed-loop characteristic equation, $\Delta(s) = 1 + KGH(s) = 0$, as the controller gain K varies from $0 \to \infty$.

Assuming that the RL contains the desired closed-loop root location, s_1, the corresponding controller gain K can be obtained from the MATLAB generated RL plot by simply clicking on it. The dialog box that appears provides additional information on the damping of the closed-loop roots, and the step overshoot of the closed-loop transfer function for the choice of K (see Figures 6.4 and 6.5 below).

In order to analytically characterize the controller gain K, assume that the loop transfer function $KGH(s)$ is given in the factored form as:

$$KGH(s) = \frac{K(s - z_1) \ldots (s - z_m)}{(s - p_1) \ldots (s - p_n)}; \quad n > m.$$

For a point s_1 to lie on the RL, and hence constitute a root of the characteristic equation, it must satisfy:

$$KGH(s_1) = \frac{K(s_1 - z_1) \ldots (s_1 - z_m)}{(s_1 - p_1) \ldots (s_1 - p_n)} = -1$$

The above relation defines two conditions, the magnitude and the angle conditions, that are given as:

Magnitude condition: $|KGH(s_1)| = \frac{K|s_1 - z_1| \ldots |s_1 - z_m|}{|s_1 - p_1| \ldots |s_1 - p_n|} = 1.$

Angle condition: $\angle KGH(s_1) = \angle(s_1 - z_1) + \cdots + \angle(s_1 - z_m)$
$- \angle(s_1 - p_1) - \cdots - \angle(s_1 - p_n) = \pm 180°.$

From the magnitude conditions, $|KGH(s_1)| = 1$, the controller gain K for a desired root location s_1 is given as: $K = \frac{|s_1 - p_1| \ldots |s_1 - p_n|}{|s_1 - z_1| \ldots |s_1 - z_m|}.$

From the angle condition, let $\theta_{z_i}, \theta_{p_i}$ denote the angles from the plant zeros and poles to the point s_1 on the RL plot, then $\sum \theta_{z_i} - \sum \theta_{p_i} = \pm 180°$ (for $K > 0$), and $\sum \theta_{z_i} - \sum \theta_{p_i} = 0°$ (for $K < 0$).

Example 6.5: Let $G(s) = \frac{1}{s(s+1)(s+2)}$, and assume that the only design requirement is to have $\zeta = 0.7$; the, we proceed with the controller design factored follows:

The closed-loop (CL) characteristic polynomial is given as:

$$\Delta(s) = s(s + 1)(s + 2) + K$$

The closed-loop (CL) characteristic polynomial is stable for $0 < K < 6$. From the RL plot (Figure 6.2 above), we may choose, e.g., $K = 0.65$ to satisfy the design condition. The resulting characteristic polynomial is factored as:

$$\Delta(s) = (s + 2.235)(s^2 + 0.765s + 0.291),$$

which has dominant CL roots at -0.38 ± 0.38 ($\zeta = 0.7$).

In the event that the RL plot does not pass through the desired root location s_1, the angle condition is used to add poles or zeros to the loop transfer function, and hence define a dynamic controller, that would force the RL to pass through s_1. The design of the dynamic controller is described below after considering the controller design specifications.

6.3 Controller Design Specifications

The controller design specifications for desired closed-loop system response are normally stated with reference to the step response of a prototype second-order system transfer function, given as:

$$T(s) = \frac{y}{r} = \frac{\omega_n^2}{s^2 + 2\zeta\omega_n s + \omega_n^2}.$$

Typical specifications include an acceptable overshoot (e.g., $OS < 5\%$) that translates into a range for the damping ratio (e.g., $0.7 < \zeta < 1$) of the dominant CL roots, and an acceptable settling time t_s, where $t_s \cong \frac{4}{\mathrm{Re}[s]}$.

These specifications are used to identify a region in the complex s-plane where the closed-loop roots should be placed. In particular, the settling time constraint defines a vertical line at: $\sigma = -\frac{4}{t_s}$, where the acceptable region lies to the left of the line. The damping ratio constraint defines radial lines at: $\theta = \pm\cos^{-1}\zeta$, where the acceptable region lies in between the lines.

As an example, assume that the design specifications are: $0.7 < \zeta < 1$ and $t_s < 2s$. Then, the s-plane region for the acceptable closed-loop root locations is bounded by the lines: $\sigma = -2$ and $\theta = \pm 45°$ (Figure 6.3).

In the event that the existing RL branches do not pass through the region containing the desired root locations, the situation calls for the design of a dynamic controller that is described next.

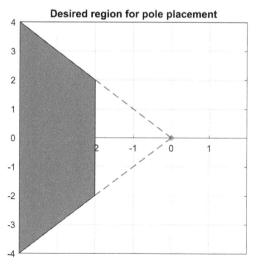

Figure 6.3 The desired region for closed-loop pole placement for $0.7 < \zeta < 1$ and $t_s < 2s$.

6.4 Dynamic Controller Design

The dynamic controller $K(s)$ (also termed as the dynamic compensator) adds one or more poles and zeros to the loop transfer function; the modified loop transfer function is given as: $K(s)G(s)H(s)$. The effect of adding poles and/or zeros to the loop transfer function can be visualized through changes to the RL plot.

The addition of a finite zero to the loop transfer function causes the RL branches to bend towards it. Hence, placing a zero to the left of the existing RL branches (as in the case of PD controller) adds dynamic stability to the closed-loop system by pulling the RL toward left.

The addition of a finite pole to the loop transfer function repels the RL branches that are close to it. The pole location is selected at the origin (as in the case of PI compensator), or close by (as in the case of phase-lag compensator), in order not to appreciably disturb the existing RL plot or compromise the existing range of K for stability.

6.4.1 Transient Response Improvement

The dynamic controller designed for transient response improvement forces the RL plot to pass through a desired closed-loop root location in the complex plane. The desired root location is selected keeping in mind the transient response specifications, as illustrated via the following example:

Example 6.6: Let $G(s) = \frac{1}{s(s+2)}$; and assume that the design requirements are to have: $\zeta = 0.7, \quad t_{\mathrm{s}} = 2\mathrm{s}$.

We first try the static gain controller. For a static controller, the CL characteristic polynomial is given as:

$$\Delta(s) = s^2 + 2s + K$$

The resulting CL roots are at: $s_{1,2} = -1 \pm \sqrt{4 - K}$, which become complex for $K > 4$.

The RL plot has two branches that break-away from the real-axis at $s = -1$, and extend along the $s = -1$ line to $-1 \pm j\infty$.

Using the static gain controller, the settling time for the CL system (for $K > 4$) is fixed at: $t_s = \frac{4}{|\mathrm{Re}(s)|} = 4\,\mathrm{s}$, i.e., the static gain controller is inadequate, and in order to realize the desired, $t_{\mathrm{s}} = 2s$, we need a dynamic (PD or phase-lead) controller.

In order to design a PD controller to meet the specifications, we define a desired root location for the closed-loop characteristic polynomial as:

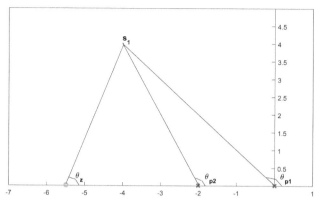

Figure 6.4 Application of the angle criteria at the desired root location s_1: $\theta_z - \theta_{p1} - \theta_{p2} = \pm 180°$.

$s_1 = -4 \pm j4$. Since the existing RL does not pass through this location, we add a zero to the loop transfer function in orderto bend the RL toward s_1.

The controller zero location is selected from the angle condition: $\sum \theta_{z_i} - \sum \theta_{p_i} = \pm 180°$.

Applying the angle condition (Figure 6.4), the angle subtended by the controller zero at the desired pole location is given as: $\theta_z - 116.6° - 135° = -180°$ or $\theta_z = 71.6°$. This translates into a real-axis zero location at -5.33; the resulting controller transfer function is given as: $K(s) = s + 5.33$. The modified loop transfer function is given as:

$$KGH(s) = \frac{K(s + 5.33)}{s(s + 2)}.$$

The closed-loop characteristic polynomial with the addition of the PD controller becomes: $\Delta(s) = s(s + 2) + K(s + 5.33)$. Indeed, for $K = 6$, the characteristic polynomial is given as: $\Delta(s) = s^2 + 8s + 32$, has dominant roots at the desired pole location $(-4 \pm j4)$ (Figure 6.5).

For noise suppression and/or performance improvement, it is advantageous to replace the PD controller with a phase-lead controller (a first-order high pass filter). This is accomplished by adding a pole located far to the left of the zero location (Figure 6.5).

Accordingly, let $K(s) = \frac{K(s+5)}{s+30}$. The resulting closed-loop characteristic polynomial is given as:

$$\Delta(s) = s^3 + 32s^2 + (60 + K)s + 5K.$$

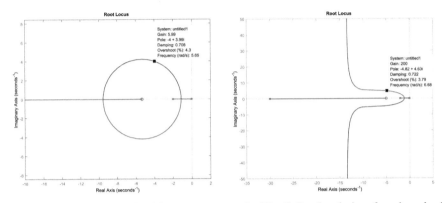

Figure 6.5 The design for PD compensator via RL (*left*); the design for phase-lead compensator (*right*).

Then, for $K = 200$, the equation has dominant CL roots at $-4.8 \pm j4.6(\zeta = 0.72)$ that satisfy the design specifications.

The MATLAB commands for the above PD and phase-lead designs are given below (Figure 6.5):

```
>> rlocus(zpk([-5.33],[0 -2],1))
>> rlocus(zpk([-5],[0 -2 -30],1))
```

Example 6.7: Let $G(s) = \frac{1}{s(s+2)(s+5)}$; and assume that the design specifications are: $\%OS \le 10\%$, $t_s \le 2$s.

A RL plot of $KG(s)$ for $K = 8.8$ shows the dominant closed-loop roots located at: $= -0.76 \pm j1.01$ ($\zeta = 0.6$, $OS \cong 10\%$), with a settling time of $t_s = 5.3$s. Thus, the static gain compensator is inadequate.

In order to meet the design specifications, let the desired dominant CL root locations be given at: $s = -2 \pm j2.5$ ($\zeta = 0.62$).

We next consider a phase-lead compensator of the form: $K(s) = \frac{K(s+z_c)}{s+p_c}$. We use the angle condition to determine the phase contribution from the controller as: $\theta_z - \theta_p = -180° + 129° + 90° + 40° \cong 79°$.

In order to design the compensator, we may arbitrarily choose a zero location: $z_c = 2$, and use the angle condition to compute $p_c \cong 15$, to obtain: $K(s) = \frac{K(s+2)}{s+15}$.

The resulting RL plot of $K(s)G(s)$, for $K = 165$, has dominant closed-loop roots at $s = -2.03 \pm j2.49$ ($\zeta = 0.63$). The closed-loop characteristic polynomial (CLCP) is given as: $s^3 + 20s^2 + 75s + 165$.

The MATLAB commands for this example are given below (Figure 6.6):

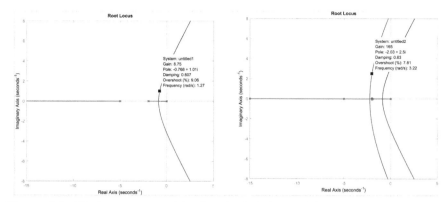

Figure 6.6 RL improvement with phase-lead design: original (*left*) and phase-lead compensated (*right*).

```
>> rlocus(zpk([],[0 -2 -5],1)), hold
>> rlocus(zpk([-2],[0 -2 -5 -15],1))
```

We note from the above example that in the case of phase-lead design, we may arbitrarily choose a location for either z_c or p_c, and then use the angle condition to determine the other.

In the above example, we chose the compensator zero to be coincident with the plant pole purely for convenience. Other zero locations are possible. However, in order to have dominant complex poles with approximately the required ζ, the compensator zero should be placed at or to the left of the plant pole at $s = -2$. In fact, the angle condition leaves little room for moving the zero around. In this case, the compensator zero may be selected in the interval: $[-2.4, -2]$.

We may also note that the projected pole-zero cancellation in the above design is only good on paper. It may not work in practice due to the component tolerances that would affect the zero location. Nonetheless, the contribution from a closed-loop pole that is located close to an open-loop zero toward the natural response of the system will be small.

6.4.2 Steady-State Error Improvement

The steady-state error characterizes the long-term behavior of the closed-loop system when a constant step input or a linearly varying ramp input is applied.

For a given plant, $G(s)$, the steady-state error to a step or ramp input is computed using the position and velocity error constants. These constants are defined in terms of the low-frequency loop gain as:

$$K_\mathrm{p} = \lim_{s \to 0} KG(s)$$

$$K_\mathrm{v} = \lim_{s \to 0} sKG(s)$$

In the case of unity gain feedback configuration, the steady-state errors to step and ramp inputs are computed as:

$$e(\infty)|_\mathrm{step} = \frac{1}{1 + K_\mathrm{p}}$$

$$e(\infty)|_\mathrm{ramp} = \frac{1}{K_\mathrm{v}}.$$

The compensator design for steady-state error improvement aims to boost K_p and/or K_v through the addition of a PI or phase-lag controller. Both compensators add a pole-zero pair to the loop transfer function.

For example, the phase-lag controller transfer function is given as:

$$K(s) = \frac{K(s + z_\mathrm{c})}{(s + p_\mathrm{c})}, \quad p_\mathrm{c} < z_\mathrm{c} \ll 1.$$

The steady-state response improvement offered by the phase-lag controller is described in terms of the position error constant, which is defined as: $K_\mathrm{p} = \lim_{s \to 0} KGH(s)$, and is modified with the addition of the phase-lag controller to: $K'_\mathrm{p} = (z_\mathrm{c}/p_\mathrm{c})K_\mathrm{p} > K_\mathrm{p}$.

The PI controller adds an integrator to the loop. Thus, we have $p_\mathrm{c} = 0$ and $K'_\mathrm{p} = \infty$; hence $e(\infty)|_\mathrm{step} = 0$.

The effect of adding a PI or phase-lag controller to the loop transfer function can be assessed on the RL plot. Assume that the controller pole and zero are located at $-p_\mathrm{c}$ and $-z_\mathrm{c}$, respectively, in such a way that $z_\mathrm{c}, p_\mathrm{c} \ll -\mathrm{Re}[s_1]$, where s_1 denotes the desired closed-loop pole location as identified on the existing RL plot.

Based on the angle condition, the angle contributed by the compensator pole-zero pair is small, i.e., $\theta_z - \theta_\mathrm{p} \approx 0°$. This is done in order not to appreciably affect the existing RL branch that is assumed to pass through the desired pole location s_1.

We may note that the closed-loop pole added close to the origin from the controller introduces a system response mode with a long settling time; however, the contribution of this mode to the overall system response will be small due to the close proximity of the controller zero.

Example 6.8: Let $G(s) = \frac{1}{s(s+2)(s+5)}$; and assume that the design specifications are: $\zeta \geq 0.6$, $e_\mathrm{ss}|_\mathrm{ramp} \leq 0.1$.

We first note that the transient response requirement can be met with a static compensator ($K = 8$), which results in the closed-loop roots at: $s = -0.78 \pm j92$ ($\zeta = 0.65$). However, since $KG(s) = \frac{K}{s(s+2)(s+5)}$, we have a velocity error constant, $K_v = K/10$, and for $K = 8$, we have $K_v = 0.8$.

Since we desire to increase K_v to 10; we may consider the following phase-lag compensator, designed to increase the low frequency gain by a factor of 12.5: $K(s) = \frac{K(s+.1)}{s+0.008}$.

The compensated system has an improved velocity error constant: $K_v = \lim_{s\to 0} sK(s)G(s) = 1.25 \, K$.

Then, for $K = 8$, we have $K_v = 10$, with $e_{ss}|_{\text{ramp}} \cong 0.1$.

The resulting CL roots are given as: $-5.43, \ -0.79 \pm j0.93$ ($\zeta \cong 0.65$).

The MATLAB commands for the phase-lag design are given below (Figure 6.7):

```
>> rlocus(zpk([],[0 -2 -5],1)), hold
>> rlocus(zpk([-.1],[0 -2 -5 -.008],1))
```

6.4.3 Lead–Lag and PID Designs

If the design specifications require simultaneous transient response and steady-state error improvements, then a lead–lag or a PID controller may be considered.

The lead–lag compensator includes a cascade of a lead stage and a lag stage. The PID compensator effectively combines a PD and a PI compensator.

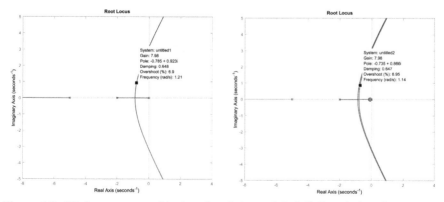

Figure 6.7 RL improvement with phase-lag design: original (*left*) and phase-lag compensated (*right*).

The steps for a lead–lag/PID design are as follows:

1. Compute the desired dominant pole location (s_i) from the transient response specifications.
2. Plot the RL for the loop transfer function to determine if a static gain compensator will suffice.
3. Design a phase-lead or PD compensator that will make the compensated RL pass through (s_i). Plot the RL for the compensated system and verify it passes through the desired location (s_i).
4. Check the RL gain at (s_i) and the relevant error constant to determine if steady-state error specifications are met. If needed, design a phase-lag or PI section to boost the error constant.
5. Plot the RL for $K(s)G(s)$ and verify that it passes through the desired location (s_i). Also check the RL gain K and verify that error constant requirements are met.

Example 6.9: Let $G(s) = \frac{1}{s(s+2)(s+5)}$; and assume that the controller design specifications are: $\%OS \leq 10\%$, $t_s \leq 2s$, $e_{ss}|_{ramp} \leq 0.1$. The steps for a lead–lag design are given as follows (Figure 6.7):

1. A RL plot of $KG(s)$ shows dominant closed-loop roots at $s = -0.76 \pm j1.01 (\zeta = 0.6)$ with $t_s = 5.3s$. Thus, static gain compensator will not suffice.
2. Let the desired root location be at: $s = -2.2 \pm j2.5$ ($\zeta = 0.65$). Use the angle condition to obtain: $\theta_z - \theta_p = -180° + 131° + 95° + 42° = 88°$.
3. Phase-lead Design: let $z_c = 2$, $p_c = 20$, $K_{lead}(s) = \frac{K(s+2)}{s+20}$. Note that this is a 'ballpark' design that approximately satisfies the angle condition above.
4. The RL plot of $K_{lead}(s)G(s)$ has, for $K = 220$, dominant CL roots at: $s = -2.15 \pm j2.49$ ($\zeta = 0.65$), and for the lead compensated system, $K_v = 2.25$.
5. Phase-lag design: we wish to improve K_v by a factor of 4.5; accordingly, let $z_c = 0.09$, $p_c = 0.02$, $K_{lag}(s) = \frac{K(s+0.09)}{s+0.02}$.
6. The RL plot of $K_{lead}(s)K_{lag}(s)G(s)$ has, for $K = 225$, dominant CL roots at $s = -2.12 \pm j2.46$ ($\zeta = 0.65$). The resulting $e_{ss}|_{ramp} \leq 0.099$ ($K_v = 10.13$).

The resulting lead-lag compensator is given as: $K(s) = K(s + 0.09)(s + 2)/(s + 0.02)(s + 20)$. Alternatively, the steps for a PID design are given as follows (Figure 6.8):

1. PD design: let $z_c = 2.05$; $K_{PD}(s) = K(s + 2.05)$.
2. The RL plot of $K_{PD}(s)G(s)$ has, for $K = 12.5$, dominant CL roots at $s = -2.4 \pm j2.5$ ($\zeta = 0.69$). For the PD compensated system, $K_v = 1.25$.
3. PI design: let $z_c = 0.05$, $p_c = 0$, $K_{PI}(s) = K(s + 0.05)/s$.
4. The RL plot of $K_{PD}(s)K_{PI}(s)G(s)$ has, for $K = 12.5$, dominant CL roots at $s = -2.4 \pm j2.5$ ($\zeta = 0.69$). The resulting $e_{ss}|_{ramp} = 0$ ($K_v = \infty$).

The resulting PID compensator is given as: $K(s + 0.05)(s + 2.05)/s$. The MATLAB commands for the above lead-lag and PID designs are given below (Figure 6.8):

```
>> G1=tf(1, [1 7 10 0]);
>> rlocus(G1), hold
>> rlocus(G1*tf([1 2],[1 20])) %lead
>> rlocus(G1,tf([1 2.090.18],[1 20.02.4])) %lead-lag
>> rlocus(G1*tf([1 2.1],[1])) %pd
>> rlocus(G1*tf([1 2.10.1025],[1])) %pid
```

6.4.4 Rate Feedback Compensation

Rate feedback signal is often available in the control systems due to frequent use of rate sensors, such as rate gyroscopes and tachometers. A rate feedback compensator improves the relative stability and transient response, similar to a PD compensator.

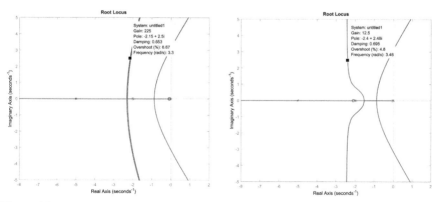

Figure 6.8 Lead–lag compensation through RL plot (*left*); PID compensation through RL plot (*right*).

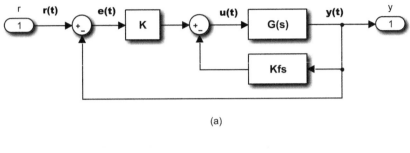

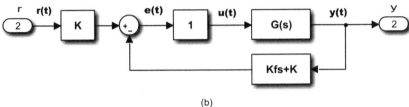

(b)

Figure 6.9   (a) Rate feedback in a unity gain feedback control system; (b) Equivalent control system diagram.

The rate feedback design (Figure 6.9a) is accomplished in two steps: the inner (minor) loop is first designed for the desired closed-loop pole locations that later serve as the plant poles in the outer loop design.

The rate feedback compensator introduces a zero at the origin in the minor loop transfer function. However, this zero does not cancel any existing plant poles at that location. Thus, despite closing the inner feedback loop through the derivative term, rate feedback does not change the system type.

A PD compensator can also replace the differentiator in the minor loop in the rate feedback design. The zero introduced by the PD compensator, similarly, does not cancel a plant pole that happens to be coincident with the compensator zero.

Example 6.10: Let $G(s) = \frac{1}{s(s+2)(s+5)}$; and assume that the design specifications are: $\%OS \le 10\%$, $t_s \le 2$ s.

We consider the rate feedback design via $H(s) = K_f s$. The inner loop transfer function is: $G(s)K_f s$. The transfer function when the inner loop is closed is given as:

$$G_{ml}(s) = \frac{G(s)K_f s}{1 + G(s)K_f s} = \frac{1}{s\left[(s+2)(s+5) + K_f\right]}.$$

The feedback gain K_f is selected via the RL design with the minor loop gain, $K_f s\, G(s) = \frac{K_f s}{s(s+2)(s+5)}$.

The minor loop RL has two branches that split at $\sigma = -3.5$ ($K_{\mathrm{f}} = 2.25$) and follow the asymptotes at $\pm 90°$. We may, for example, choose $K_{\mathrm{f}} = 15$ for CL roots at: $s = -3.5 \pm j3.57$ ($\zeta \cong 0.7$).

Once the minor loop is closed, the resulting loop transfer function for the outer loop design is given as:

$$KG_{\mathrm{ml}}(s) = \frac{K}{s\left[s^2 + 10s + 25\right]}.$$

The RL for this stage starts from the above pole locations (Figure 6.10). On the updated RL, we may choose $K = 34$ for the resulting closed-loop roots at: $s = -2.21 \pm j2.95$ ($\zeta = 0.6$), with $t_{\mathrm{s}} \cong 1.8$ s.

Finally, we note that it is possible, through block diagram manipulation, to combine the rate feedback controller with the cascade controller (Figure 6.9(b)). The reduced system has a single feedback loop, and the resulting closed-loop characteristic polynomial is given as:

$$\Delta(s) = s\left[(s + 2)(s + 5) + K_f\right] + K.$$

The controller design involves choosing both K and K_f. As an example, choosing $K_f = 15$, $K = 35$, realizes the above design.

The MATLAB commands for rate feedback design are given below (Figure 6.10):

```
>> G1=tf(1, [1 7 10 0]);
>> G2=tf([1 0], 1);
>> rlocus(G1*G2), hold
>> axis([-6 1 -6 6])
>> rlocus(feedback(G1, 15*G2))
```

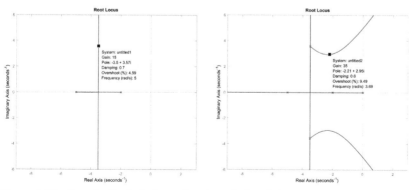

Figure 6.10 Rate feedback design: minor loop design (*left*); outer loop design (*right*).

6.4.5 Controller Design Comparison

The controller design choices presented in this chapter include the following: static compensator, phase-lag, phase-lead, lead–lag, PID, and the rate compensator.

We now compare the performance of these compensators for the plant $G(s) = \frac{1}{s(s+2)(s+5)}$. Accordingly, we plot the step responses of the compensated systems in each case (Figure 6.11).

The following table compares the overshoot, settling time, and steady-state error for the six compensators (Table 6.1). The rise time is defined as the first time when the step response equals: $y(t_r) = 1.0$.

We make the following observations from the table:

1. The PID compensator has the fastest rise time followed by the Lead-lag design.
2. The rate feedback offers the shortest settling time followed by phase-lead design.
3. The phase-lag compensator has highest overshoot (20%).
4. The rate feedback compensator has the lowest overshoot (1%) and offers the best overall performance.

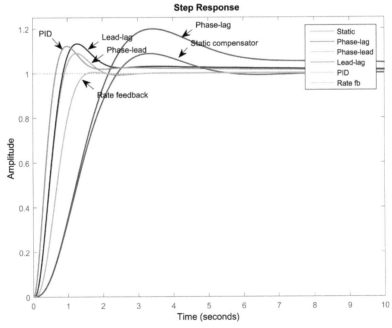

Figure 6.11 A comparison of the step responses for the various compensator designs.

Table 6.1 A comparison of compensator designs for $G(s) = \left(\frac{1}{s(s+2)(s+5)} \right)$

Compensator	Transfer Function $K(s)$	Closed-loop Pole Locations	Rise Time (s)	Percentage Overshoot	Settling Time (s)
Static	8.75	$-0.77 \pm j1.0$	2.42	9%	5.0
Phase-lag	$\dfrac{8.75(s+.1)}{s+.008}$	$-0.72 \pm j0.95$	2.17	20%	20
Phase-lead	$\dfrac{175(s+2)}{s+15}$	$-2.0 \pm j2.64$	0.91	9%	1.8
Lead-lag	$\dfrac{175(s+2)(s+.1)}{(s+15)(s+.008)}$	$-1.95 \pm j2.58$	0.87	13%	6.0
PID	$\dfrac{16.75(s^2+2.1s+.205)}{s}$	$-2.45 \pm j3.2$	0.66	12%	3.0
Rate feedback	$15s+35$	$-2.21 \pm j2.95$	1.56	1%	1.4

6.4.6 Controller Design with MATLAB SISO Tool

The MATLAB Control System Toolbox provides a utility for single-input single-output (SISO) controller design. The design tool is accessed by typing 'sisotool' on the command line interface, or by entering the Control System Designer App (2017 release).

For the above example, we may initialize the design tool as:

```
>> sisotool(tf(1,[1 2 0]))
```

Subsequently, we can add compensator poles and zeros under the compensator editor tab, and observe the step response under the analysis plots tab. Please refer to the MATLAB Control Systems Toolbox documentation for further details.

6.5 Controller Realization

Dynamic compensators of the phase-lead/phase-lag type or PD, PI, PID types may be realized through electronic circuits built with operational amplifiers.

An operational amplifier in the inverting configuration has the following input-output transfer function:

$$\frac{V_0(s)}{V_i(s)} = -\frac{Z_f(s)}{Z_i(s)}$$

where $Z_i(s)$ is the input impedance and $Z_f(s)$ is the impedance in the feedback path.

For controller realization, these impedances comprise RC networks in series and/or parallel configuration, where

Series RC circuit: $Z_{\text{ser}}(s) = R + \frac{1}{Cs} = \frac{RCs+1}{Cs}$

Parallel RC circuit: $Z_{\text{par}}(s) = \frac{R/Cs}{R+\frac{1}{Cs}} = \frac{R}{RCs+1}$

6.5.1 Phase-Lead/Phase-Lag Compensators

A phase lead/lag compensator is formed by parallel RC circuits in the input and feedback paths; then,

$$K(s) = -\frac{Z_f(s)}{Z_i(s)} = -\frac{R_f}{R_i}\frac{(R_iC_is + 1)}{(R_fC_fs + 1)}$$

The compensator thus has a zero at: $z_c = \frac{1}{R_iC_i}$, and a pole at: $p_c = \frac{1}{R_fC_f}$.

We may choose $R_iC_i > R_fC_f$ for phase-lead, and $R_iC_i < R_fC_f$ for the phase-lag compensator, where we note that some of the component values may need to be chosen arbitrarily.

For output sign correction, a resistive op-amp circuit with static gain: $\frac{V_0}{V_i} = -\frac{R_f}{R_i} = -1$ can be used.

Example 6.11: Let $K(s) = \frac{5(s+1)}{s+10}$; then, we may choose: $R_i = 200\ \text{K}\Omega$, $R_f = 1\ \text{M}\Omega$, $C_i = 5\ \mu\text{F}$, $C_f = 0.1\ \mu\text{F}$.

6.5.2 PD, PI, PID Compensators

These compensators may be realized by combining the following impedances in series/parallel:

$$G_{\text{PD}}(s) = -\frac{R_f}{Z_{\text{par}}(s)} = -\frac{R_f}{R_i}(R_iC_is + 1)$$

$$G_{\text{PI}}(s) = -\frac{1/C_fs}{Z_{\text{par}}(s)} = -\frac{1}{R_iC_f}\frac{(R_iC_is + 1)}{s}$$

$$G_{\text{PID}}(s) = -\frac{Z_{f-\text{ser}}(s)}{Z_{i-\text{par}}(s)} = -\frac{1}{R_iC_f}\frac{(R_iC_is + 1)(R_fC_fs + 1)}{s}.$$

In the case of PID compensator, first choosing the component values realizes the following PID gains:

$$k_p = \frac{R_f}{R_i} + \frac{C_i}{C_f}, \quad k_i = R_fC_i, \quad k_d = \frac{1}{R_iC_f}$$

Example 6.12: Let $G_{\text{PID}}(s) = \frac{(s+0.1)(s+10)}{s}$; then, we may choose: $R_i = 10\ \text{M}\Omega$, $R_f = 1\ \text{M}\Omega$, $C_i = 1\ \mu\text{F}$, $C_f = 0.1\ \mu\text{F}$.

Skill Assessment Questions

Link to the answers:
http://www.riverpublishers.com/book_details.php?book_id=449

6.1 Consider the model of a DC motor given as: $G(s) = \frac{5000}{s(s+10)(s+100)}$.
Assume that the motor is connected in unity gain feedback configuration.

   a. Sketch the root locus (RL) for the motor model.
   b. Choose a controller gain K to achieve $\zeta \cong 0.7$.
   c. Find the range of K for stability from the RL plot.

6.2 Let $G(s) = \frac{s+3}{s(s+1)(s+2)}$; assume that the system is connected in unity gain feedback configuration.

   a. Plot the root locus for the system and find the range of K for stability.
   b. Design a phase-lead controller for unity gain feedback to meet the following specifications: $\omega_n \cong 2\sqrt{2}$, $\zeta \cong 0.7$.

6.3 Consider the plant in Question 6.2 with the phase-lead controller.

   a. Add a phase-lag controller to boost the velocity error constant to $K_v > 10$.
   b. Plot the ramp response to verify $e(\infty)|_{ramp} < 0.1$.
   c. Realize the controller using op-amp circuits.

6.4 Consider the simplified model of a flexible beam given as: $G(s) = \frac{100}{s^2+s+100}$. Assume that the beam is connected in unity gain feedback configuration.

   a. Design a PID compensator for the model (use MATLAB 'pidtune' command)
   b. Realize the controller using op-amp circuits.

6.5 Consider the model of the DC motor in Question 6.1.

   a. Design a rate feedback controller to achieve: $\omega_n \geq 50\frac{rad}{s}$, $\zeta \geq 0.6$.
   b. Plot the step response of the closed-loop system.

6.6 Consider the model of human postural dynamics described as an inverted pendulum, given as: $G(s) = \Omega^2/(s^2 - \Omega^2)$, where $\Omega = \sqrt{10}$.

    a. Design a Phase-lead controller for the model to achieve: $\omega_n \geq 5\frac{\text{rad}}{\text{s}}, \xi \geq 0.8$.

    b. Add a phase-lag controller to achieve smaller than 2% steady-state error to a step input.

7

Sampled-Data Systems

Learning Objectives

1. Obtain a model of the sampled-data system that includes a sampler and zero-order hold.
2. Obtain the pulse transfer function of the sampled-data system using the z-transform.
3. Determine the stability of the closed-loop sampled-data system.

In the contemporary control systems technology, data acquisition card (DAQ) is used to sense, sample and process variables of interest, while the process controller is implemented as a software routine on a programmable logic controller (PLC), microcontroller, or digital signal processor (DSP). The systems that involve sampled variables and software based controllers are analyzed using techniques designed for the sampled-data systems.

The sampled-data systems contain a clock-driven element in the feedback loop (Figure 7.1). The DAQ includes an analog-to-digital converter (ADC) that samples, quantizes, and stores the variables. The output of the DAQ is a sequence of numbers that represent variable values at multiples of sampling period T.

The digital controller output is similarly represented as a sequence given as $u(kT)$, $k = 0, 1, \ldots$, where $u(kT)$ represents the sampled values of a continuous signal $u(t)$. Before applying it to the process under control, the controller output is converted to a continuous-time signal via a zero-order hold (ZOH), i.e., a device that holds its output constant for one time period.

The choice of the sampling period is an important consideration in designing digital control systems. In general, the sampling frequency should be selected 4–10 times higher than the system bandwidth, approximately given as the inverse of the smallest time constant of the system. A lower sampling frequency causes degradation in the phase margin, adversely affecting relative stability, because of the presence of ZOH.

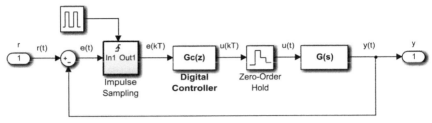

Figure 7.1   The closed-loop sampled-data system.

7.1 Models of Sampled-Data Systems

To model the sampled-data systems, we consider an ideal sampler that samples a physical signal $r(t)$ every T seconds, and generates a series of impulses with amplitudes $r(kT), \ k = 0, 1, \ldots$. The sampler output represents multiplication of $r(t)$ with a train of impulses, and is mathematically represented as $r^*(t)$, where

$$r^*(t) = \sum_0^\infty r(kT)\delta(t - kT).$$

We apply the Laplace transform to the sampled signal $r^*(t)$ to obtain:

$$r^*(s) = \sum_0^\infty r(kT)e^{-skT}.$$

Next, we define $z = e^{sT}$ as a derived complex variable whose value explicitly depends on the sampling time T. The result is called the z-transform of the sequence $r(kT)$, and is given as:

$$r(z) = z[r(kT)] = \sum_0^\infty r(kT)z^{-k}.$$

The domain of the z-transform is a real or complex-valued number sequence, $r(kT)$; the resulting $r(z)$ represents a complex function.

As an example, the z-transform of a sampled unit step signal is computed as follows:

$$u(kT) = \{1, \ 1, \ \ldots\}$$

$$u(z) = \sum_0^\infty z^{-k} = \frac{1}{1 - z^{-1}}; \ |z^{-1}| < 1.$$

We note that the convergence of the above geometric series is dependent on the condition: $|z^{-1}| < 1$, i.e., the region of convergence (ROC) of $u(z)$ is confined to the inside of a unit circle in the complex z-plane.

Note, also, that though the definition of z-transform involves the Laplace transform variable ($z = e^{sT}$), its actual computation has no reference to s.

Example 7.1: Let $r(t) = e^{-at}u(t)$; then

$$r(kT) = \{1,\ e^{-aT}, e^{-2aT}, \ldots\}$$

$$r(z) = \sum_0^\infty e^{-akT} z^{-k} = \frac{1}{1 - e^{-aT}z^{-1}};\quad |e^{-aT}z^{-1}| < 1$$

Example 7.2: Let $r(t) = e^{j\omega t}u(t)$; then

$$r(kT) = \{1, e^{j\omega T}, e^{2j\omega T}, \ldots\}$$

$$r(z) = \sum_0^\infty e^{jk\omega T} z^{-k} = \frac{1}{1 - e^{j\omega T}z^{-1}};\quad |e^{j\omega T}z^{-1}| = |z^{-1}| < 1.$$

We note that in each case, the z-transform representation includes the specified region of convergence (ROC). In fact the z-transform description is incomplete without the ROC being specified.

The z-transform of complex signals is obtained by using its properties of linearity, differentiation, and translation, etc. (see References).

The following z-transforms of the sinusoidal functions are obtained by applying the Euler's identity to z-transform of $e^{jk\omega T}$ and separating the real and imaginary parts.

$$\sin(k\omega T) \overset{z}{\leftrightarrow} \frac{\sin(\omega T)\, z}{z^2 - 2\cos(\omega T) + 1}$$

$$\cos(k\omega T) \overset{z}{\leftrightarrow} \frac{z(z - \cos(\omega T))}{z^2 - 2\cos(\omega T) + 1}$$

$$e^{-akT}\sin(k\omega T) \overset{z}{\leftrightarrow} \frac{e^{-aT}\sin(\omega T)\, z}{z^2 - 2\cos(\omega T)\, e^{-aT} + e^{-2aT}}$$

$$e^{-akT}\cos(k\omega T) \overset{z}{\leftrightarrow} \frac{z(z - e^{-aT}\cos(\omega T))}{z^2 - 2\cos(\omega T)\, e^{-aT} + e^{-2aT}}.$$

7.1.1 Zero-Order Hold

The zero-order hold (ZOH), representing the digital-to-analog converter (DAC), is a device that holds its output constant for one time period T, i.e.,

$$r(t) = r((k-1)T) \quad \text{for} \quad (k-1)T \le t < kT.$$

The output of the ZOH for an arbitrary input is a stair-case type signal, as easily seen by selecting the input signal to be a unit ramp: $r(t) = tu(t)$.

The impulse response of ZOH is given as: $\delta_{zoh}(t) = 1, \ 0 < t < 1$.

Applying the Laplace transform to $\delta_{\text{ZOH}}(t)$, the impulse transfer function of ZOH is given as:

$$G_{\text{ZOH}}(s) = \frac{1}{s} - \frac{e^{-sT}}{s} = \frac{1 - e^{-sT}}{s}$$

The frequency response of ZOH is computed as:

$$G_{\text{ZOH}}(\omega) = \frac{1 - e^{-j\omega T}}{j\omega} = T\frac{\sin(\omega T/2)}{\omega T/2}e^{-j\omega T/2}$$

The frequency response of the ZOH resembles a *sinc* function with a delay term that contributes additional phase when the ZOH is placed inside the feedback loop. The added phase reduces the available phase margin (the amount of phase in degrees that can be added to the control loop before compromising stability), adversely impacting stability of the closed-loop system.

The phase margin (PM) is read from the Bode phase plot at the gain crossover frequency ω_{gc}, i.e., the frequency at which the Bode magnitude plot has a gain of unity (0 dB) (see Chapter 11).

Since the phase margin (PM) is reduced by an amount $\omega_{gc}T/2$, the desired minimum sampling time T can be computed from acceptable degradation in the PM. For example, if we want to limit the PM degradation to $5°$ (0.087 rad), the sampling time must be selected as: $T \le \frac{0.175}{\omega_{\text{gc}}}$.

7.2 The Pulse Transfer Function

To analyze a continuous-time plant driven by a digital controller via the ZOH, we use the z-transform to obtain a transfer function description of the plant cascaded with a ZOH. The resulting pulse transfer function provides the input-output description of the plant at the sampling intervals.

The pulse transfer function of a continuous-time plant described by transfer function $G(s)$, is defined as:

$$G(z) = z \left[\frac{1 - e^{-sT}}{s} G(s) \right] = (1 - z^{-1}) z \left[\frac{G(s)}{s} \right].$$

The computation of pulse transfer function is illustrated through the following examples.

Example 7.3: Let $G(s) = \frac{a}{s+a}$; then

$$G(z) = (1 - z^{-1}) z \left[\frac{a}{s(s+a)} \right] = (1 - z^{-1}) z \left[\frac{1}{s} - \frac{1}{s+a} \right].$$

From the z-transform tables (see references), $G(z) = \frac{z-1}{z} \left(\frac{z}{z-1} - \frac{z}{z-e^{-aT}} \right) = \frac{1-e^{-aT}}{z-e^{-aT}}$.

Example 7.4: Let $G(s) = \frac{a}{s(s+a)}$; then

$$G(z) = (1 - z^{-1}) z \left[\frac{a}{s^2(s+a)} \right] = (1 - z^{-1}) z \left[\frac{1}{s^2} - \frac{1/a}{s} + \frac{1/a}{s+a} \right]$$

Using the z-transform table,

$$G(z) = \frac{z-1}{z} \left[\frac{Tz}{(z-1)^2} - \frac{1}{a} \left(\frac{z}{z-1} - \frac{z}{z-e^{-aT}} \right) \right]$$

$$= \frac{T(z - e^{-aT}) - \frac{1}{a(z-1)(z-e^{-aT})}}{(z-1)(z-e^{-aT})}.$$

From the above examples, we make the following observations:

1. The order of the pulse transfer function is the same as that of the continuous-time plant.
2. The poles of the pulse transfer function are related to the plant poles through $z_i = e^{s_i T}$.

7.2.1 Pulse Transfer Function in MATLAB

The pulse transfer function of a continuous-time system is conveniently obtained in MATLAB by using the 'c2d' command and specifying a sampling time. In addition to the ZOH (default), the command allows a choice of input methods for converting a continuous-time object to discrete time.

The input argument to the 'c2d' command is a Dynamic System Object (DSO) created in the Control Systems Toolbox using multiple input commands. For example, the 'tf' command creates a DSO in the transfer function form. The MATLAB commands to obtain the pulse transfer function are invoked as follows:

```
>> G=tf(num, den);
>> Gz=c2d(G, T)
```

Example 7.5: Let $G(s) = \frac{1}{s+1}$, $T = 1s$; then, the pulse transfer function is obtained from MATLAB as: $G(z) = \frac{0.632}{z-0.368}$.

Example 7.6: Let $G(s) = \frac{1}{s(s+1)}$, $T = 1s$; then, the pulse transfer function is obtained from MATLAB as: $G(z) = \frac{0.368z+0.264}{(z-1)(z-0.368)}$.

7.3 Closed-Loop Sampled-Data Systems

We consider a unity gain feedback control system, where a continuous-time plant is driven by a digital controller through a ZOH. We further assume that a static controller, represented by a scalar gain K, is employed. Accordingly, the closed-loop pulse transfer function is given as:

$$T(z) = \frac{Y(z)}{R(z)} = \frac{KG(z)}{1 + KG(z)}.$$

The closed-loop characteristic polynomial is given as: $\Delta(z) = 1 + KG(z)$.

   Using the closed-loop pulse transfer function, we can compute the output of the sampled-data system to a given input sequence.

7.3.1 Step Response

The step response of the closed-loop sampled-data system is its response to a step input sequence: $r\{kT\} = \{1, 1, 1 \ldots\}$; $r(z) = \frac{1}{1-z^{-1}}$. The step response is illustrated by the following examples:

Example 7.7: Let $G(s) = \frac{1}{s+1}$; then, $G(z) = \frac{1-e^{-T}}{z-e^{-T}}$, and the closed-loop pulse transfer function is given as:

$$T(z) = \frac{K(1 - e^{-T})}{z - e^{-T} + K(1 - e^{-T})}.$$

Let $T = 1$ s; then, $T(z) = \frac{0.632K}{z+0.632K-0.368}$. Next, assume a unit step sequence: $r(kT) = 1$, $r(z) = \frac{1}{1-z^{-1}}$; then, for $K = 1$, the output is given in z-domain as:

$$Y(z) = \frac{0.632z}{(z-1)(z+0.268)}.$$

In order to take the inverse z-transform of the output, we expand $Y(z)/z$ in partial fractions to obtain:

$$Y(z) = 0.5 \left(\frac{z}{z-1} - \frac{z}{z+0.268} \right).$$

Then, by taking the inverse z-transform, the sampled output sequence is given as: $y(kT) = 0.5(1 - (-0.268)^k)$, $k = 0, 1, \ldots$, i.e., the output at the sampling intervals is given in terms of the sequence: $\{0,\ 0.634,\ 0.464,\ \ldots\}$

We may want to compare the step response of the sampled-data system with that of an equivalent continuous-time system. Accordingly, let $G(s) = \frac{1}{s+1}$, $T(s) = \frac{K}{s+1+K}$, $K = 1$, and $r(t) = u(t)$; then, the continuous-time system output is given as: $y(t) = \frac{1}{2}(1 - e^{-2t})$, $t > 0$.

The step response of the sampled-data and the continuous-time systems are plotted in MATLAB using the following commands (Figure 7.2).

```
>> G=tf(1,[1 1])
>> Gz=c2d(G, 1)
>> step(feedback(Gz,1)), hold
>> step(feedback(G,1)), hold
>> legend('sampled-data','continuous-time')
```

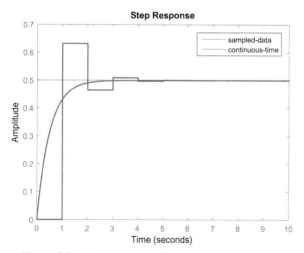

Figure 7.2 First-order sampled-data system response.

We make the following observations based on this comparison:

1. A first-order continuous-time system translates into a first-order sampled-data system.
2. The step response of first-order continuous-time system has no overshoot; however, the equivalent discrete system has an overshoot. This happens due to the response term $(-0.268)^k$.
3. The expression for $y(kT)$ describes the system output at the sampling intervals alone, i.e., we will need to interpolate between sampling times in order to plot the output in continuous-time.

Example 7.8: Let $G(s) = \frac{1}{s(s+1)}$; we use a sampling time $T = 1$s to obtain:
$G(z) = \frac{0.368z+0.264}{(z-1)(z-0.368)}$.

The closed-loop pulse transfer function is given as:

$$T(z) = \frac{(0.368z + 0.264)K}{(z - 1)(z - 0.368) + (0.368z + 0.264)K}$$

Assume a controller gain $K = 1$; then, the closed-loop pulse transfer function is: $T(z) = \frac{0.368z+0.264}{z^2-z+0.632}$.

Let $R(z) = \frac{z}{z-1}$ (unit step); then, the output response is given as: $Y(z) = \frac{z(0.368z+0.264)}{(z-1)(z^2-z+0.632)}$.

In order to apply the inverse z-transform, we may use the long division to write:

$$Y(z) = 0.368z^{-1} + z^{-2} + 1.4z^{-3} + 1.4z^{-4} + 1.15z^{-5} + \dots$$

The output at sampling intervals $(T = 1$s$)$ is given by the following sequence:

$$y(kT) = \{0,\ 0.368,\ 1,\ 1.4,\ 1.4,\ 1.15,\ \dots\}$$

For comparison, the closed-loop transfer function for a continuous-time plant is given as: $T(s) = \frac{1}{s^2+s+1}$.

Its step response is given by: $y(s) = \frac{1}{s(s^2+s+1)} = \frac{1}{s} - \frac{1}{s^2+s+1}$, or $y(t) = (1 - 1.15e^{-0.5t}\sin 0.866t)u(t)$. The MATLAB commands for this example are given as follows (Figure 7.3):

```
>> G=tf(1,[1 1 0])
>> Gz1=c2d(G, 1)
>> Gz2=c2d(G, 0.5)
>> step(feedback(Gz1,1)), hold
```

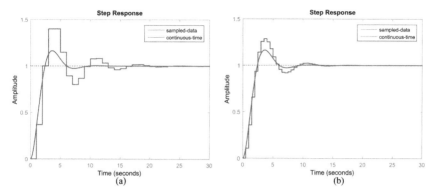

Figure 7.3 Second-order sampled-data system response: T = 1s (*left*); T = 0.5s (*right*).

```
>> step(feedback(G,1))
>> legend('sampled-data','continuous-time')
>> figure
>> step(feedback(Gz2,1)), hold
>> step(feedback(G,1))
>> legend('sampled-data','continuous-time')
```

We note that while the continuous-time system step response with $\zeta = 0.5$ has a 16.3% overshoot (Figure 7.3a), the corresponding sampled-data system response has about 45% overshoot (when interpolated between $t = 3$s and $t = 4$s). In addition, the settling time of the sampled-data system is twice that of the continuous-time system.

The high overshoot in the case of sampled-data system occurs due to the presence of the ZOH that reduces the available PM by an amount: $\Delta\phi_m = \frac{\omega_{gc}T}{2}$.

For this example, the continuous-time system has a PM of $52°$ at $\omega_{gc} = 0.786 \frac{rad}{s}$. The drop in PM due to the introduction of ZOH is: $\Delta\phi_m = \frac{\omega_{gc}T}{2} = 22.5°$. In comparison, if the sampling time is reduced to $(T = 0.5$s$)$, the reduction in the PM is $11°$, which results in about 30% overshoot (Figure 7.3b).

7.3.2 Steady-State Error

The output error, $e(z)$, in response to a given reference input $r(z)$, in the case of a unity feedback sampled-data system is computed as:

$$e(z) = r(z)(1 - T(z)).$$

The steady-state error is computed via the application of the final value theorem (FVT) in the z-domain, which is given as:

$$\lim_{k \to \infty} e(k) = \lim_{z \to 1} (z - 1)e(z).$$

We can define the following error constants for the sampled-data systems:

$$K_p = \lim_{z \to 1} G(z)$$

$$K = \lim_{z \to 1} \frac{(z - 1)}{T} G(z)$$

Then, the steady-state errors to step and ramp inputs are computed as:

$$e_{ss}|_{step} = \frac{1}{1 + K_p}; \quad e_{ss}|_{ramp} = \frac{1}{K}.$$

Example 7.9: Let $G(z) = \frac{0.368z + 0.264}{(z-1)(z-0.368)}$ $(T = 1s)$; then $K_p = \infty$, $K = 1$. Accordingly, $e_{ss}|_{step} = 0$; $e_{ss}|_{ramp} = 1$.

7.4 Stability of Sampled-Data Systems

Recall that the BIBO stability in the case of continuous-time systems requires that the closed-loop system poles (roots of the characteristic polynomial, $\Delta(s)$) are confined to the OLHP, where $s = j\omega$ defines the stability boundary.

In the case of sampled-data systems, the z-plane stability boundary is derived as: $z = e^{j\omega T} = 1\angle \omega T$, which maps the $j\omega$-axis to the unit circle in the complex z-plane, and the OLHP to the inside of the unit circle: $|z| < 1$.

Accordingly, the closed-loop sampled-data system is stable if and only if the pulse characteristic polynomial $\Delta(z)$ has its roots inside the unit circle, i.e., if z_i is root of $\Delta(z)$, then the stability requirement is: $|z_i| < 1$.

The stability of the sampled-data systems can be characterized in various ways that are described next.

7.4.1 Unit Pulse Response

The unit pulse response of a sampled-data system, described by pulse transfer function $G(z)$, is its response to a unit pulse $r(k) = \delta(k)$. Given an input-output description of the system in the time-domain, the unit pulse response can be computed through iteration.

The unit pulse response can be used to characterize the stability of the sampled-data system as follows: first, assume that $\Delta(z)$ has distinct roots: z_i, $i = 1, \ldots, n$, then its unit pulse response contains system response modes: $\phi_i(k) = (z_i)^k$. These terms die out with time if $|z_i| < 1$.

Next, assume that $\Delta(z)$ has complex roots of the form $re^{\pm j\theta}$; then, the unit pulse response contains system modes: $\phi_i(k) = \{r^k \cos k\theta, \ r^k \sin k\theta\}$. These modes die out with time if $|r| < 1$, i.e., if the closed-loop z-domain poles are inside the unit circle.

7.4.2 Schur–Cohn Stability Test

The Schur–Cohn test is an analytical stability test that determines if all roots of a polynomial $A(z)$ are inside the unit circle. The test is applied in the form of a computational algorithm, given as follows:

Initialize: $A_N(z) = \sum_{k=0}^{N} a_k z^{-k}, \ a_N(k) = a_k$

Define: $A_m(z) = \sum_{k=0}^{m} a_k z^{-k}, \ a_m(0) = 1$

$$B_m(k) = z^{-m} A_m(z^{-1}) = \sum_{k=0}^{N} a_m(m - k) z^{-k}$$

$$K_m = a_m(m), \ m = 1, \ldots, N$$

Recursion: $A_{m-1}(z) = \frac{A_m(z) - K_m B_m(z)}{1 - K_m^2}, \ m = N, \ N - 1, \ldots, 1$

Test: $A(z)$ has roots $|z| < 1$ if and only if $|K_m| < 1$ for $m = 1, \ldots, N$

For example, in the case of a second order polynomial: $A_2(z) = 1 + a_1 z^{-1} + a_2 z^{-2}$, we have

$$K_2 = a_2(2) = a_2, \ B_2(z) = a_2 + a_1 z^{-1} + z^{-2},$$

$$A_1(z) = \frac{A_2(z) - a_2 B_2(z)}{1 - a_2^2} = 1 + \frac{a_1}{1 + a_2} z^{-1}.$$

So $K_1 = a_1(1) = \frac{a_1}{1 + a_2}$; and the stability conditions are: $|a_2| < 1$, $\left|\frac{a_1}{1 + a_2}\right| < 1$ or $|1 + a_2| > |a_1|$.

Example 7.10: Let $A(z) = z + 0.632K - 0.368$; then $A(z)$ is stable for $(1 - 0.368) > 0.632K$, or $K < 1$.

7.4.3 The Jury's Test

When adapted to real polynomials, the Schur–Cohn test results in a criterion similar to the Ruth's test, and is known as the Jury's test.

Assume that the z-domain polynomial to be investigated is given as: $A(z) = a_0 z^n + a_1 z^{n-1} + \cdots + a_{n-1} z + a_n; a_0 > 0$. Next, we build the Jury's table, given as:

$$
\begin{vmatrix}
a_n & a_{n-1} & \cdots & a_0 \\
a_0 & a_1 & \cdots & a_n \\
b_{n-1} & b_{n-2} & \cdots & b_0 \\
b_0 & b_1 & \cdots & b_{n-1} \\
c_{n-2} & c_{n-1} & \cdots & c_0 \\
c_0 & c_1 & \cdots & c_{n-2} \\
& & \vdots &
\end{vmatrix}
$$

The coefficients of the third and subsequent rows are computed as:

$$
b_k = \begin{vmatrix} a_n & a_{n-1-k} \\ a_0 & a_{k+1} \end{vmatrix}, \quad k = 0, \ldots, n-1
$$

$$
c_k = \begin{vmatrix} b_{n-1} & b_{n-2-k} \\ b_0 & b_{k+1} \end{vmatrix}, \quad k = 0, \ldots, n-2, \text{ etc.}
$$

The necessary conditions for polynomial stability are: $A(1) > 0$, $(-1)^n A(-1) > 0$.

The sufficient conditions for stability, given by the Jury's test, are:
$$a_0 > |a_n|, \ |b_{n-1}| > |b_0|, \ |c_{n-2}| > |c_0|, \ldots \ (n-1 \text{ conditions})$$

Example 7.11: In the case of a second order polynomial: $A(z) = z^2 + a_1 z + a_2$, the Jury's table is given as:

$$
\begin{vmatrix}
a_2 & a_1 & 1 \\
1 & a_1 & a_2 \\
b_1 & b_0 & \\
b_0 & b_1 &
\end{vmatrix}
$$

The resulting necessary conditions are: $1 + a_1 + a_2 > 0$, $1 - a_1 + a_2 > 0$. And, the sufficient conditions are: $|a_2| < 1$, $|1 + a_2| > |a_1|$.

We note that the above conditions are similar to those obtained from the Schur–Cohn test.

7.4.4 Stability through Bilinear Transform

The bilinear transform (BLT) provides a mapping from the s-domain to the z-domain and vice versa. In order to develop the BLT, we use the first-order Pade' approximation for $z = e^{sT}$ to obtain:

$$z = \frac{e^{sT/2}}{e^{-sT/2}} \cong \frac{1 + sT/2}{1 - sT/2}, \quad s = \frac{2}{T} \frac{z - 1}{z + 1}$$

Bilinear transform can be used to determine the stability of the closed-loop sampled-data system as follows: first, since T has no impact on stability determination, we may use $T = 2$ for simplicity.

Next, in order to distinguish the characteristic polynomial obtained through BLT from the original $\Delta(s)$, a new frequency-domain complex variable w is introduced. The resulting BLT is given as:

$$z = \frac{1 + w}{1 - w}, \quad w = \frac{z - 1}{z + 1}$$

Application of the above BLT to $\Delta(z)$ returns the characteristic polynomial $\Delta(w)$, whose stability is then determined through the application of Hurwitz criterion.

In particular, consider a second-order polynomial: $\Delta(z) = z^2 + a_1 z + a_2$. The application of BLT, ignoring the denominator term, results in:

$$\Delta(w) = \Delta(z)|_{z = \frac{1+w}{1-w}} = (1 - a_1 + a_2)w^2 + 2(1 - a_2)w + 1 + a_1 + a_2$$

By applying the Hurwitz criterion, we obtain the following stability conditions for $\Delta(z)$:

$$a_2 + a_1 + 1 > 0$$
$$a_2 - a_1 + 1 > 0$$
$$1 - a_2 > 0$$

The resulting stability region is graphed in the $a_1 - a_2$ plane (Figure 7.4).

Example 7.11: Let $\Delta(z) = z^2 + z + K$; then,

$$\Delta(w) = \Delta(z)|_{z = \frac{1+w}{1-w}} = Kw^2 + 2(1 - K)w + 1 + K$$

The application of the Hurwitz criteria to $\Delta(w)$ reveals $0 < K < 1$ for stability.

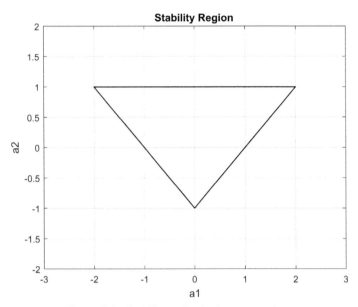

Figure 7.4   Stability region in the a_1-a_2 plane.

Example 7.12: Let $G(s) = \frac{1}{s(s+1)}$, $T = 1s$; then, the pulse transfer function is given as: $G(z) = \frac{0.368z+0.264}{(z-1)(z-0.368)}$, and the characteristic polynomial is:

$$\Delta(z) = z^2 + (0.368K - 1.368)z + 0.264K + 0.368$$

The w-polynomial is obtained through BLT as:

$$\Delta(w) = (2.736 - 0.104K)w^2 + (0.632 - 0.264K)2w + 0.632K$$

The application of the Hurwitz criteria to $\Delta(w)$ gives $0 < K < 2.394$ for stability.

Skill Assessment Questions

Link to the answers:
http://www.riverpublishers.com/book_details.php?book_id=449

7.1 Consider the simplified model of a DC motor, given as: $G(s) = \frac{1}{s(s+10)}$.

    a. Choose a sampling time T and obtain the pulse transfer function for the DC motor.

    b. Determine the range of K for the stability of the closed-loop characteristic polynomial.

    c. Obtain the first few terms of the output sequence for a unit pulse input sequence.

7.2 Consider the simplified model of a flexible beam, given as: $G(s) = \frac{100}{s^2+s+100}$.

    a. Obtain the pulse transfer function for the beam (use $T = 0.01$ s).

    b. Determine the range of K for the stability of the closed-loop characteristic polynomial.

    c. Obtain the first few terms of the output sequence for a unit input sequence.

7.3 Consider a system model described as: $G(s) = \frac{s+3}{s(s+1)(s+2)}$.

    a. Use MATLAB to obtain the pulse transfer function for the system (assume $T = 0.1$ s).

    b. Determine the range of K for the stability of the closed-loop characteristic polynomial.

    c. Obtain the first few terms of the output sequence for a unit pulse input sequence.

7.4 Consider the simplified model of an automobile given as: $G(s) = \frac{28s+120}{s^2+7s+14}$.

    a. Choose a sampling time T and obtain a pulse transfer function for the automobile.

    b. Determine the range of K for the stability of the closed-loop characteristic polynomial.

    c. Obtain the first few terms of the output sequence for a unit pulse input sequence.

7.5 Consider the discrete-time model of an automobile developed in Question 7.4.

  a. Obtain the closed-loop pulse transfer function for $K = 0.5\,K_{\text{max}}$
  b. Determine the steady-state error to a unit step input.
  c. Plot and compare the step responses of the continuous and discrete systems.

7.6 Consider the model of human postural dynamics described as an inverted pendulum, given as: $G(s) = \Omega^2/(s^2 - \Omega^2)$, where $\Omega = \sqrt{10}$.

  a. Obtain a pulse transfer function for the model (use $T = 0.1$ s).
  b. Obtain the first few terms of the impulse response.

8

Digital Controller Design

Learning Objectives

1. Obtain the discrete-time equivalent of an analog controller by emulation.
2. Emulate the analog PID controller for computer implementation.
3. Perform root locus design of the digital controller for systems described by their pulse transfer function.

This chapter discusses the digital controller design techniques for continuous-time plants. We begin with the analog controller emulation methods, followed by computer implementation of the analog PID controller. Later, we discuss the root locus design of the digital control systems. The state-space design of the digital systems will be taken up later (in Chapter 10).

The controller emulation is a practical and time-saving approach to obtain an approximate digital controller that emulates an already designed analog controller for a continuous-time plant. Assuming a high enough sampling rate, the approximate digital controller obtained by emulation gives comparable performance to the analog controller it mimics.

The root locus design of the digital controller is based on the pulse transfer function of a continuous-time system that describes its input-output relationship that is valid at the sampling intervals. The rules for plotting the root locus for the discrete-time systems are the same as those for the continuous-time systems, the only difference being the desired z-plane characteristic polynomial should have its roots confined to the unit circle.

8.1 Controller Emulation

Assuming that a continuous-time controller with satisfactory output response has already been designed, a corresponding digital controller for computer implementation can be obtained by emulation. Popular methods for controller emulation are discussed below.

8.1.1 Controller Emulation Using Impulse Invariance

Impulse invariance technique is commonly used in digital signal processing to transform analog filter designs to their equivalent digital filters. However, impulse invariance works well only for band-limited filters, i.e., it may not be used with PD, PI, or PID type controllers.

Assume that the filter transfer function is written in partial fraction form as: $H(s) = \sum_{k=1}^{n} \frac{A_k}{s-s_k}$.

The filter impulse response is given as: $h(t) = \sum_{k=1}^{n} A_k e^{s_k t}$.

The sampled impulse response is: $h(kT) = T \sum_{k=1}^{n} A_k e^{s_k kT}$.

Using z-transform, the filter pulse transfer function is:

$$H(z) = T \sum_{k=1}^{n} \frac{A_k}{1 - p_k z^{-1}}, \quad p_k = e^{s_k T}.$$

The MATLAB command for controller emulation via impulse invariance is:

```
>> Gz=c2d(G, T, 'impulse')
```

The input argument G above is a Dynamic System Object created in the MATLAB Control System Toolbox, and T is the sampling time.

Example 8.1: A phase-lead controller for the continuous-time plant: $G(s) = \frac{2}{s(s+1)(s+2)}$ was earlier designed as: $K(s) = \frac{14.58(s+0.36)}{s+5.25} = 14.58 \left(1 - \frac{4.89}{s+5.25}\right)$. Then, an equivalent digital controller ($T = 0.1$s) is obtained as:

$$K(z) = 14.58 \left(1 - \frac{0.489}{s - 0.592}\right) = \frac{7.45(z - 0.957)}{z - 0.592}.$$

The update rule for the digital controller implementation with input e_k and output u_k is given as:

$$u_k = 0.592 u_{k-1} + 7.45(e_k - 0.957 e_{k-1}).$$

8.1.2 Controller Emulation Using Pole-Zero Matching

In the impulse invariance method, the plant poles are mapped to their discrete-time equivalents in the z-plane. We can additionally map the zero locations using a similar transform: $z_i = e^{s_i T}$.

In pole-zero matching, if the analog transfer function has n poles and m finite zeros, to be matched, then another $n - m$ zeros at $z = 1$ are added to the pulse transfer function. Additionally, the gain of the filter is matched to that of the analog filter at some frequency of interest. For example, the DC gain may be matched in the case of a low-pass filter.

The MATLAB command for controller emulation through pole-zero matching is:

```
>> Gz=c2d(G, T, 'matched')
```

Example 8.2: A phase-lead controller for the plant: $G(s) = \frac{2}{s(s+1)(s+2)}$ was previously designed as: $K(s) = \frac{14.58(s+0.36)}{s+5.25}$.

Then, a pole-zero matching digital controller $(T = 0.1s)$ is given as: $K(z) = \frac{K(z+e^{0.36T})}{z+e^{5.25T}}$.

If the DC gain is matched, the equivalent digital controller is given as: $K(z) = \frac{11.548(z-0.965)}{z-0.592}$.

Alternatively, we may want to match the filter gain at $\omega_m = \sqrt{\omega_z \omega_p}$, to obtain: $K(z) = \frac{5.177(z-0.965)}{z-0.592}$.

The update rule for digital controller implementation is given as: $u_k = 0.592u_{k-1} + 5.177(e_k - 0.965e_{k-1})$.

8.1.3 Controller Emulation Using Bilinear Transform

The bilinear transform (or Tustin's method) can be used to approximately convert an analog controller into a digital controller. This emulation method is effective at high enough sampling rates.

In controller emulation using bilinear transform, the sampling frequency is selected at least 20 times the bandwidth of the compensated system, i.e., $f_s \geq \frac{10\omega_B}{\pi}$ or $T \leq \frac{\pi}{10\omega_B}$.

Using the bilinear transform, the equivalent digital controller is obtained as: $K(z) = K(s)|_{s=\frac{2z-1}{Tz+1}}$.

The MATLAB command for controller emulation through BLT is:

```
>> c2d(G, T, 'tustin')
```

Example 8.3: A phase-lead controller for the plant: $G(s) = \frac{2}{s(s+1)(s+2)}$ was previously designed as: $K(s) = \frac{14.58(s+0.36)}{s+5.25}$.

The compensated system has $\omega_B \cong 1.8 \frac{rad}{s}$. Accordingly, let $T = 0.1$ s, then a matching digital controller is given as: $K(z) = \frac{11.756(z-0.965)}{z-0.584}$.

The update rule for digital controller implementation is given as: $u_k = 0.584u_{k-1} + 11.56(e_k - 0.965e_{k-1})$.

8.1.4 Controller Emulation Using ZOH

For controller emulation via zero-order hold (ZOH), the analog controller is first realized as a state-space model; next, it is transformed into its discrete-time equivalent through ZOH; finally, the resulting digital controller is converted into its pulse transfer function $G(z)$. The following example illustrates this method:

Example 8.4: As in the previous example, a phase-lead controller for the plant: $G(s) = \frac{2}{s(s+1)(s+2)}$ is given as: $K(s) = \frac{u(s)}{e(s)} = \frac{14.58(s+0.36)}{s+5.25}$.
Using MATLAB, we convert the controller in the state-space form as:

$$\dot{x} = -5.25x + 8.845e, \quad u = -8.06x + 14.58e.$$

The discrete-time state-space controller model $(T = 0.1 \text{ s})$ is given as:

$$x_{k+1} = 0.592x_k + 0.688e_k, \quad u_k = -8.06x_k + 14.58u_k.$$

The corresponding pulse transfer function is given as: $K(z) = \frac{14.58\,(z-0.972)}{z-0.592}$.
The update rule for digital controller implementation is given as: $u_k = 0.592u_{k-1} + 14.58(e_k - 0.972e_{k-1})$.

The MATLAB command for the controller emulation via ZOH (default) is:

```
>> c2d(G, T)
```

8.1.5 Comparison of Controller Emulation Methods

The following plot (Figure 8.1) compares the step response of the closed-loop system using impulse invariance, pole-zero matching, bilinear transform (Tustin's Method), and the ZOH methods. In the figure, the left-hand plot in each case was drawn in MATLAB using the pulse transfer function $G(z)$. The right-hand plot was simulated in Simulink, using digital controller in cascade with the continuous-time plant $G(s)$.

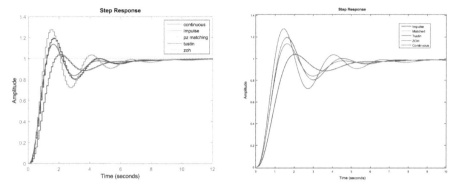

Figure 8.1 Step response comparison for controller emulation: MATLAB simulation with the pulse transfer functions (*left*); Simulink simulation of continuous-time plant with the digital controllers (*right*).

As seen from the plots, the bilinear transform and pole-zero matching methods produce comparable step responses with about 20% overshoot, compared to about 28% overshoot for the ZOH method. The impulse invariance method has a much lower 4% overshoot. For comparison, the continuous-time closed-loop system has about 15% overshoot.

The MATLAB commands for the above simulation are given below (Figure 8.1):

```
>> G=tf(2, [1 3 2 0]);
>> Gc=zpk(-.36,-5.25,14.58);
>> Gz=c2d(G,.1);
>> Gci=zpk(.957,.592,7.45,-1)
>> Gcm=c2d(Gc,.1,'matched');
>> Gct=c2d(Gc,.1,'tustin');
>> Gcz=c2d(Gc,.1,'zoh');
>> T=feedback(Gc*G,1)
>> Ti=feedback(Gci*Gz,1);
>> Tm=feedback(Gcm*Gz,1);
>> Tt=feedback(Gct*Gz,1)
>> Tz=feedback(Gcz*Gz,1)
>> step(T,Ti,Tm,Tt,Tz)
>> legend('continuous','impulse', 'pz matching',
    'tustin','zoh')
```

8.2 Emulation of Analog PID Controller

The PID controller is a general-purpose controller that is popular in the industrial process control applications. A PID controller provides closed-loop stability and robustness against model mismatches. The industrial PID controller is commonly implemented on a digital computer, a microprocessor, or a PLC.

An analog PID controller with input $e(t)$ and output $u(t)$ is described by the input-output relation:

$$u(t) = k_{\mathrm{p}}e(t) + k_{\mathrm{d}}\frac{de(t)}{dt} + k_i \int e(t)dt.$$

To derive an update rule for computer implementation of the PID controller, we use numerical differentiation via backward difference method, and numerical integration via the Euler's method. Therefore,

For the differentiator, let $u_k = \frac{1}{T}(e_k - e_{k-1})$

For the integrator, let $u_k = u_{k-1} + Te_k$.

The equivalent z-domain transfer function for the PID controller is given as:

$$K(z) = k_{\mathrm{p}} + k_{\mathrm{d}}\left(\frac{z-1}{Tz}\right) + k_{\mathrm{i}}\left(\frac{Tz}{z-1}\right).$$

The corresponding update rule is obtained as follows, where v_k represents the integrator output:

$$u_k = k_{\mathrm{p}}e_k + \frac{k_{\mathrm{d}}}{T}(e_k - e_{k-1}) + k_i v_k$$

$$v_k = v_{k-1} + Te_k.$$

Example 8.5: An analog PID controller for the plant model: $G(s) = \frac{1}{s(s+2)(s+5)}$ was designed earlier as: $K(s) = \frac{12.5(s+0.05)(s+2.05)}{s} = 26.25 + 12.5s + \frac{1.28}{s}$. An equivalent digital controller is given as:

$$K(z) = 26.25 + 12.5\left(\frac{z-1}{Tz}\right) + 1.28\left(\frac{Tz}{z-1}\right)$$

$$= \frac{151.378(z-0.995)(z-0.83)}{z(z-1)}.$$

The update rule for the computer implementation of the PID controller ($T = 0.1s$) is given as:

$$u_k = 26.25e_k + 125(e_k - e_{k-1}) + 1.28v_k$$

$$v_k = v_{k-1} + 0.1e_k.$$

For comparison, two equivalent digital PID controllers were also obtained.

The PID equivalent obtained through pole-zero matching is given as:
$K(z) = \frac{138.52(z-0.995)(z-0.815)}{(z-1)}$.

The PID equivalent obtained through bilinear transform is given as:
$K(z) = \frac{276.31(z-0.995)(z-0.814)}{(z-1)(z+1)}$.

The following plot (Figure 8.2) shows a comparison of the step response for the three PID controllers. The pulse transfer function for the plant was obtained through the ZOH method.

A corresponding Simulink plot shows the step response of the analog plant, when controlled by the digital PID controllers. Since Simulink could only simulate proper transfer functions, the pole-zero matching controller was modified as: $K(z) = \frac{138.52(z-0.995)(z-0.815)}{(z-1)(z+1)}$.

The Simulink plot shows that the forward difference method results in a smoother response as compared to emulation by pole-zero matching and BLT, though the latter have lower overshoots.

The MATLAB code for this example is given below (Figure 8.2):

```
>> G=zpk([],[0 -2 -5],1)
>> Gc=zpk([-.05, -2.05],0,12.5);
>> Gz=c2d(G,.1)
>> Gcz=tf([151.3781 -276.25 125],[1 -1 0],.1)
>> Gcz1=c2d(Gc,.1,'matched');
>> Gcz2=c2d(Gc,.1,'tustin');
>> T=feedback(Gc*G,1);
>> Tz=feedback(Gcz*Gz,1);
>> Tz1=feedback(Gcz1*Gz,1);
>> Tz2=feedback(Gcz2*Gz,1);
>> step(Tz,Tz1,Tz2,T)
>> legend('forward diff','matched','bilinear',
   'continuous')
```

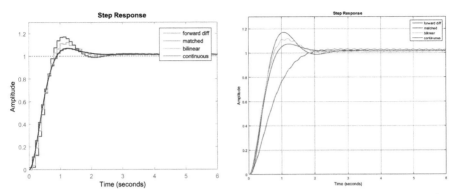

Figure 8.2   Comparison of the step response for PID controller: MATLAB simulation with pulse transfer function (*left*); Simulink simulation with the analog plant (*right*).

8.3 Root Locus Design of Digital Controllers

Recall that the root locus plot for a given loop transfer function is a locus of the roots of the closed-loop characteristic polynomial with variation in the controller gain. The root locus can be plotted in the s-plane or in the z-plane. The rules for plotting root locus in the z-plane are similar to those for the s-plane.

To plot the root locus in the z-plane, the closed-loop characteristic polynomial is given as: $\Delta(z) = 1 + KG(z)$. To ensure stability, the polynomial should have its roots inside the unit circle.

8.3.1 Design for a Desired Damping Ratio

Assuming that the controller design specifications are stated in terms of a desired damping ratio ζ, we consider the z-plane root locus design based on ζ.

To proceed further, we consider a prototype second-order analog transfer function: $T(s) = \frac{\omega_n^2}{s^2 + 2\zeta\omega_n s + \omega_n^2}$. The s-plane root locations for this transfer function are given as: $s = \sigma \pm j\omega = -\zeta\omega_n \pm \omega_n\sqrt{1 - \zeta^2}$.

The constant ζ lines are characterized by the relation: $\frac{\sigma}{\omega} = \frac{\zeta}{\sqrt{1-\zeta^2}}$. In the s-plane, these are radial lines at angles defined by $\theta = \cos^{-1}\zeta$.

To plot the constant ζ lines in the z-plane, we consider the equivalence: $z = e^{Ts} = e^{\sigma T}e^{j\omega T}$, where $\sigma = \frac{\zeta}{\sqrt{1-\zeta^2}}\omega$. The resulting plot, when ωT ranges from 0 to π, gives the constant ζ lines in the z-plane.

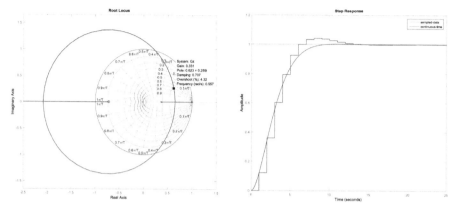

Figure 8.3 Root locus design in the z-plane (*left*); step response of the closed-loop system (*right*).

For example, let $\zeta = \frac{1}{\sqrt{2}}$; then $\sigma = \omega$, and the constant ζ line is defined by $z = e^{-\omega T} \angle \pm \omega T$, where ωT ranges from 0 to π. The closed-loop roots crossing this line can then be selected from the root locus plot.

Alternatively, we can use the MATLAB 'zgrid' command to obtain the constant ζ contours in the z-plane, as shown in the following example:

Example 8.6: Let $G(z) = \frac{0.368z + 0.264}{(z-1)(z-0.368)}$; then, the closed-loop characteristic polynomial is given as:

$$\Delta(z) = z^2 + (0.368K - 1.368)z + 0.264K + 0.368.$$

The following commands are used to plot the z-plane root locus in MATLAB (Figure 8.3):

```
>> Gz=tf([.368 .264],[1 -1.368 .368], -1)
>> rlocus (Gz), zgrid
```

From the RL plot, the controller gain where the z-plane RL crosses $\zeta = 0.7$ line is found as: $K = 0.33$.

The resulting closed-loop characteristic polynomial is: $\Delta(z) = z^2 - 1.246z + 0.455$, with roots at: $z = 0.623 \pm j0.258$. The step response of the closed-loop system is plotted as (Figure 8.3).

The continuous-time system with the same controller gain $K = 0.33$ has $\zeta = 0.87$ with $< 1\%$ overshoot in the step response (Figure 8.3).

The z-plane grid in Figure 8.3 additionally shows constant frequency (ω_n) lines. These lines are helpful in selecting a suitable sampling time (T)

for a given natural frequency. Accordingly, the desirable region for z-plane root locations is bounded as: $0.1\pi \leq w_nT \leq 0.5\pi$, $\xi \geq 0.6$. A sampling frequency higher than $10w_n$ can also be selected but requires faster processing.

8.3.2 Settling Time and Damping Ratio

The settling time and the damping ratio of the dominant closed-loop roots are often used as quality metrics in the design of continuous-time systems. These metrics may be deduced from the z-plane pole locations by making the following comparison with reference to the prototype second-order system:

$$re^{\pm j\theta} = z = e^{Ts} = e^{-\zeta w_nT}e^{\pm jw_dT}, \quad w_d = w_n\sqrt{1 - \zeta^2}.$$

Separating the above relation into its real and imaginary part gives two equations: $\ln r = -\zeta w_nT$ and $\theta = w_dT$, which can be solved to obtain:

$$\zeta = -\ln r/\sqrt{\ln^2 r + \theta^2}$$

$$w_n = \sqrt{\ln^2 r + \theta^2}/T$$

$$\tau = \frac{1}{\zeta w_n} = -\frac{T}{\ln r}.$$

Example 8.7: Let $T(z) = \frac{0.368z+0.264}{z^2-z+0.632}$; so that the closed-loop roots in the z-plane are given at: $z = 0.5\pm j0.618 = 0.795\,e^{\pm j0.89}$. Then, from the above relations, $\zeta = 0.25$, $w_n = 0.919$, $\tau = 4.36$ s.

Skill Assessment Questions

Link to the answers:
http://www.riverpublishers.com/book_details.php?book_id=449

8.1 An analog PID controller is given by: $G_c(s) = 20 + 10s + \frac{1}{s}$.

   a. Obtain an equivalent digital PID controller via emulation (use $T = 0.1$ s).

   b. Provide an update rule for controller implantation on digital computer.

8.2 A lead-lag compensator for an inertial mass $G(s) = \frac{1}{s^2}$ was designed as: $G_c(s) = \frac{20(s+.1)(s+1)}{(s+.01)(s+6)}$. Let $T = 0.05$ s; obtain the equivalent digital compensator via the following methods:

   a. Pole-zero matching

   b. Zero-order-hold

   c. Bilinear transform

   Plot and compare the closed-loop step response of the compensators.

8.3 Consider the simplified model of a flexible beam given as: $G(s) = \frac{100}{s^2+s+100}$.

   a. Design an PID controller for the model (use MATLAB 'pidtune' command)

   b. Discretize the model using the MATLAB 'c2d' command (use $T = 0.001$ s).

   c. Discretize the controller using the forward difference method (use $T = 0.001$ s).

   d. Plot the step response of the closed-loop system.

8.4 Consider the simplified model of an automobile given as: $G(s) = \frac{28s+120}{s^2+7s+14}$.

   a. Discretize the model using the MATLAB 'c2d' command (use $T = .01$ s).

   b. Use root locus technique to design a digital controller to achieve: $\zeta \cong 0.7$.

   c. Plot the step response of the closed-loop system.

8.5 Consider the plant transfer function $G(s) = \frac{s+3}{s(s+1)(s+2)}$.

    a. Discretize the model using the MATLAB 'c2d' command (use $T = 0.5$ s).

    b. Use root locus technique to design a digital controller to achieve: $\zeta \cong 0.7$.

    c. Plot the step response of the closed-loop system.

8.6 Consider the model of human postural dynamics described as an inverted pendulum, given as: $G(s) = \Omega^2/(s^2 - \Omega^2)$, where $\Omega = \sqrt{10}$. Assume that a Phase-lead controller for the model was designed as: $K(s) = 7(s + 3.16)/(s + 13.16)$.

    a. Discretize the model using the MATLAB 'c2d' command (use $T = 0.1$ s).

    b. Discretize the controller using MATLAB 'c2d' command (use $T = 0.1$ s with 'tustin' option).

    c. Plot the step response of the closed-loop system.

9

Control System Design in State-Space

Learning Objectives

1. Perform pole placement design using full state feedback for a given state variable model.
2. Transform a given state variable model into its controller form.
3. Design a tracking controller for the state variable model.

The state-space model of a dynamic system comprises of time derivatives of a set of state variables. The state variables are often the natural variables associated with the energy storage elements present in the system. Examples are the capacitor voltages and inductor currents in the electrical circuits, and the displacement and velocity of the inertial elements in the case of mechanical systems.

The controller design for a given state variable model involves feeding back all (or a selection of) the state variables. The controller design thus involves design of multiple feedback loops.

The pole placement design through full state feedback refers to the selection of n feedback gains for placing the n roots of the closed-loop characteristic equation at the desired locations in the complex plane. The design is facilitated by first transforming the model into the controller form.

Reference waveform for the output variable may be provided for tracking system design that involves the use of integral control to reduce the steady-state error to zero.

Compared to the root locus design that allows selective placement of the closed-loop poles of the system, arbitrary pole placement is made possible through full state feedback design.

In the following, we assume that the system to be controlled is of single-input single-output (SISO) type.

107

9.1 Pole Placement with Full State Feedback

Let $x(t)$ denote a vector of state variables, $u(t)$ denote the input, and $y(t)$ denote the output, then the state-space model of a SISO systems is written as:

$$\dot{x}(t) = Ax(t) + bu(t)$$

$$y(t) = c^T x(t).$$

In the above, A is the system matrix, b is the input distribution matrix (a column vector), and c^T is the output distribution matrix (a row vector).

Using full state feedback, the controller output (input to the plant) is generated as:

$$u = -k^T x + r,$$

where $k^T = [k_1, \ k_2, \ \ldots, k_n]$ is a vector of n feedback gains to be selected, one for each of the state variables, and r is a scalar reference input, which may be time varying.

Using the full state feedback, the closed-loop system dynamics are described as:

$$\dot{x}(t) = (A - bk^T)x(t) + br.$$

The design problem, then, is to select the feedback gains k for the state variable vector x, such that the closed-loop system matrix $A - bk^T$ has a characteristic polynomial that aligns with a desired polynomial, i.e.,

$$\left| sI - A + bk^T \right| = \Delta_{\text{des}}(s)$$

The above equation represents an nth order polynomial equation; by equating the coefficients on both sides of the equation, we can solve for the n feedback gains: $k_i, \ i = 1, \ldots, n$.

The desired characteristic polynomial for pole placement design can be selected with desired root locations that meet the time-domain design specifications, as shown in the following example.

Example 9.1: Pole placement design of a DC motor
The state and output equations for a small DC motor model are:

$$\frac{d}{dt} \begin{bmatrix} i_a \\ \omega \end{bmatrix} = \begin{bmatrix} -1000 & -50 \\ 5 & -10 \end{bmatrix} \begin{bmatrix} i_a \\ \omega \end{bmatrix} + \begin{bmatrix} 1000 \\ 0 \end{bmatrix} V_a$$

$$\omega = \begin{bmatrix} 0 & 1 \end{bmatrix} \begin{bmatrix} i_a \\ \omega \end{bmatrix}.$$

The DC motor model has an open-loop characteristic polynomial $\Delta(s) = s^2 + 1010s + 10250$, with roots at -10.25, -999.75 that correspond to motor time constants ($\tau_e \cong 1$ ms, $\tau_m \cong 100$ ms).

Assume that we wish to improve the motor transient response by moving the dominant pole to -50; the remaining pole can be left at its current location or selected at -1000. Thus, a suitable choice for the desired polynomial is: $\Delta_{des}(s) = s^2 + 1050s + 50000$.

The controller, with $r = 0$, is defined as: $V_a = -k^T x$, where $k^T = [k_1, k_2]$.

The resulting closed-loop system model is given as:

$$A - bk^T = \begin{bmatrix} -1000(1 + k_1) & -50(1 + 20k_2) \\ 5 & -10 \end{bmatrix}$$

The closed-loop characteristic polynomial is obtained as:

$$\left| sI - A + bk^T \right| = (s + 10)\left[s + 1000(1 + k_1)\right] + 250(1 + 20k_2).$$

By comparing the closed-loop polynomial with the $\Delta_{des}(s)$, we obtain the following feedback gains: $k_1 = 0.04$, $k_2 = 7.87$.

9.1.1 Pole Placement in MATLAB

The MATLAB Control System Toolbox 'place' command is used for the pole placement design. The command is invoked with the system and input matrices, and returns the feedback gain vector that places the eigenvalues of the system matrix, equivalently, the roots of the closed-loop characteristic polynomial, at the desired root locations.

The MATLAB commands for pole placement for the DC motor example are given below:

```
>> A=[-1000 -50; 5 -10]; B=[1000; 0];
>> k=place(A, B, [-50, -1000])
```

9.2 Controller Form Pole Placement Design

The control system design in state-space gets easier if the system and input matrices are in their controller form, as defined by the following structure:

$$A = \begin{bmatrix} 0 & 1 & 0 & \cdots \\ 0 & 0 & 1 & \cdots \\ \vdots & \vdots & \ddots & 1 \\ -a_n & -a_{n-1} & \cdots & -a_1 \end{bmatrix}, \quad b = \begin{bmatrix} 0 \\ 0 \\ \vdots \\ 1 \end{bmatrix}.$$

We note that in controller form the coefficients of the characteristic polynomial appear in reverse order in the last row of A matrix.

Using full state feedback through $u = -k^T x + r$, the closed-loop system matrix is given as:

$$A - bk^T = \begin{bmatrix} 0 & 1 & 0 & \cdots \\ 0 & 0 & 1 & \cdots \\ \vdots & \vdots & \ddots & 1 \\ -a_n - k_1 & -a_{n-1} - k_{n-1} & \cdots & -a_1 - k_n \end{bmatrix}.$$

The closed-loop characteristic polynomial includes the controller gains, and is given as:

$$\Delta(s) = s^n + (a_1 + k_n)s^{n-1} + \cdots + a_n + k_1$$

Assume that the desired characteristic polynomial is defined as:

$$\Delta_{\text{des}}(s) = s^n + \bar{a}_1 s^{n-1} + \cdots + \bar{a}_{n-1}s + \bar{a}_n.$$

The feedback gains, obtained by comparing the polynomial coefficients, are:

$$k_1 = \bar{a}_n - a_n, \quad k_2 = \bar{a}_{n-1} - a_{n-1}, \ldots, \quad k_n = \bar{a}_1 - a_1.$$

Since the state variables in the controller form include the output and its derivatives, pole placement through full state feedback may be considered a generalization of the proportional-derivative (PD) controller.

Example 9.2: The mass–spring–damper system

Consider the mass–spring–damper model, where the following parameter values are assumed: $m = 1, b = 1, k = 10$. The resulting state-space model is given as:

$$\frac{d}{dt}\begin{bmatrix} x \\ v \end{bmatrix} = \begin{bmatrix} 0 & 1 \\ -10 & -1 \end{bmatrix}\begin{bmatrix} x \\ v \end{bmatrix} + \begin{bmatrix} 0 \\ 1 \end{bmatrix} f$$

$$x = \begin{bmatrix} 1 & 0 \end{bmatrix}\begin{bmatrix} x \\ v \end{bmatrix}.$$

The characteristic polynomial of the mass-spring-damper system can be written by inspection as the system is already in the controller form, and is given as: $\Delta(s) = s^2 + s + 10$.

The system has low damping, and, in order to improve the damping, the desired characteristic polynomial is selected as: $\Delta_{des}(s) = s^2 + 4s + 10$.

The desired feedback gains can be written by inspection, and are given as: $k^T = \begin{bmatrix} 0 & 3 \end{bmatrix}$.

9.2.1 Linear Transformation to the Controller Form

We first note that given a transfer function model, the controller form state-space structure can be realized by choosing the output variable and its time derivatives as state variables (see Chapter 3).

Alternatively, a given state variable model can be transformed into its controller form representation via a linear state transformation, $z = Px$.

A necessary condition to find a linear transformation to convert a given state-space model into controller form is that the system model is controllable, which holds if the following controllability matrix is of full rank:

$$M_C = [b, \; Ab, \; \ldots, \; A^{n-1}b].$$

We note that for the SISO systems, the controllability matrix is $n \times n$.

The controllability matrix is guaranteed to be of full rank if the transfer function description has the same order, n, as the length of the state vector in the state variable representation of the system.

Next, assume that the controllability matrix is of full-rank, indicating that an equivalent controller form representation is attainable. Let the controllability matrix for the desired controller form representation be given as: M_{CF}. We note that M_{CF} involves the coefficients of the characteristic polynomial and can be written by inspection.

Then, the matrix that transforms the given state-space model into its controller form representation is given as:

$$Q = P^{-1} = M_C M_{CF}^{-1}.$$

After transforming into the controller form, we can perform pole placement using new state variables as: $u = -k'^T z$. The corresponding feedback gains for the original system are given as: $k^T = k'^T P$.

Example 9.3: The DC motor model

The state and output equations for a small DC motor model are:

$$\frac{d}{dt}\begin{bmatrix} i_a \\ \omega \end{bmatrix} = \begin{bmatrix} -1000 & -50 \\ 5 & -10 \end{bmatrix}\begin{bmatrix} i_a \\ \omega \end{bmatrix} + \begin{bmatrix} 1000 \\ 0 \end{bmatrix}V_a$$

$$\omega = \begin{bmatrix} 0 & 1 \end{bmatrix}\begin{bmatrix} i_a \\ \omega \end{bmatrix}$$

The controllability matrix for the model is given as: $M_C = \begin{bmatrix} b, & Ab \end{bmatrix} = \begin{bmatrix} 10^3 & -10^6 \\ 0 & 5 \times 10^3 \end{bmatrix}.$

This matrix is of full rank. The transfer function for the motor model is:
$G(s) = \frac{5000}{s^2 + 1010s + 10250}.$

From the transfer function description, we obtain the accompanying controller form realization as:

$$\dot{x} = \begin{bmatrix} 0 & 1 \\ -10250 & -1010 \end{bmatrix}x + \begin{bmatrix} 0 \\ 1 \end{bmatrix}V_a$$

$$\omega = \begin{bmatrix} 5000 & 0 \end{bmatrix}x.$$

The controllability matrix for the controller form representation is given as:
$M_{CF} = \begin{bmatrix} 0 & 1 \\ 1 & -1010 \end{bmatrix}.$

Then, the state transformation matrix for the DC motor model is computed as: $P^{-1} = \begin{bmatrix} 10000 & 1000 \\ 5000 & 0 \end{bmatrix}$, $P = \begin{bmatrix} 0 & 0.0002 \\ 0.001 & -0.002 \end{bmatrix}$. Indeed, $\bar{A} = PAQ$ gives the system matrix in the controller form.

For the pole placement design of the DC motor, let the desired characteristic polynomial be given as: $\Delta_{des}(s) = s^2 + 1100s + 10^5.$

Then, the feedback gains in the transformed coordinates are given as: $k'^T = \begin{bmatrix} 89750 & 90 \end{bmatrix}.$

The corresponding feedback gains in the original coordinates are: $k^T = k'^T P = \begin{bmatrix} 0.09 & 17.77 \end{bmatrix}.$

Finally, we note that the same result can be obtained by using the following MATLAB commands:

```
>> A=[-1000 -50; 5 -10]; B=[1000; 0];
>> k=place(A, B, [-100, -1000])
```

9.3 Tracking System Design

A tracking system is designed to maintain zero steady-state error with respect to a reference input, i.e., its output follows the reference input. In order to design a tracking system, let the control law be defined as: $u = -\mathbf{k}^T x + k_r r$, where k_r is a feedforward gain for the reference input $r(t)$.

To simplify the design, we may assume that the state vector includes the output variable. In particular, let $y = x_1$; then, the control law can be written as: $u = k_r r - k_1 y - k_2 x_2 - \cdots - k_n x_n$.

The gain k_r may be selected as $k_r = k_1$, so that $r - y = 0$ in the steady-state; alternately, it can be selected to have a unity DC gain for the closed-loop transfer function $T(s)$, i.e., to make $T(0) = 1$.

Example 9.4: The mass–spring–damper system
Consider the mass–spring–damper model in the sate-space, described as:

$$\frac{\mathrm{d}}{\mathrm{d}t} \begin{bmatrix} x \\ v \end{bmatrix} = \begin{bmatrix} 0 & 1 \\ -10 & -1 \end{bmatrix} \begin{bmatrix} x \\ v \end{bmatrix} + \begin{bmatrix} 0 \\ 1 \end{bmatrix} f$$

$$x = \begin{bmatrix} 1 & 0 \end{bmatrix} \begin{bmatrix} x \\ v \end{bmatrix}.$$

Let the control law be given as: $f = -k_1 x - k_2 v + k_r r$.
Then, the closed-loop system is described as:

$$\frac{\mathrm{d}}{\mathrm{d}t} \begin{bmatrix} x \\ v \end{bmatrix} = \begin{bmatrix} 0 & 1 \\ -10 - k_1 & -1 - k_2 \end{bmatrix} \begin{bmatrix} x \\ v \end{bmatrix} + \begin{bmatrix} 0 \\ k_r \end{bmatrix} r.$$

The transfer function of the above closed-loop system is given as: $T(s) = \frac{k_r}{s^2 + (k_2+1)s + (k_1+10)}$.

For a desired characteristic polynomial: $\Delta_{\mathrm{des}}(s) = s^2 + 4s + 10$, the feedback gains for the mass–spring–damper model were earlier computed as: $\mathbf{k}^T = \begin{bmatrix} 0 & 3 \end{bmatrix}$. The accompanying feedforward gain for the tracking system is now selected as: $k_r = 10$, to achieve $T(0) = 1$.

9.3.1 Tracking PI Control

In a more general state-space controller design, if elimination of steady-state error to a step input is desired, then an integral controller needs to be added to the loop (Figure 9.1). The integral controller integrates the error signal; thereby the error is forced to zero in the steady-state.

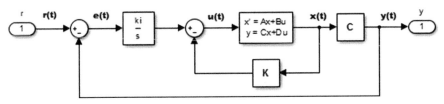

Figure 9.1   The tracking PI control model.

The proportional-integral (PI) type control law for tracking system design using full-state feedback is defined as:

$$u = -k^T x + k_i \int (r - y)\mathrm{d}t$$

The input to the integrator is the error: $e = r - y$, where r represents a reference input.

Let the integrator output be denoted as x_a; then, the integrator differential equation is given as: $\dot{x}_a = r - y = r - c^T x$. The integrator thus adds a new state variable, its output x_a, to the state-space model.

The augmented system, including the integrator, has $n+1$ state variables, and is described as:

$$\begin{bmatrix} \dot{x} \\ \dot{x}_a \end{bmatrix} = \begin{bmatrix} A & 0 \\ -c^T & 0 \end{bmatrix} \begin{bmatrix} x \\ x_a \end{bmatrix} + \begin{bmatrix} b \\ 0 \end{bmatrix} u + \begin{bmatrix} 0 \\ 1 \end{bmatrix} r.$$

Then, full-state feedback controller for the augmented system is given as:

$$u = -\begin{bmatrix} k^T & -k_i \end{bmatrix} \begin{bmatrix} x \\ x_a \end{bmatrix},$$ where k_i represents the integral gain.

Addition of the above controller to the augmented system results in the closed-loop system described as:

$$\begin{bmatrix} \dot{x} \\ \dot{x}_a \end{bmatrix} = \begin{bmatrix} A - bk^T & bk_i \\ -c^T & 0 \end{bmatrix} \begin{bmatrix} x \\ x_a \end{bmatrix} + \begin{bmatrix} 0 \\ 1 \end{bmatrix} r$$

The characteristic polynomial of the above system is given as:

$$\Delta(s) = s \left| sI - A + bk^T \right| + c^T bk_i = 0,$$

where I is an identity matrix of order n.

Next, we may choose a desired characteristic polynomial of $(n+1)$ order, and perform the regular pole placement design using the augmented system. The integrator pole may be selected keeping in view the desired settling time of the system, as shown in the following examples.

Example 9.5: The mass–spring–damper system

Consider the mass-spring-damper model above. In order to perform integral control, the augmented state-space model is given as:

$$\frac{d}{dt}\begin{bmatrix} x \\ v \\ x_a \end{bmatrix} = \begin{bmatrix} 0 & 1 & 0 \\ -10 & -1 & 0 \\ -1 & 0 & 0 \end{bmatrix}\begin{bmatrix} x \\ v \\ x_a \end{bmatrix} + \begin{bmatrix} 0 \\ 1 \\ 0 \end{bmatrix}u + \begin{bmatrix} 0 \\ 0 \\ 1 \end{bmatrix}r.$$

The control law is given as: $u = -k_1 x - k_2 v + k_i \int (r - x)\mathrm{d}t$.

The closed-loop system characteristic polynomial is given as: $\Delta(s) = s^3 + (k_2 + 1)s^2 + (k_1 + 10)s + k_i$.

Let the $(n + 1)$th order desired characteristic polynomial be given as: $\Delta_{\mathrm{des}}(s) = (s + 1)(s^2 + 4s + 10)$.

Then, by comparing polynomial coefficients, the controller gains are found as: $k_1 = 4$, $k_2 = 4$, $k_i = 10$.

Alternatively, the controller gains for $\Delta_{\mathrm{des}}(s) = (s + 2)(s^2 + 4s + 10)$ are given as: $k_1 = 8$, $k_2 = 5$, $k_i = 20$.

The MATLAB commands used for computing the controller gains and plotting the step response are:

```
>> A=[0 1; -10 -1]; B=[0; 1];
>> k=place([A 0*B;-1 0 0], [B;0], [-1 -2+j*sqrt(6)
     -2-j*sqrt(6)])
>> step(ss([A 0*B;-1 0 0]-[B;0]*k, [0 0 1]',
     [1 0 0],0))
```

The step response of the closed-loop system is compared for two sets of PI gains (Figure 9.2).

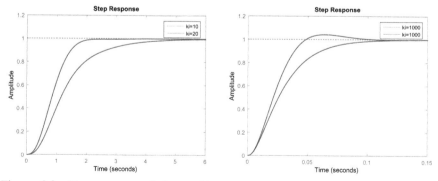

Figure 9.2 Step response of the tracking system: mass-spring-damper model (*left*); DC motor model (*right*).

Example 9.6: The DC motor model

The state variable model for a small DC motor is given as:

$$\frac{d}{dt}\begin{bmatrix} i_a \\ \omega \end{bmatrix} = \begin{bmatrix} -1000 & -50 \\ 5 & -10 \end{bmatrix}\begin{bmatrix} i_a \\ \omega \end{bmatrix} + \begin{bmatrix} 1000 \\ 0 \end{bmatrix}V_a$$

$$\omega = \begin{bmatrix} 0 & 1 \end{bmatrix}\begin{bmatrix} i_a \\ \omega \end{bmatrix}$$

The control law for the tracking PI controller is given as: $u = -k_1 i_a - k_2\omega + k_i \int (r - \omega)dt$.

The augmented system model for pole placement design using integral control is given as:

$$\frac{d}{dt}\begin{bmatrix} i_a \\ \omega \\ x_a \end{bmatrix} = \begin{bmatrix} -1000 & -50 & 0 \\ 5 & -10 & 0 \\ 0 & -1 & 0 \end{bmatrix}\begin{bmatrix} i_a \\ \omega \\ x_a \end{bmatrix} + \begin{bmatrix} 1000 \\ 0 \\ 0 \end{bmatrix}u + \begin{bmatrix} 0 \\ 0 \\ 1 \end{bmatrix}r.$$

In order to choose a desired characteristic polynomial, the pole of the PI controller is placed at –50 for an acceptable settling time. Accordingly, let $\Delta_{des}(s) = (s + 50)(s + 100)(s + 1000)$. The resulting feedback gains are computed as: $k_1 = 0.14$, $k_2 = 28.7$, $k_i = -1000$.

The closed-loop system with the above gains has a settling time of about 0.12 s. Alternatively, we may chose a characteristic polynomial with complex poles as $\Delta_{des}(s) = (s + 1000)(s^2 + 100s + 5000)$. The resulting feedback gains are given as: $k_1 = 0.09, k_2 = 18.77, k_i = -1000$.

We may try to increase the integral gain in order to reduce the settling time; however, a high integral gain may result in complex dominant poles displaying low damping and an oscillatory step response.

The step response of the DC motor model for two values of the integral gain is compared (Figure 9.2).

Finally, the transfer function model of the DC motor is obtained as: $G(s) = \frac{5000}{s^2+1010s+1250}$. Let an integral compensator be defined as: $K(s) = \frac{k_i}{s}$. Then, root techniques can be used to find an output feedback controller for the DC motor model. A corresponding root locus plot with integrator in the loop shows $\zeta \cong 0.22$ for an integral gain of $k_i = 100$ (Figure 9.3).

The MATLAB commands for the DC motor example are given below:

```
>> A=[-1000 -50;5 -10]; B=[1000 0]'; C=[0 1];
>> place([A [0 0]';-C 0],[B;0],[-50 -100 -1000])
```

```
>> step(ss([A 0*B;0 -1 0]-[B;0]*k, [0 0 1]',
   [0 1 0],0))
>> figure, rlocus(ss(A,B,[0 1],0)*tf(1, [1 0]))
```

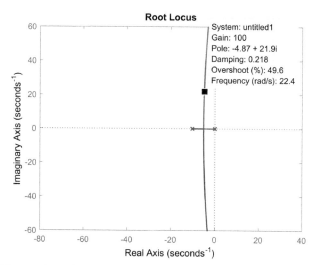

Figure 9.3 The root locus for the DC motor model with integrator in the loop. An integrator gain of 100 results in oscillatory response with low damping.

Skill Assessment Questions

Link to the answers:
http://www.riverpublishers.com/book_details.php?book_id=449

9.1 The linearized model of an inverted pendulum is described as: $\dfrac{d}{dt}\begin{bmatrix} \theta \\ \omega \end{bmatrix} =$
$\begin{bmatrix} 0 & 1 \\ \frac{g}{l} & 0 \end{bmatrix}\begin{bmatrix} \theta \\ \omega \end{bmatrix} + \begin{bmatrix} 0 \\ 1 \end{bmatrix}u$, where $l = 1m$, $g = 9.8\frac{m}{s^2}$. Design a full state
feedback controller to achieve ITAE specifications for $\omega_n = 5\frac{rad}{s}$.

9.2 The model of a small armature-controlled DC motor is given as:
$\dfrac{d}{dt}\begin{bmatrix} i_a \\ \omega \end{bmatrix} = \begin{bmatrix} -R/L & -k_b/L \\ k_t/J & -b/J \end{bmatrix}\begin{bmatrix} i_a \\ \omega \end{bmatrix} + \begin{bmatrix} 1/L \\ 0 \end{bmatrix}V_a.$
The following parameter values are assumed: $R = 1, L = .01, J = .01,$
$b = .1, k_t = k_b = 0.02$. Design a full state feedback controller for the
motor to achieve: $t_s \le 20\ ms$, $\zeta \ge 0.8$.

9.3 The simplified model of airplane longitudinal dynamics using angle of
attack (α) and pitch rate (q) as state variables is described as:

$$\frac{d}{dt}\begin{bmatrix} \alpha \\ q \end{bmatrix} = \begin{bmatrix} \frac{Z_\alpha}{V} & 1 \\ M_\alpha & M_q \end{bmatrix}\begin{bmatrix} \alpha \\ q \end{bmatrix} + \begin{bmatrix} \frac{Z_E}{V} \\ M_E \end{bmatrix}\delta_E.$$

Assume the following parameter values: $\frac{Z_\alpha}{V} = -1, \frac{Z_E}{V} = -.1, M_\alpha = -7,$
$M_q = -.5, M_E = -5.$

 a. Design a full state feedback controller to achieve: $\omega_n \ge 7\frac{rad}{s}$,
 $\zeta \ge 0.7$.
 b. Design a tracking PI controller to track the angle of attack
 reference. Plot the step response.

9.4 Consider an inverted pendulum over cart model described in terms of
cart position (y) and pendulum angle (θ) via the following equations:
$(M + m)\ddot{y} + ml\ddot{\theta} = f$, $m\ddot{y} + ml\ddot{\theta} - mg\theta = 0$. Assume the following
parameter values: $m = 1$ kg; $l = 1$ m; $M = 10$ kg; $g = 9.8$. Design a
full state feedback controller to achieve: $t_s \le 1s$, $\%OS \le 2\%$.

9.5 The simplified model of a flexible beam is given as: $\dfrac{d}{dt}\begin{bmatrix} x \\ v \end{bmatrix} =$
$\begin{bmatrix} 0 & 1 \\ -w_n^2 & -2\zeta w_n \end{bmatrix}\begin{bmatrix} x \\ v \end{bmatrix} + \begin{bmatrix} 0 \\ w_n \end{bmatrix} u$, where $w_n^2 = 10^4$, $\zeta = .005$.

Design a tracking P1 controller for the beam to achieve: $w_n \geq 100\frac{rad}{s}$, $\%OS \leq 5\%$, $e(\infty)|_{step} = 0$.

9.6 Consider the model of human postural dynamics described as an inverted pendulum, given as: $G(s) = \Omega^2/(s^2 - \Omega^2)$, where $\Omega = \sqrt{10}$.

a. Obtain a state-space model of the plant.

b. Design an pole placement controller to achieve: $w_n \geq 5\frac{rad}{s}$, $\xi \geq 0.85$.

c. Design a tracking PI controller to track the postural command θ_{ref} with no error.

10

Digital Controller Design in State-Space

Learning Objectives

1. Describe a sampled-data system in the state variable form.
2. Obtain solution to the discrete state equations.
3. Perform digital controller design in the state-space.

The digital controller design of sampled-data systems described by their pulse transfer functions was addressed earlier (in Chapter 7). In this chapter, we will discuss the controller design for the state variable models of sample-data systems.

The discrete-time state-space description of a system is obtained via discretizing the continuous-time state equations with a zero-order-hold (ZOH), and is valid at the sampling intervals. A solution to the discrete-state equations is easily obtained by iteration.

Pole placement design, similar to the continuous-time case, can be performed on discrete state variable models. The desired characteristic polynomial in the discrete case should have its roots inside the unit circle in order to ensure stability of the closed-loop system.

A peculiarity in the case of discrete systems is the deadbeat controller design that places all roots of the closed-loop characteristic polynomial at the origin, and thus ensures that the system response reaches the steady state in exactly n iterations.

10.1 Sampled-Data Systems in State-Space

System models developed in state variable form can be converted to their discrete-time equivalent system models by considering the effect of ZOH at the input. In particular, the output of ZOH is held constant for one time period.

To develop this approach, let the analog system model be given as:

$$\dot{x}(t) = Ax(t) + Bu(t)$$

$$y(t) = Cx(t).$$

In the above equations, A is the system matrix, B is the input matrix, and C is the output matrix. In the case of SISO system models, we may substitute $B = b$, $C = c^{\mathrm{T}}$.

Recall, from Chapter 3, that the time-domain solution to the state equation is given as:

$$x(t) = e^{At}x_0 + \int_0^\tau e^{A(t-\tau)}Bu(\tau)\mathrm{d}\tau.$$

Since the plant input, generated by the ZOH, is step-wise constant over each time period T, we may assume that system state is available at $(k-1)T$ and compute the system state at $t = kT$ via the convolution integral; the result is given as:

$$x_k = e^{At}x_{k-1} + \int_{(k-1)T}^{kT} e^{AT}Bu_k\mathrm{d}\tau = e^{At}x_{k-1}$$

$$+ \int_0^T e^{AT}\mathrm{d}\tau Bu_k = A_dx_k + B_du_k,$$

Then, given a continuous-time state variable model, the corresponding discrete-time state variable model is obtained as:

$$x_{k+1} = A_dx_k + B_du_k, \quad y_k = Cx_k.$$

The system and input matrices appearing in the discrete model are defined as:

$$A_{\mathrm{d}} = e^{AT}, \quad B_d = \int_0^T e^{A\tau}\mathrm{d}\tau B$$

The expression for B_d may be further simplified (assuming matrix A is invertible) as:

$$B_d = \left[\int_0^T \left(I + A\tau + \frac{A^2\tau^2}{2!} + \ldots \right) \mathrm{d}\tau \right]$$

$$B = \left(IT + \frac{AT^2}{2!} + \dots \right) B = A^{-1}(e^{AT} - I)B.$$

Discretization of state variable models is illustrated via the following example.

Example 10.1: The DC motor model
The state-space model of a small DC motor is given as:

$$\frac{d}{dt} \begin{bmatrix} i \\ \omega \end{bmatrix} = \begin{bmatrix} -1000 & -50 \\ 5 & -10 \end{bmatrix} \begin{bmatrix} i \\ \omega \end{bmatrix} + \begin{bmatrix} 1000 \\ 0 \end{bmatrix} V$$

$$\omega = \begin{bmatrix} 0 & 1 \end{bmatrix} \begin{bmatrix} i \\ \omega \end{bmatrix}.$$

Let $T = 0.02$ s; then, the system and input matrices for the discrete model are computed as:

$$A_{\mathrm{d}} = \begin{bmatrix} -0.0002 & -0.0412 \\ 0.0041 & 0.8148 \end{bmatrix}, \quad B_{\mathrm{d}} = \begin{bmatrix} 0.9959 \\ 0.0863 \end{bmatrix}.$$

The resulting discrete state-space model is given as:

$$\begin{bmatrix} i_{k+1} \\ \omega_{k+1} \end{bmatrix} = \begin{bmatrix} -0.0002 & -0.0412 \\ 0.0041 & 0.8148 \end{bmatrix} \begin{bmatrix} i_k \\ \omega_k \end{bmatrix} + \begin{bmatrix} 0.9959 \\ 0.0863 \end{bmatrix} V_k$$

$$y_k = \begin{bmatrix} 0 & 1 \end{bmatrix} \begin{bmatrix} i_k \\ \omega_k \end{bmatrix}.$$

We may note that in the above example, the sampling time $T = 0.02$ s was selected keeping in view the dominant motor time constant: $\tau_m \cong 0.1$ s.

The discretized state variable model can be obtained via the MATLAB 'c2d' command, as follows:

```
>> A=[-1000 -50; 5 -10]; B=[1000; 0];
>> [Ad,Bd]=c2d(A,B,.02)
```

10.2 Solution to the Discrete State Equations

The discrete state equations can be solved via iteration by assuming an initial state vector and an input sequence as follows: let the discrete-time state-equation be given as:

$$\boldsymbol{x}_{k+1} = \boldsymbol{A}_{\mathrm{d}}\boldsymbol{x}_k + \boldsymbol{B}_{\mathrm{d}}u_k$$

$$y_k = \boldsymbol{C}\boldsymbol{x}_k.$$

An iterative solution is developed as:

$$\boldsymbol{x}_1 = \boldsymbol{A}_{\mathrm{d}}\boldsymbol{x}_0 + \boldsymbol{B}_{\mathrm{d}}u_0$$

$$\boldsymbol{x}_2 = \boldsymbol{A}_{\mathrm{d}}^2\boldsymbol{x}_0 + \boldsymbol{A}_{\mathrm{d}}\boldsymbol{B}_{\mathrm{d}}u_0 + \boldsymbol{B}_{\mathrm{d}}u_1$$

$$\vdots$$

$$\boldsymbol{x}_n = \boldsymbol{A}_{\mathrm{d}}^n\boldsymbol{x}_0 + \sum_{k=0}^{n-1} \boldsymbol{A}_{\mathrm{d}}^{n-1-k}\boldsymbol{B}_{\mathrm{d}}u_k.$$

The state transition matrix in the discrete case is defined as: $\Phi(k) = \boldsymbol{A}_{\mathrm{d}}^k$.
In terms of the state transition matrix, the state vector evolves as:

$$\boldsymbol{x}_n = \Phi(n)\boldsymbol{x}_0 + \sum_{k=0}^{n-1} \Phi(n-1-k)\boldsymbol{B}_{\mathrm{d}}u_k.$$

Example 10.2: The DC motor model
The discrete state-space model of a small dc motor is described by:

$$\boldsymbol{A}_d = \begin{bmatrix} -0.0002 & -0.0412 \\ 0.0041 & 0.8148 \end{bmatrix}, \quad \boldsymbol{B}_d = \begin{bmatrix} 0.9959 \\ 0.0863 \end{bmatrix}, \quad \boldsymbol{C} = \begin{bmatrix} 0 & 1 \end{bmatrix}$$

Assume that the initial conditions are zero, and the input is a unit-step
sequence: $u_k = \{1, 1, \ldots\}$; then, the output sequence is iteratively computed
in MATLAB as follows:

```
>> A=[-1000 -50; 5 -10]; B=[1000; 0]; C=[0 1];
>> [Ad,Bd]=c2d(A,B,.02);
>> x=[0 0]';
>> for k=1:10, x=[x Ad*x(:,end)+Bd]; end
>> y=C*x
```

The results of the iteration are given as:

$\{0, 0.0863, 0.1608, 0.2214, 0.2708, 0.3110, 0.3438, 0.3705, 0.3922, 0.4100, 0.4244\}$

10.3 Pulse Transfer Function from the State Equations

The pulse transfer function $G(z)$ of the sampled-data system can be obtained from the discrete state-space description via the application of z-transform, which gives:

$$z\boldsymbol{x}(z) - z\boldsymbol{x}_0 = \boldsymbol{A}\boldsymbol{x}(z) + \boldsymbol{B}u(z)$$

The above equation is solved for zero initial conditions to obtain:

$$\boldsymbol{x}(z) = (z\boldsymbol{I} - \boldsymbol{A})^{-1}\boldsymbol{B}u(z)$$

$$y(z) = \boldsymbol{C}(z\boldsymbol{I} - \boldsymbol{A})^{-1}\boldsymbol{B}u(z) = G(z)u(z).$$

Thus, $G(z) = \boldsymbol{C}(z\boldsymbol{I} - \boldsymbol{A})^{-1}\boldsymbol{B}$.

The state transition matrix for the sampled-data system is obtained by taking the inverse z-transform of $(z\boldsymbol{I} - \boldsymbol{A})^{-1}$, and is given as:

$$\phi(k) = z^{-1}\left\{ (z\boldsymbol{I} - \boldsymbol{A})^{-1} \right\} = \boldsymbol{A}^k.$$

The unit pulse response the sampled-data system is obtained as:

$$g_k = \boldsymbol{C}\boldsymbol{A}^{k-1}\boldsymbol{B}, \ \ k \geq 0.$$

The pulse transfer function is obtained in the MATLAB Control System Toolbox by invoking the 'c2d' command, by first defining a Dynamic System Object via, e.g., the 'ss' command, and selecting a sampling time T. The MATLAB commands for this purpose are given below:

```
>> G=ss(A,B,C,0);
>> Gz=c2d(G,T);
>> tf(Gz)
```

10.4 Digital Controller Design via Pole Placement

Given a discrete-time system model in state variable form, and a desired pulse characteristic polynomial $\Delta_{\text{des}}(z)$, the digital controller gains for full state feedback design can be obtained via pole placement. The pole placement design in the case of digital systems is similar to that of the continuous-time systems.

To illustrate the design process, let the discrete-time SISO system model be given as:

$$x_{k+1} = A_d x_k + b_d u_k$$
$$y_k = c^T x_k.$$

The digital feedback controller for the state variable system model is given as:

$$u_k = -k^T x_k.$$

Where k^T represents a (row) vector of feedback gains. These gains are solved by equating the coefficients of the characteristic polynomial with the desired polynomial:

$$\Delta(z) = |zI - A_d| = \Delta_{des}(z).$$

The $\Delta_{des}(z)$ may be selected as a stable polynomial (in z), whose roots lie inside the unit circle, and satisfy the given damping and/or settling time requirements. If the desired s-plane root locations are known, the corresponding z-plane locations can be obtained as: $z = e^{Ts}$.

The controller design process is illustrated via the following example.

Example 10.3: The discrete-time model of a small DC motor is given as ($T = 0.02$ s):

$$\begin{bmatrix} i_{k+1} \\ \omega_{k+1} \end{bmatrix} = \begin{bmatrix} -0.0002 & -0.0412 \\ 0.0041 & 0.8148 \end{bmatrix} \begin{bmatrix} i_k \\ \omega_k \end{bmatrix} + \begin{bmatrix} 0.9959 \\ 0.0863 \end{bmatrix} V_k$$

$$y_k = \begin{bmatrix} 0 & 1 \end{bmatrix} \begin{bmatrix} i_k \\ \omega_k \end{bmatrix}.$$

The desired s-plane roots for small DC motor model were earlier selected as: $s = -100, -1000$. The corresponding z-plane roots are given as: $z = e^{-2}, e^{-20}$. The desired characteristic polynomial is given as: $\Delta_{des}(z) = z(z - 0.135)$.

The desired feedback gains $k^T = [k_1, k_2]$, computed by using the MATLAB 'place' command, are given as: $k_1 = 0.0376$, $k_2 = 7.4359$.

The update rule for the digital controller is given as:

$$u_k = 0.0376 i_k + 7.4359 \omega_k.$$

The step response of the closed-loop system is shown below (Figure 10.1). The step response of the continuous-time system is also shown, where the continuous-time response has been scaled to match the discrete system response.

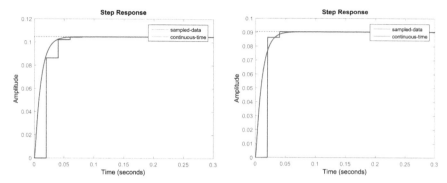

Figure 10.1 The step response of the small DC motor model: digital controller matching continuous controller design (*left*); the deadbeat controller (*right*).

The MATLAB script for this example is given as follows:

```
>> A=[-1000 -50; 5 -10]; B=[1000 0]'; C=[0 1];
>> [Ad,Bd]=c2d([-1000 -50; 5 -10],[1000;0],.02);
>> K=place(A,B, [-100 -1000]);
>> Acl=A-B*K;T0=-C*(Acl\B);
>> Kd=place(Ad,Bd, exp([-100 -1000]*.02));
>> Adl=Ad-Bd*Kd; Td0=C*((I2-Adl)\Bd);
>> step(ss(Adl,Bd,C,0,.02), Td0/T0*ss(Acl,B,C,0))
```

10.4.1 Deadbeat Controller Design

A closed-loop digital system is deadbeat if all closed-loop poles are placed at the origin $(z = 0)$. A deadbeat system has the remarkable property that its response reaches steady-state in n-steps. This property makes it appealing to choose the desired characteristic polynomial as $\Delta_{des}(z) = z^n$ or $T(z) = z^{-n}$.

To design the deadbeat controller, let the closed-loop pulse transfer function $T(z)$ be defined as:

$$T(z) = \frac{K(z)G(z)}{1 + K(z)G(z)}.$$

The above equation is solved for $K(z)$ to obtain:

$$K(z) = \frac{1}{G(z)} \frac{T(z)}{1 - T(z)}.$$

Let $T(z) = z^{-n}$; then, $K(z) = \frac{1}{G(z)(z^n-1)}$

Example 10.4: Let $G(s) = \frac{1}{s+1}$; then $G(z) = \frac{1-e^{-T}}{z-e^{-T}}$

and we obtain the deadbeat controller as: $K(z) = \frac{z-e^{-T}}{(1-e^{-T})(z-1)}$.

Example 10.5: The DC motor model

The discrete-time DC motor model, obtained via the MATLAB 'c2d' command for $T = 0.02$ s is given as:

$$\begin{bmatrix} i_{k+1} \\ \omega_{k+1} \end{bmatrix} = \begin{bmatrix} -0.0002 & -0.0412 \\ 0.0041 & 0.8148 \end{bmatrix} \begin{bmatrix} i_k \\ \omega_k \end{bmatrix} + \begin{bmatrix} 0.9959 \\ 0.0863 \end{bmatrix} V_k$$

$$y_k = \begin{bmatrix} 0 & 1 \end{bmatrix} \begin{bmatrix} i_k \\ \omega_k \end{bmatrix}.$$

The desired characteristic polynomial is selected as: $\Delta_{\text{des}}(z) = z^2$.

In this case, the MATLAB 'place' command cannot be used due to repeated poles. However, direct calculations using $u_k = -[k_1, k_2] x_k$ results in the following closed-loop characteristic polynomial:

$$\Delta(z) = z^2 + (0.9959k_1 + 0.0863k_2 - 0.8146)z - 0.815k_1 + 0.0041k_2$$
$$+ 1.69 \times 10^{-9}.$$

Then, by equating the second and third term in the polynomial to zero, we obtain the feedback controller gains as: $k_1 = 0.045$, $k_2 = 8.9173$. The update rule for the digital controller is given as:

$$u_k = 0.045i_k + 8.917\omega_k.$$

The step response of the deadbeat controller shows no overshoot (Figure 10.1). The step response of the continuous-time system (scaled to match the discrete system response) is also shown in the figure.

The MATLAB script for this example is given as follows:

```
>> Kd=[0.045 8.9173];
>> Adl=Ad-Bd*Kd; Td0=C*((I2-Adl)\Bd);
>> step(ss(Adl,Bd,C,0,.02), Td0/T0*ss(Acl,B,C,0))
```

Skill Assessment Questions

Link to the answers:
http://www.riverpublishers.com/book_details.php?book_id=449

10.1 Consider the linearized model of an inverted pendulum given as:

$$\frac{d}{dt}\begin{bmatrix} \theta \\ \omega \end{bmatrix} = \begin{bmatrix} 0 & 1 \\ \frac{g}{l} & 0 \end{bmatrix}\begin{bmatrix} \theta \\ \omega \end{bmatrix} + \begin{bmatrix} 0 \\ 1 \end{bmatrix} u, \text{ where } l = 1 \text{ m}, \ g = 9.8 \ \tfrac{m}{s^2}.$$

    a. Choose a sampling time T and discretize the model.

    b. Obtain the solution to discrete state equations by iteration.

10.2 Consider the simplified model of airplane longitudinal dynamics given
as: $\dfrac{d}{dt}\begin{bmatrix} \alpha \\ q \end{bmatrix} = \begin{bmatrix} \frac{Z_\alpha}{V} & 1 \\ M_\alpha & M_q \end{bmatrix}\begin{bmatrix} \alpha \\ q \end{bmatrix} + \begin{bmatrix} \frac{Z_E}{V} \\ M_E \end{bmatrix}\delta_E. \ \text{Let } \dfrac{Z_\alpha}{V} = -1,$

$\dfrac{Z_E}{V} = -0.1, \ M_\alpha = -7, \ M_q = -0.5, \ M_E = -5.$

    a. Choose a sampling time T and obtain a discrete-time model for the airplane.

    b. Obtain the two pulse transfer functions for the airplane model.

    c. Obtain the output sequences for angle-of-attack α and pitch rate q in response to a unit pulse at δ_E.

10.3 The model of a small armature-controlled DC motor is given as:

$$\frac{d}{dt}\begin{bmatrix} i_a \\ \omega \end{bmatrix} = \begin{bmatrix} -R/L & -k_b/L \\ k_t/J & -b/J \end{bmatrix}\begin{bmatrix} i_a \\ \omega \end{bmatrix} + \begin{bmatrix} 1/L \\ 0 \end{bmatrix} V_a. \ \text{Let } R = 1, L = $$

$0.01, J = 0.01, b = 0.1, \ k_t = k_b = 0.02.$

    a. Discretize the model using the MATLAB 'c2d' command (use $T = 0.01$ s).

    b. Design a state feedback controller to place the poles at: z_1, $z_2 = e^{-1}$.

    c. Plot the step response for the motor speed ω.

    d. Design a deadbeat controller and plot the step response for ω.

10.4 Consider the simplified model of a flexible beam given as: $\dfrac{d}{dt}\begin{bmatrix} x \\ v \end{bmatrix} =$

$\begin{bmatrix} 0 & 1 \\ -w_n^2 & -2\zeta w_n \end{bmatrix}\begin{bmatrix} x \\ v \end{bmatrix} + \begin{bmatrix} 0 \\ 1 \end{bmatrix} u$, where $w_n^2 = 10^4$, $\zeta = .005$.

    a. Discretize the model using $T = 1$ ms.

    b. Design an pole placement controller to achieve: $\zeta \geq 0.5$, $t_s \leq$ 20 ms.

    c. Plot the step response of the system.

10.5 Consider the model of airplane longitudinal dynamics in Question 10.2 above.

    a. Discretize the model using the MATLAB 'c2d' command (use $T = 0.05$ s).

    b. Design a state feedback controller to achieve: $w_n \geq 5\frac{\text{rad}}{\text{s}}$, $\zeta \geq 0.8$.

    c. Plot the step response of the state variables.

    d. Design a deadbeat controller and plot the step response of state variables.

10.6 Consider the model of human postural dynamics described as an inverted pendulum, given as: $G(s) = \Omega^2/(s^2 - \Omega^2)$, where $\Omega = \sqrt{10}$.

    a. Obtain a state-space model and discretize it using MATLAB 'c2d' command (use $T = 0.1$ s).

    b. Design a state feedback controller to achieve: $w_n \geq 5\frac{\text{rad}}{\text{s}}$, $\xi \geq 0.8$. Plot the step response.

    c. Design a deadbeat controller for the model and plot the step response.

11

Compensator Design via Frequency Response Modification

Learning Objectives

1. Characterize frequency response of a transfer function model using Bode and polar plots.
2. Performance criteria for controller design in the frequency-domain.
3. Design static gain compensator for the plant using frequency response methods.
4. Design phase-lead and phase-lag compensators using frequency response methods.

In this final chapter we will discuss frequency response based design of compensators for control systems.

Frequency response methods of compensator design predate the root locus method and the time-domain (state variable) methods. These methods work with the frequency response of the system, which maybe empirically obtained without resorting to the mathematical system model. The frequency response is graphically represented through Bode and/or Nyquist/polar plots.

The frequency response design seeks to impart a certain degree of relative stability to the control loop, as measured by the available gain and phase margins on the frequency response plots. The closed-loop stability may also be ascertained via the Nyquist criterion that relates to the encirclements of the $-1 + j0$ point on the Nyquist plot by the frequency response curve (see references for details).

Control systems design via the frequency response method requires knowledge of the frequency response $G(j\omega)$. Strictly speaking, the knowledge of the plant transfer function $G(s)$ is not needed. In fact, in the absence of a mathematical model, the plant transfer function can be identified from empirical measurements of the frequency response $G(j\omega)$.

In the following, we first describe the plotting of frequency response and the associated quality metrics. Later, we will discuss the frequency response modification through adding the phase-lead, phase-lag, lead–lag, PD, PI, and PID compensators to the control loop. The closed-loop frequency response may be visualized on the Nichol's chart.

The frequency response of the system can be graphed in multiple ways. The two most common representations are: (i) the Bode plot, and (ii) the Polar plot. These are described next.

11.1 The Bode Plot

Let the plant transfer function be given as $G(s)$; then, its frequency response function is given as $G(j\omega)$.

For a particular value of ω, the $G(j\omega)$ is a complex number, which may be described in terms of its magnitude and phase as $G(j\omega) = |G(j\omega)|e^{j\phi(\omega)}$. Then, as ω varies from 0 to ∞, both magnitude and phase may be plotted as functions of ω, and are called the Bode magnitude and phase plots.

It is customary to plot the Bode magnitude plot as $|G(j\omega)|_{dB} = 20 \log_{10}|G(j\omega)|$. The phase $\phi(\omega)$ is plotted in degrees.

11.1.1 Bode Plot of First Order Factors

Bode plots of low-order transfer functions can be sketched by hand as shown by the following examples:

We begin with an integrator: $G(s) = \frac{1}{s}$, so that $G(j\omega) = \frac{1}{j\omega} = \frac{1}{\omega}\angle -90°$. Then, $|G(j\omega)| = \frac{1}{\omega}$, $|G(j\omega)|_{dB} = -20 \log_{10}\omega$, and $\phi(\omega) = -90°$ (Figure 11.1).

When plotted on a semi-log paper, i.e., when $\log_{10}\omega$ is linearly plotted along the horizontal axis, the $-20 \log_{10}\omega$ appears as a straight line with a slope of -20 dB/decade (referring to an order of magnitude change in the frequency ω).

Next, consider a first-order transfer function: $G(s) = \frac{a}{s+a}$, $G(j\omega) = \frac{a}{a+j\omega} = \frac{a}{\sqrt{a^2+\omega^2}}\angle -\tan^{-1}(\frac{\omega}{a})$. Then, $|G(j\omega)|_{dB} = 20 \log_{10} a - 10 \log_{10}(a^2 + \omega^2)$. Further, at low and high frequencies,

For $\omega \ll a$, $|G(j\omega)|_{dB} = 0$ dB, and

For $\omega \gg a$, $|G(j\omega)|_{dB} = -20 \log_{10}\omega$.

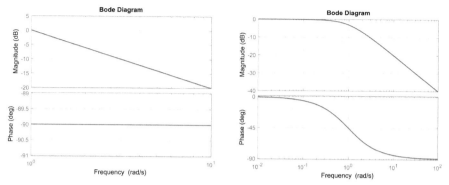

Figure 11.1 Bode plot for an integrator (*left*), and the first-order lag (*right*).

The Bode plot of a first-order $G(s)$ is characterized by two asymptotes, one along the 0 dB line, and the other sloping downward at -20 dB/decade for large ω. These asymptotes intersect at the corner frequency $\omega = a$, where $G(ja) = \frac{1}{1+j\omega}$, $|G(ja)|_{dB} = -3$ dB.

A freehand sketch can be used to join the two asymptotes passing through the -3 dB point (Figure 11.1).

The phase plot $\phi(\omega) = -\tan^{-1}(\frac{\omega}{a})$ is characterized by the following points, and can be easily sketched: $\phi(0) = 0°$, $\phi(.1a) = -5.7°$, $\phi(a) = -45°$, $\phi(10a) = -84.3°$, and $\phi(\infty) = -90°$.

The Bode magnitude plot of a first-order zero of the form $(s + a)$ is a reflection of the first-order pole plot. Specifically, it has a positive slope of 20 dB/decade for large ω. The corresponding phase plot is given as: $\phi(\omega) = \tan^{-1}(\frac{\omega}{a})$ and is a reflection of the phase plot for the first-order pole.

11.1.2 Bode Plot of Second Order Factors

Second order factors of the form: $G(s) = \frac{a^2+b^2}{(s+a)^2+b^2}$ with complex roots $s = -a \pm jb$ have a low-frequency gain of unity (0 dB), and the pair of poles contribute a -40 dB/dec slope to the magnitude plot for large ω.

In the case of second-order factors, the exact shape of the magnitude plot as well as the phase plot depends on the damping ratio defined as: $\zeta = \frac{a}{\sqrt{a^2+b^2}}$. A low value of ζ results in an overshoot in the magnitude plot at the corner frequency: $\omega = \sqrt{a^2 + b^2}$, accompanied by a sharp drop in phase $\phi(\omega)$, a characteristic of the tuned circuits.

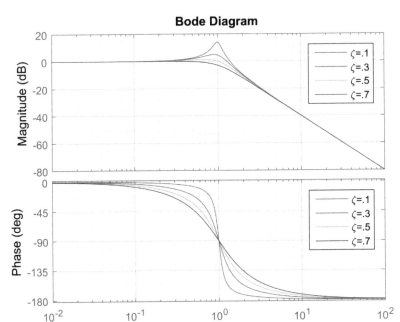

Figure 11.2   Bode plots of second-order factors.

The second-order system model is alternately expressed in terms of its natural frequency ω_n and damping ratio ζ as: $G(s) = \frac{\omega_n^2}{s^2+2\zeta\omega_n s+\omega_n^2}$, $G(j\omega) = \frac{1}{1-\omega^2/\omega_n^2+j2\zeta\omega/\omega_n}$.

The resonant frequency at which the magnitude plot peaks, is given as: $\omega_r = \omega_n\sqrt{1-2\zeta^2}$, $\zeta < \frac{1}{\sqrt{2}}$; the magnitude peak in the frequency response is computed as: $M_{p\omega} = \frac{1}{2\zeta\sqrt{1-\zeta^2}}$, $\zeta < \frac{1}{\sqrt{2}}$.

For example, for $G(s) = \frac{10}{s^2+2s+10}$, $\omega_n = \sqrt{10}\ \frac{\text{rad}}{\text{sec}}$, $\zeta = \frac{1}{\sqrt{10}}$; $\omega_r = 8\ \frac{\text{rad}}{\text{sec}}$, and $M_{p\omega} = 1.67$ or 4.44 dB.

Bode plots for second-order factors with various ζ values are plotted below (Figure 11.2).

11.1.3 The Composite Bode Plot

The composite Bode plot of $G(s)$ is an aggregation of the individual plots contributed by its first-order (for real poles and zeros) and second-order factors (for complex poles and zeros).

Assume that the system transfer function is given in the factored form:

$$G(s) = \frac{K(s-z_1)\ldots(s-z_m)}{s^l(s-p_{l+1})\ldots(s-p_n)}.$$

For computing the frequency response, it is convenient to express the transfer function as:

$$G(s) = \frac{K' \prod_{i=1}^{m}(1+s/z_i)}{s^l \prod_{i=1}^{n-l}(1+s/p_i)}; \quad K' = \frac{K \prod_{i=1}^{m} z_i}{\prod_{i=1}^{n} p_i}.$$

Then, the frequency response function is given as:

$$G(j\omega) = \frac{K' \prod_{i=1}^{m}(1+j\omega/z_i)}{(j\omega)^l \prod_{i=1}^{n-l}(1+j\omega/p_i)}.$$

The composite Bode magnitude plot combines the individual magnitudes in dB:

$$|G(j\omega)|_{dB} = 20 \log K' + \sum_{i=1}^{m} 20 \log \left|1 + \frac{j\omega}{z_i}\right|$$

$$- \sum_{i=1}^{n} 20 \log \left|1 + \frac{j\omega}{p_i}\right| - (20l)\log \omega.$$

The composite phase plot aggregates the phase angles in degrees:

$$\angle G(j\omega) = \sum_{i=1}^{m} \angle \left(1 + \frac{j\omega}{z_i}\right) - \sum_{i=1}^{n} \angle \left(1 + \frac{j\omega}{p_i}\right) - l(90°).$$

For $l = 0$, the low-frequency gain (or the DC gain) is given as: $20 \log_{10} K' = 20 \log_{10} \frac{K \prod_{i=1}^{m} z_i}{\prod_{i=1}^{n} p_i}$.

For $l > 0$, the magnitude plot at low frequencies has a slope of $-20l$ dB/dec.

The slope of the composite Bode magnitude plot for large ω is: $20(n-m)$, where $n - m$ represents the pole excess of the plant transfer function. The phase angle for large ω is given as: $\phi(\omega) = 90°(n - m)$.

11.2 The Polar Plot

The polar plot is a graph of $G(j\omega)$ in the complex plane with ω as a parameter. For a particular value of ω, $G(j\omega)$ is a complex number with magnitude and phase, where the latter restricted to $[0, 2\pi]$. The locus of all such points, as ω varies from $0 \to \infty$, constitutes the polar plot of $G(j\omega)$.

The shape of the polar plot depends on the locations of poles and zeros of $G(j\omega)$, and can be visualized by computing its magnitude and phase at low and high frequencies as explained below.

First, assume that $G(s)$ has no poles at the origin. Then, at low frequencies, $G(j0) \cong K' \angle 0°$, and at high frequencies, $|G(j\infty)| \to 0$, $\angle G(j\infty) = -90°(n-m)$, where $n-m$ represents the pole excess of $G(s)$.

If $G(s)$ includes a single pole at the origin, then at low frequencies, the polar plot has a high magnitude, $|G(j0)| \to \infty$, at a phase angle of $\phi(0) = -90°$.

Similarly, in the case that $G(s)$ has l poles at the origin, $\phi(0) = -90°(l)$. The high-frequency behavior of $G(j\omega)$ remains the same as described above.

The Nyquist plot of $G(s)$ additionally includes the polar plot of $G(j\omega)$ when ω varies from $-\infty$ to 0. The polar plot of $G(j\omega)$ for the negative frequencies is a reflection about the real-axis of the positive frequency plot.

The Nyquist plot is obtained in the MATLAB Control System Toolbox using the following command:

```
>> nyquist(tf(num,den))
```

The following examples illustrate the shape of the polar plot for transfer functions of various orders.

Example 11.1: Let $G(s) = \frac{1}{s+1}$; then, $G(j\omega) = \frac{1}{1+j\omega} = \frac{1}{\sqrt{1+\omega^2}} \angle -\tan^{-1}\omega$

Low frequency: $G(j0) = 1 \angle 0°$

High frequency: $G(j\infty) = 0 \angle -90°$

The resulting polar plot is a semi-circle in the fourth quadrant (Figure 11.3).

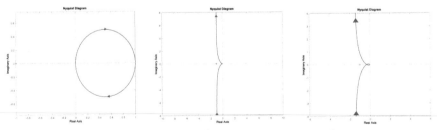

Figure 11.3   Nyquist plot for $G(s) = \frac{1}{(s+1)}$ (*left*); for $G(s) = \frac{1}{s(s+1)}$ (*middle*); and, for $G(s) = \frac{1}{s(s+1)(s+2)}$ (*right*).

Example 11.2: Let $G(s) = \frac{1}{s(s+1)}$; then, $G(j\omega) = \frac{1}{j\omega(1+j\omega)} = \frac{1}{\omega}\frac{1}{\sqrt{1+\omega^2}}$ $\angle -90° - \tan^{-1}\omega$

Low frequency: $G(j0) = \infty \angle -90°$

High frequency: $G(j\infty) = 0 \angle -180°$

The resulting polar plot is a curve in the third quadrant (Figure 11.3).

Example 11.3: Let $G(s) = \frac{2}{s(s+1)(s+2)}$; then, $G(j\omega) = \frac{2}{j\omega(1+j\omega)(2+j\omega)} = \frac{1}{\omega}\frac{1}{\sqrt{1+\omega^2}}\frac{2}{\sqrt{2+\omega^2}} \angle -90° - \tan^{-1}\omega - \tan^{-1}2\omega$.

Low frequency: $G(j0) = \infty \angle -90°$

High frequency: $G(j\infty) = 0 \angle -270°$

The resulting polar plot is a curve that begins along the negative $j\omega$-axis as $\phi(0) = -90°$, crosses the negative real-axis, and approaches the origin from the positive $j\omega$-axis as $\phi(\infty) = 90°$ (Figure 11.3).

11.3 Relative Stability

In frequency-domain methods, the relative stability (of the system impulse response) is described in terms of the gain and phase margins as defined below. For stability determination, we may assume that the plant transfer function $G(j\omega)$ is minimum-phase, i.e., it has poles and zeros located in the left-half plane (LHP).

Gain Margin. The maximum amount of loop gain that can be added to the control loop without compromising stability.

Phase Margin. The maximum amount of phase that can be added to the control loop without compromising stability.

11.3.1 Relative Stability on Frequency Response Plots

The phase margin (PM) on the Bode phase plot is indicated as: PM $= -180° + \phi(\omega_{gc})$, where ω_{gc} is the gain crossover frequency, i.e., the frequency at which the Bode magnitude plot crosses the 0 dB line.

The gain margin (GM) on the Bode plot is indicated at the phase crossover frequency ω_{pc}, i.e., the frequency at which $\phi(\omega_{pc}) = 180°$, and is given as: GM $= -|G(j\omega_{pc})|_{dB}$.

We note that the gain margin is equal to the maximum gain on the root locus (RL) plot before the closed-loop roots cross the $j\omega$-axis.

Alternatively, the gain and phase margins can be obtained from the polar plot, which enters the unit circle at the gain crossover frequency ω_{gc}, and crosses the negative real-axis at the phase cross-over frequency ω_{pc}.

Let the polar plot cross the negative real-axis at $g \angle -180°$, then the gain margin is given as: $\mathrm{GM} = g^{-1}$.

Let the polar plot enter the unit circle at $1 \angle -\phi$, then the phase margin is given as: $\mathrm{PM} = 180° - \phi$.

For plants with # poles ($n \leq 2$), the polar plot does not cross the negative real-axis; hence, the $\mathrm{GM} = \infty$. Also, in such case the RL plot does not crossover to the right half-plane (RHP).

The plants with # poles ($n - m \geq 3$) have a finite gain margin $0 < \mathrm{GM} < \infty$. Also, their root locus plot, for large K, crosses the stability boundary into the RHP. If the polar plot crosses the negative real-axis to the left of -1 point, then $\mathrm{GM} < 0$, and the closed-loop system is unstable.

In the case of minimum-phase plants, the closed-loop stability depends on the polar plot keeping a finite distance from the $-1 + j0$ point. The minimum distance of the polar plot from the $-1 + j0$ point is called the minimum return difference and is given by the minimum value of $\max_{\omega} |1 + KGH(j\omega)|$. The minimum return difference is a measure of the robustness of the control loop to parameter variations in the model.

The Nyquist plot of $KGH(j\omega)$ (for $-\infty < \omega < \infty$) represents a closed curve in the complex plane. The celebrated Nyquist criterion defines the closed-loop stability in terms of encirclements of the $-1 + j0$ point by the $KGH(j\omega)$ plot on the complex plane (see References for details).

The MATLAB 'margin' command can be used to obtain the GM and PM, along with gain and phase crossover frequencies, on the Bode plot:

```
>> margin(tf(num,den)); grid
```

11.3.2 Phase Margin and the Transient Response

The phase margin of a minimum-phase system transfer function is related to the damping ratio of the dominant poles of the system. The relationship is approximately given as: $\zeta = 0.01\phi_m$, where ϕ_m represents the phase margin (PM). Thus, in order to have a $\zeta = 0.7$, we may design the system for a $PM \cong 70°$ (65° to be exact).

The phase margin is also related to settling time of the step response. In particular, for a second order transfer function with complex poles, the settling time t_s and the phase margin ϕ_m are related by:

$$\tan \phi_{\mathrm{m}} = \frac{2\zeta\omega_{\mathrm{n}}}{\omega_{gc}} = \frac{8}{t_s\omega_{gc}}$$

The transient response performance is more effectively judged by the closed-loop bandwidth ω_B of the system. Assume that the closed-loop system response, given by $T(j\omega)$, has unity DC gain, i.e., let $T(j0) = 0$ dB. Then, ω_B is defined at the frequency where $|T(j\omega_B)| = -3$ dB.

The bandwidth is related to the rise time t_r of the system step response, i.e., the time when the response first reaches 100% (for $\zeta < 1$). Specifically, $\omega_B t_r \approx 1$, or $t_r \cong 1/\omega_B$. In the case of second-order system the rise time can also be approximated as: $t_r \cong \frac{3\zeta}{w_{\mathrm{n}}}$ around ($\xi = 0.7$).

11.3.3 Sensitivity

Sensitivity is a measure of the change in the closed-loop system characteristics to incremental changes in the system parameters. The sensitivity of the closed-loop transfer function to changes in the plant transfer function is a function of frequency, and is given as:

$$S_G^T(j\omega) = \frac{\partial T/T}{\partial G/G} = \frac{1}{1 + KG_c(j\omega)G(j\omega)}$$

In order for the sensitivity to be small over a frequency band, the loop gain $KG_c(j\omega)G(j\omega)$ must be high. In general, increasing the loop gain decreases the stability margins, i.e., a trade-off between the two exists. Accordingly, the control system designers generally aim for high loop gain at low frequencies.

11.4 Frequency Response Design

The Bode plot serves as a design tool for the frequency response design method in control systems. The performance indices used to evaluate the frequency response design are the GM and the PM of the loop transfer function $KGH(j\omega)$.

For stability, both the GM and the PM are required to be positive. Additionally, the PM should be adequate to ensure good relative stability and acceptable transient response. The GM should be adequate to achieve robustness and low sensitivity to parameter variations.

In the following, we assume that the plant and compensator transfer functions are given by $G(s)$ and $G_c(s)$, respectively. The sensor transfer function is $H(s)$. The characteristic equation of the closed-loop system is given as:

$$\Delta(s) = 1 + G_c GH(s) = 0$$

We assume a unity gain feedback system $(H(s) = 1)$ unless stated otherwise. We further assume that the plant transfer function has # poles $(n \geq 3)$ so that the $0 < \text{GM} < \infty$. Recall that for a given $G(s)$, the GM, PM, ω_{gc}, and ω_{pc} can be obtained from the MATLAB 'margin' command.

In the frequency response design, the choice of compensators includes the gain compensator, phase-lag and phase-lead compensators, and PD, PI, and PID compensators. These are described as follows:

11.4.1 Gain Compensation

In the case of gain compensation, $G_c(s) = K$. The gain compensation raises the loop gain is K, i.e., the Bode magnitude plot is shifted up by $20 \log_{10} K$. The resulting change in ω_{gc} affects the PM. Let the new gain crossover frequency be ω'_{gc}; then, the new phase margin is given as: PM $= 180° - \phi(\omega'_{gc})$.

The phase margin is reduced (for $K > 1$), and increased for (for $K < 1$). For $(K > 1)$, the system bandwidth increases, which improves the transient response by reducing the settling time.

Example 11.4: Let $G(s) = \frac{2}{s(s+1)(s+2)}$; the resulting gain and phase margins are found from MATLAB, and are given as: GM $= 3$ (9.54 dB), $\omega_{pc} = 1.41 \frac{\text{rad}}{\text{s}}$; PM $= 32.6°$, $\omega_{gc} = 0.749 \frac{\text{rad}}{\text{s}}$.

Assume that we want to increase the phase margin to 50° using gain compensation. Then, from the Bode plot, $\phi(j0.49) = 130°$, and $|G(j0.49)| = 5$ dB. Thus, the required gain K to achieve a PM $= 50°$ is given as: -5 dB or $K = 0.56$. Note that the loop gain is reduced in order to increase the PM.

The design is verified by plotting the frequency response for the loop transfer function: $KG(s) = \frac{1.12}{s(s+1)(s+2)}$, which shows $PM = 50°$ at $\omega_{gc} = 0.49 \frac{\text{rad}}{\text{s}}$ (Figure 11.4).

The MATLAB commands for the gain compensator example are given as:

```
>> G=tf(2, [1 3 2 0]);
>> margin(G), grid
>> margin (0.56*G), grid
```

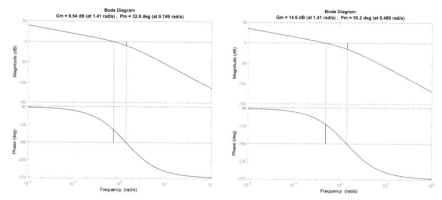

Figure 11.4 Gain compensation for desired phase margin: uncompensated system (*left*); compensated system (*right*).

11.4.2 Phase-Lag Compensation

In the frequency response design, the phase-lag compensator is used to improve the phase margin (a measure of transient response) and/or the DC gain (a measure of steady-state response).

The phase-lag compensator transfer function is given as:

$$G_c(s) = \frac{K(1 + s/\omega_z)}{1 + s/\omega_p}, \quad G_c(j\omega) = \frac{K(1 + j\omega/\omega_z)}{(1 + j\omega/\omega_p)}, \quad \omega_z > \omega_p$$

First assume that $K = 1$. Then, the compensator response at low and high frequencies is given as:

Low frequency: $G(j0) = 0$ dB $\angle 0°$

High frequency ($\omega \to \infty$): $G(j\omega) = -20 \log_{10}(\frac{\omega_z}{\omega_p}) \angle 0°$

Since the high frequency gain of the compensator is less than one, the addition of the compensator to the feedback loop moves the gain crossover frequency to the lower value that increases the phase margin. The phase-lag compensator thus acts as a low pass filter.

Next, we assume that $K = \omega_z/\omega_p$; then, the compensator gain at low and high frequencies is given as:

Low frequency: $G(j0) = 20 \log_{10} K \angle 0°$

High frequency ($\omega \to \infty$): $G(j\omega) = 0$ dB$\angle 0°$

Thus, addition of the compensator to the feedback loop will boost the relevant error constant by a factor of $K = \omega_z/\omega_p$.

Due to the destabilizing effect of the phase-lag, the compensator pole and zero locations are selected such that: $\omega_p < \omega_z < 0.1\omega_{gc}$, where ω_{gc} is the gain crossover frequency.

Assume that a phase margin of ϕ_m is desired; then, the phase-lag compensator design follows these steps:

1. Adjust the DC gain K for the loop transfer function $KGH(s)$ to satisfy the low-frequency requirements.
2. On the Bode plot of $KGH(j\omega)$, select the frequency ω_1 so that $\angle KGH(j\omega_1) = -180° + \phi_m + 5°$. Then, ω_1 serves as the new gain crossover frequency. The five degree safety margin serves to compensate for the estimated: $G_c(j\omega_1) \cong -5°$.
3. Select pole and zero frequencies as: $\omega_z = 0.1\omega_1$, $\omega_p = \omega_z/|KGH(j\omega_1)|$.
4. Draw the Bode plot for the compensated system and verify that the design requirements are met.

Example 11.5: Let $G(s) = \frac{2}{s(s+1)(s+2)}$; $PM = 32.6°$. Assume that we desire $PM = 50°$, and $e_{ss}|_{ramp} < 0.1$, i.e., $K_v > 10$.

The phase-lag compensator design steps are:

1. Choose $KG(s) = \frac{22}{s(s+1)(s+2)}$ for $K_v = 11$. Draw the Bode plot for $KG(j\omega)$.
2. From the Bode plot, we choose $\omega_1 = 0.4 \text{ rad/s}$, to obtain $\angle KG(j\omega_1) = -123°$; then, $|KG(j\omega_1)| = 24.9$ (28 dB)
3. Choose $\omega_z = 0.04$, $\omega_p = 0.0016$, $G_c(s) = \frac{11(1+s/0.04)}{1+s/0.0016}$
4. The Bode plot of $G_cG(j\omega)$ has $\omega_{gc} = 0.4 \text{ rad/s}$ and a phase margin of $PM = 51°$ (Figure 11.5)

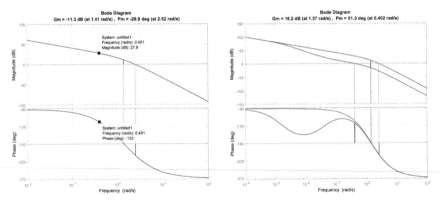

Figure 11.5   Phase-lag compensation: magnitude compensated plot (*left*); phase-lag compensated plot (*right*).

The MATLAB commands for the phase-lag design example are given as:

```
>> G=tf(2, [1 3 2 0]);
>> bode(11*G), grid, hold
>> margin(11*G*tf([1/.04 1],[1/.0016 1]))
```

11.4.3 Phase-Lead Compensation

In the frequency response design, the phase-lead compensator is used to improve the closed-loop bandwidth, leading to transient response improvements. The phase-lead compensator is given as:

$$G_c(s) = \frac{K(1 + s/\omega_z)}{1 + s/\omega_p}, \quad G_c(j\omega) = \frac{K(1 + j\omega/\omega_z)}{(1 + j\omega/\omega_p)}, \quad \omega_z < \omega_p$$

Assume that $K = 1$; then, the compensator response at low and high frequencies is given as:

Low frequency: $G_c(j0) = 0\ \text{dB}\angle 0°$

High frequency $(\omega \to \infty)$: $G_c(j\omega) = 20\log_{10}(\frac{\omega_p}{\omega_z})\angle 0°$.

Since the high frequency gain of the compensator is greater than one, the addition of the compensator to the feedback loop moves the gain crossover frequency to the right, increasing the bandwidth. The phase-lead compensator thus acts as a high pass filter.

For maximum effectiveness, the pole and zero of the phase-lead compensator should be placed in the vicinity of the gain crossover frequency ω_{gc}. The compensator contributes maximum phase lead of θ_m at a frequency $\omega_m = \sqrt{\omega_z\omega_p}$. The compensator transfer function at ω_m is given as:

$$G_c(\omega_m) = \sqrt{\frac{\omega_p}{\omega_z}}\angle\theta_m, \quad \tan\theta_m = \frac{1}{2}\left(\sqrt{\frac{\omega_p}{\omega_z}} + \sqrt{\frac{\omega_z}{\omega_p}}\right).$$

Practically, the maximum achievable value of θ_m is about 70°, which corresponds to $\sqrt{\frac{\omega_p}{\omega_z}} \cong 6$.

The phase-lead design first selects a gain crossover frequency ω_{gc} to meet the desired bandwidth and/or the settling-time requirement. Let the desired settling-time be t_s, then the desired $\omega_{gc} = \frac{8}{t_s\tan\phi_m}$.

In addition the phase contribution from the compensator should be positive, i.e.,

$$\theta = -180° + \phi_m - \angle GH(j\omega_{gc}) > 0°.$$

The loop gain at the desired gain crossover frequency is constrained as: $G_cGH(j\omega_{gc}) = 1\angle\theta$, or

$$\left(1 + j\frac{\omega_{gc}}{\omega_z}\right)|GH(j\omega_{gc})| = \left(1 + j\frac{\omega_{gc}}{\omega_p}\right)(\cos\theta + j\sin\theta).$$

The above equation, when separated into real and imaginary parts and solved for ω_z and ω_p, gives:

$$\omega_p = \frac{\sin\theta}{\cos\theta - |GH|}\omega_{gc}, \quad \omega_z = \frac{|GH|\sin\theta}{1 - |GH|\cos\theta}\omega_{gc}.$$

The above equations are used to compute the pole and zero locations of the phase-lead compensator.

A simplified method for phase-lead design is also available, and is given as follows: let $\alpha = \frac{\omega_p}{\omega_z}$; then, $\sin\theta_m = \frac{\alpha-1}{\alpha+1}$ or $\alpha = \frac{1+\sin\theta_m}{1-\sin\theta_m}$. Since $\omega_m = \sqrt{\omega_z\omega_p}$, we obtain: $\omega_z = \frac{\omega_m}{\sqrt{\alpha}}$; $\omega_p = \omega_m\sqrt{\alpha}$.

The phase-lead design can be carried out using either the regular or the simplified method and involves the following steps:

1. Choose a desired crossover frequency ω_{gc} as the greater of $\frac{8}{t_s\tan\phi_m}$ and ω_1 where $\angle KGH(j\omega_1) = -180° + \phi_m + 5°$.
2. Compute compensator phase $\theta = \phi_m - \angle G(j\omega_{gc}) - 180°$. Also compute $|GH(j\omega_{gc})|$.
3. Solve for ω_z and ω_p using the regular or the simplified method.
4. Inspect the Bode plot of the compensated system and verify that the design requirements have been met.

Example 11.5: Let $G(s) = \frac{2}{s(s+1)(s+2)}$; $PM = 32.6°$. Assume that we desire: $\phi_m = 50°$, $t_s = 4$ s.

The phase-lead compensator design steps are:

1. From the settling time requirement, $\omega_{gc} = \frac{2}{\tan\phi_m} = 1.68 \frac{rad}{s}$; $G(j1.68)$ $= 0.23 \angle -189°$.
2. Compute $\theta = 50° + 9° = 59° \cong 60°$.
3. The solution to the regular compensator design equations gives: $\omega_z = 0.38 \frac{rad}{s}$, $\omega_p = 5.43 \frac{rad}{s}$, $G_c(s) = \frac{14.29(s+0.38)}{s+5.43}$. The Bode plot of the compensated system has $\phi_m = 50.4°$ at $\omega_{gc} = 1.69 \frac{rad}{s}$; (Figure 11.6).
4. The simplified method with ($\theta_m = 60°$, $\alpha \cong 13.92$) returns the compensator: $G_c(s) = \frac{13.92(s+0.45)}{s+6.26}$. The Bode plot of the compensated system in this case has $\phi_m = 56.6°$ at $\omega_{gc} = 1.5 \frac{rad}{s}$.

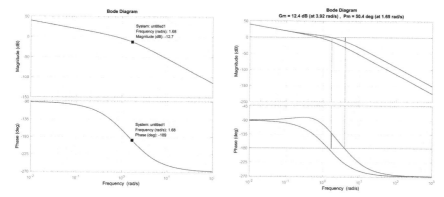

Figure 11.6 Phase-lead compensation: uncompensated plot (*left*); phase-lead compensated (*right*).

We may note that the use of simplified method results in a lower ω_{gc}, which results in comparitively higher settling time. Thus, when using the simplified design method, as a precaution, a safety margin may be added to the settling-time requirement.

The MATLAB commands for the phase-lead design example are:

```
>> G=tf(2, [1 3 2 0]);
>> bode(G), grid, hold
>> margin(G*tf([1/.38 1],[1/5.43 1]))
```

11.4.4 Lead-Lag Compensation

A lead-lag compensator combines the phase-lead and phase-lag sections, where the phase-lead section improves the bandwidth and the phase margin, and the phase-lag section improves the phase margin and the DC gain. Since both lead and lag sections can contribute to the phase margin improvement, the desired PM improvement can be distributed among the two sections.

The lead-lag compensator transfer function is given as:

$$G_c(s) = KG_{c-\text{lead}}(s)G_{c-\text{lag}}(s).$$

The steps to design a lead-lag compensator are as follows:

1. Choose static gain K to meet the steady-state error requirement.
2. Design the phase-lag section to meet part of the phase margin requirement.
3. Design phase-lead section to meet the bandwidth/settling time requirement.

Example 11.6: Let $G(s) = \frac{2}{s(s+1)(s+2)}$; and, assume that we require: $\phi_m = 50°$, $t_s = 4s$, and $e_{ss}|_{ramp} < 0.1$.

The design steps are as follows:

1. Choose $K = 11$ to meet the e_{ss} requirement. Draw the Bode plot for $KG(s)$. The plot has $\omega_{gc} = 2.5$ rad/s and PM $= -30°$.
2. In the phase-lag design, we aim to raise the PM $\approx 40°$. From the plot, let $\omega_1 = 0.5$ rad/s, such that KG $(j0.5) = 18.6$ $(25.8$ dB$)\angle-130°$.
3. Complete the phase-lag design with: $\omega_z = 0.05$, $\omega_p = 0.003$.
4. Draw the Bode plot for $KG_{c-lag}(j\omega)$. The plot shows a PM of $40°$.
5. Choose $\omega_{gc} = 1.7$ rad/s to meet the t_s requirement. Then, from the Bode magnitude plot, $KG_{c-lag}(j1.7) = 0.15$ $(-16.5$ dB$)\angle-191°$.
6. Compute the compensator phase angle $\theta = 62°$, and solve the phase-lead compensator equations to get: $\omega_z = 0.24$, $\omega_p = 4.7$.
7. The Bode plot for the compensated system has PM $= 50.2°$ at $\omega = 1.71$ rad/s (Figure 11.7).

The MATLAB commands for the lead-lag design example are:

```
>> G=tf(2, [1 3 2 0]);
>> margin(11*G*tf([1/.05 1],[1/.003 1])), grid,
   hold
>> Gc=tf([1/.05 1],[1/.003 1])*tf([1/.24 1],
   [1/4.7 1]);
>> margin(11*G*Gc)
```

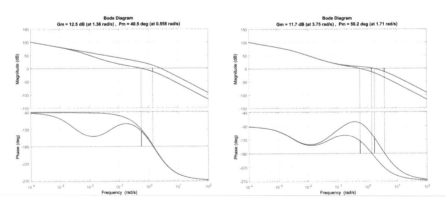

Figure 11.7 Lag-lead compensation: phase lag compensated system (*left*); lag-lead compensated system (*right*).

11.4.5 PI Compensator

The PI compensator increases system type, i.e., it increases the number of poles at the origin and is used for steady-state error improvement.

The PI compensator transfer function is given as: $G_c(s) = k_p + \frac{k_i}{s} = \frac{k_i(1+s/\omega_z)}{s}$, $G_c(j\omega) = \frac{k_i(1+j\omega/\omega_z)}{j\omega}$ where $\omega_z = \frac{k_i}{k_p}$.

The PI compensator is normally designed for a high frequency gain of one, i.e, by choosing $k_p = 1$. Additional gain compensation, similar to the phase-lag design, can be used to increase the phase margin to some desired value. The zero location for the PI compensator is arbitrarily selected close to the origin as shown in the following example.

Example 11.7: Let $G(s) = \frac{2}{s(s+1)(s+2)}$; PM $= 32.6°$. The design specifications are: PM $= 50°$, and $e_{ss}|_{\text{ramp}} < 0.1$, i.e., $K_v > 10$.

The PI compensator design steps are as follows:

1. We have $-180° + \phi_m + 5° = -125°$. The Bode plot for $G(j\omega)$ has, for $\omega = 0.42\frac{rad}{s}$, $G(j0.42) = 2.11$ (6.6 dB)$\angle-125°$. Thus, a gain compensation of -6.6 dB is needed.
2. For the PI design, let $\omega_z = 0.04$, $G_c(s) = \frac{s+.04}{s}$.
3. The Bode plot of $0.465\, G_cG(j\omega)$ has $\omega_{gc} = 0.42\ rad/s$ and a phase margin of PM $= 49.8°$ (Figure 11.8).

The MATLAB commands for the PI design example are:

```
>> G=tf(2, [1 3 2 0]);
>> bode(G), grid, hold
>> margin(G*.04*.465*tf([25 1],[1 0]))
```

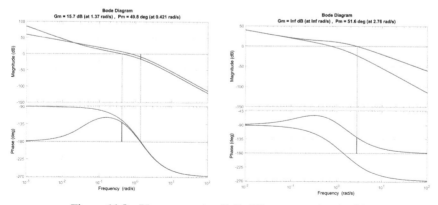

Figure 11.8 PI compensation (*left*); PD compensation (*right*).

11.4.6 PD Compensator

The PD compensator adds a first-order zero to the numerator of the loop transfer function, which increases the bandwidth and hence improves the transient response.

Let ω_{gc} denote the desired gain crossover frequency; then, from the requirement: $|G_c GH(j\omega_{gc})| = 1$, the zero location may be selected as:

$$\omega_z = \frac{|GH|}{\sqrt{1 - |GH|^2}} \omega_{gc}$$

The desired gain crossover ω_{gc} is selected as the greater of $\frac{8}{t_s \tan \phi_m}$ and ω_1, where ω_1 is found from the Bode plot for $\angle KGH(j\omega_1) = -180° + \phi_m - 90° + 5°$; the $90°$ term in this expression is the phase added by the PD compensator, and $5°$ is a safety margin.

Example 11.8: Let $G(s) = \frac{2}{s(s+1)(s+2)}$; PM $= 32.6°$. The design specifications are: $\phi_m = 50°$, $t_s = 4s$.

1. From the settling time requirement, $\omega'_{gc} = \frac{2}{\tan \phi_m} \cong 1.7 \frac{\text{rad}}{\text{s}}$; $G(j1.7) = 0.23 \angle{-190°}$. From the angle requirement, $\omega_1 = 2.8 \frac{\text{rad}}{\text{s}}$; $G(j2.8) = 0.07 \angle{-215°}$.
2. From the design equation, the compensator zero location is given as: $\omega_z = 0.2 \frac{\text{rad}}{\text{s}}$, $G_c(s) = (s + 0.2)$.
3. The Bode plot of the compensated system has $\phi_m = 51.6°$ at $\omega_{gc} = 2.76 \frac{\text{rad}}{\text{s}}$ (Figure 11.8).

The MATLAB commands for the phase-lead design example are:

```
>> G=tf(2, [1 3 2 0]);
>> bode(G), grid, hold
>> margin(G*tf([1/.2 1],[1]))
```

11.4.7 PID Compensator

The PID compensator represents a combination of the PD and PI designs. As a rule, PI section is designed first, and can be used to realize a part of or the entire phase margin improvement desired.

The bode plot of the PI compensated system is plotted, and followed by the PD design, which imparts the desired transient response improvement.

Example 11.9: Let $G(s) = \frac{2}{s(s+1)(s+2)}$; PM $= 32.6°$. The design specifications are: $\phi_m = 50°$, $t_s = 4$ s.

1. The Bode plot for $G(j\omega)$ has $G(j0.8) = 0.9 \angle -150°$.
2. PI design: let $\omega_z = 0.05$, $G_{c\_PI}(s) = 0.9(s + .05)/s$.

The Bode plot of $G_{c-PI}G(j\omega)$ has $\omega_{gc} = 0.7$ rad/s and $G(j0.7) = 1 \angle -148°$.

1. From the settling time requirement, $\omega_{gc} = \frac{2}{\tan \phi_m} \cong 1.7 \frac{rad}{s}$; $G(j1.7) = 0.23 \angle -190°$.

From the phase angle requirement, $\omega_1 = 2.5 \frac{rad}{s}$; $G(j2.5) = 0.1 \angle -210°$.

1. Let the desired $\omega_{gc} = 2.5 \frac{rad}{s}$; from the design formula, the compensator is zero selected as: $\omega_z = 0.2 \frac{rad}{s}$, $G_{c-PD}(s) = 5(s + 0.2)$
2. $G_c(s) = 4.5(s + 0.05)(s + 0.2)/s$.

The Bode plot of the compensated system has $\phi_m = 53.4°$ at $\omega_{gc} = 2.58 \frac{rad}{s}$ (Figure 11.9).

The MATLAB code for PID design is as follows:

```
>> G=tf(2, [1 3 2 0]);
>> bode(G), grid, hold
>> margin(G*.045*tf([20 1],[1 0]))
>> margin(G*.045*tf([20 1],[1 0])*tf([5 1],1))
```

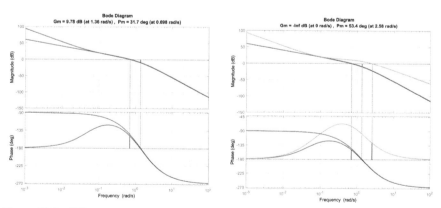

Figure 11.9 PID compensator design: PI compensation (*left*), followed by PD compensation (*right*).

11.5 Closed-Loop Frequency Response

The closed-loop frequency response for a plant $G(j\omega)$ driven by a static gain controller K in a unity gain feedback configuration $(H(s) = 1)$ is computed as:

$$T(j\omega) = \frac{KG(j\omega)}{1 + KG(j\omega)}$$

To characterize the closed-loop frequency response, let $G(j\omega) = X + jY$; then,

$$T(j\omega) = \frac{K(X + jY)}{1 + KX + jKY} = Me^{j\phi}.$$

The magnitude relation for $T(j\omega)$ can be rearranged as:

$$\left(X + \frac{M^2}{M^2 - 1}\right)^2 + Y^2 = \frac{M^2}{(M^2 - 1)^2}.$$

Since this represents the equation of a circle, the constant magnitude locus on the closed-loop frequency plot constitutes a series of constant M circles.

The phase relationship, similarly, reveals an equation for constant phase circles, given as:

$$\left(X + \frac{1}{2}\right)^2 + \left(Y - \frac{1}{2N}\right)^2 = \frac{1}{4} + \left(\frac{1}{2N}\right)^2; \quad N = \tan\phi.$$

When $G(j\omega)$ is super imposed onto constant M and constant N contours, it reveals the magnitude peak $M_{p\omega}$ of the closed-loop frequency response $T(j\omega)$.

The MATLAB Control System Toolbox plots the constant M, N contours on the Nyquist plot by turning on the 'grid' following the plotting of frequency response:

```
>> nyquist(tf(num,den)); grid
```

11.5.1 The Nichol's Chart

An alternate way to visualize the frequency response, $G(j\omega)$, is to plot the magnitude in dB along the vertical axis, and the phase is degrees along the horizontal axis. The resulting magnitude-phase plot is known as the Nichol's chart.

The 'grid' command can be similarly used with the Nichol's chart to plot the constant M, N contours to aid in the visualization of the closed-loop frequency response.

```
>> nichols(tf(num,den)); grid
```

Example 11.10: Let $G(s) = \frac{10}{s(s+1.86)}$, $G(j\omega) = \frac{10}{-\omega^2+j1.86\omega}$; then the closed-loop frequency response has a peak $M_{p\omega} = 5$ dB, as can be observed on both the Nyquist plot and the Nichol's chart (Figure 11.10).

The corresponding MATLAB commands are:

```
>> nyquist(tf([10],[1 1.86 0])),grid
>> nichols(tf([10],[1 1.86 0])),grid
```

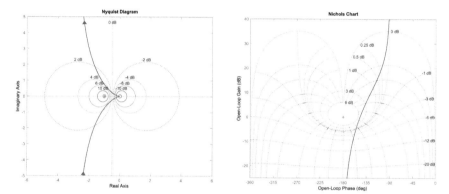

Figure 11.10 Closed-loop frequency response on the Nyquist plot (*left*), and the Nichol's chart (*right*).

Skill Assessment Questions

Link to the answers:
http://www.riverpublishers.com/book_details.php?book_id=449

11.1 Consider a plant with transfer function $G(s) = \frac{10}{s^2+2s+10}$. Use gain compensation to obtain a $\phi_m \geq 65°$.

11.2 Consider the plant in question 11.1. Modify the compensator to achieve: $KGH(0) \geq 5$.

11.3 Consider a plant with transfer function $G(s) = \frac{5(3s+1)}{s(s^2+2s+10)}$. Design a phase-lead compensator to achieve: $t_s \leq 1$ s, $\phi_m \geq 50°$.

11.4 Consider the simplified model of a flexible beam given as: $G(s) = \frac{10^4}{s^2+10s+10^4}$. Design a lead-lag compensator to achieve: $\omega_B \geq 100 \frac{rad}{s}$, $\phi_m \geq 45°$.

11.5 Consider the simplified model of an automobile given as: $G(s) = \frac{28s+120}{s^2+7s+14}$. Design a lead-lag/PID compensator to achieve: $\omega_B \geq 10 \frac{rad}{s}$, $K_p = \infty$, $K_v = 20$, $M_p \leq 1$ dB (closed-loop frequency response peak). State the gain and phase margins achieved.

11.6 Consider the model of human postural dynamics described as an inverted pendulum, given as: $G(s) = \Omega^2/(s^2 - \Omega^2)$, where $\Omega = \sqrt{10}$. Design a lead-lag/PID compensator for the model to achieve: $t_s \leq 2$ s, $\phi_m \geq 50°$.

Appendix

The Laplace Transform

The Laplace transform of a causal time function $f(t)$ is a complex function $f(s)$, defined as:

$$\mathcal{L}[f(t)] = f(s) = \int_0^\infty f(t)e^{-st}\mathrm{d}t,$$

where $s = \sigma + j\omega$ is a complex variable. In case of discontinuities, the lower limit is changed to 0^-.

Laplace transform of simple functions may be computed by hand. For example:

$$\mathcal{L}[u(t)] = \int_0^\infty e^{-st}\mathrm{d}t = \frac{1}{s}$$

$$\mathcal{L}[e^{-at}u(t)] = \int_0^\infty e^{-(s+a)t}\mathrm{d}t = \frac{1}{s+a}$$

$$\mathcal{L}[e^{j\omega t}u(t)] = \int_0^\infty e^{j\omega t}e^{-st}\mathrm{d}t = \frac{1}{s - j\omega} = \frac{s + j\omega}{s^2 + \omega^2}.$$

Laplace transform of complex functions is obtained by invoking its properties:

1. Linearity: $\mathcal{L}[a_1 f_1(t) + a_2 f_2(t)] = a_1 f_1(s) + a_2 f_2(t)$
2. Time shift: $\mathcal{L}[f(t - T)] = e^{-sT} f(s)$
3. Scaling: $\mathcal{L}[f(at)] = f(\frac{s}{a})$
4. Differentiation: $\mathcal{L}[\frac{\mathrm{d}f(t)}{\mathrm{d}t}] = sf(s) - f(0)$
5. Integration: $\mathcal{L}[\int_0^t f(\tau)\mathrm{d}\tau] = \frac{f(s)}{s}$
6. Initial value theorem: $f(0) = \lim_{s \to \infty} sf(s)$

7. Final value theorem: $f(\infty) = \lim_{s \to 0} s f(s)$.

8. Convolution: $\mathcal{L}[f(t) \times g(t)] = F(s)G(s)$.

A dynamic system model is often obtained in terms of an ODE. A potent advantage of using Laplace transform as an analysis tool is that it transforms an ODE into an algebraic equation that can be solved by algebraic methods. For example,

$$\frac{dy(t)}{dt} + ay(t) = u(t) \xrightarrow{\mathcal{L}} (s+a)y(s) = u(s) + y(0).$$

The inverse Laplace of a complex function may be obtained by using partial fraction expansion (PFE) followed by the use of Laplace transform tables. For example:

$$\mathcal{L}^{-1}\left[\frac{a}{s(s+a)}\right] = \mathcal{L}^{-1}\left[\frac{1}{s} - \frac{1}{s+a}\right] = (1 - e^{-at})u(t).$$

The Laplace transform is also used to obtain the input-output transfer function of a differential equation model. For example, for a first-order ODE model: $\frac{dy(t)}{dt} + ay(t) = u(t)$, the transfer function is found as: $\frac{y(s)}{u(s)} = \frac{1}{s+a}$. Using the transfer function, the step response of the system is computed as: $y(s) = \frac{1}{s+a}\left(\frac{1}{s}\right)$, or $y(t) = \frac{1}{a}(1 - e^{-at})u(t)$.

The $\mathcal{Z}$-Transform

The $\mathcal{Z}$-transform of a sequence $x(k)$ is defined as:

$$\mathcal{Z}[x(k)] = X(z) = \sum_{k} = -\infty^{\infty} x(k)z^{-k}$$

where z is a complex variable. To obtain a unique $F(z)$ the region of convergence (ROC) for $\mathcal{Z}$-transform must be specified.

We will assume that the sequence $x(k)$ is obtained by sampling a continuous-time causal signal $x(t) \to x(kT)$, so that the lower summation limit can be taken as $k = 0$. Then, the respective $\mathcal{Z}$ and Laplace transform variables are related as: $z = e^{sT}$.

The $\mathcal{Z}$-transforms of sampled values of common signals are given as:

$$\mathcal{Z}[u(kT)] = \frac{1}{1 - z^{-1}}; |z^{-1}| < 1$$

$$\mathcal{Z}[e^{(-akT)}] = \frac{1}{1 - e^{-aT}z^{-1}}; |e^{-aT}z^{-1}| < 1$$

$$\mathcal{Z}[e^{(j\omega kT)}] = \frac{1}{1 - e^{j\omega T}z^{-1}}; |z^{-1}| < 1$$

The $\mathcal{Z}$-transform properties include:

1. Linearity: $\mathcal{Z}[a_1 f_{1(k)} + a_2 f_2(k)] = a_1 F_1(z) + a_2 F_2(z)$
2. Time-shift: $\mathcal{Z}[f(k - n_0)] = z^{-n_0} F(z)$
3. Scaling: $\mathcal{Z}[a^k f(k)] = F(\frac{z}{a})$
4. Time differentiation: $\mathcal{Z}[f(k) - f(k-1)] = (1 - z^{-1})F(z)$
5. Time reversal: $\mathcal{Z}[f(-n)] = F(z^{-1})$
6. Final value theorem: $\lim_{k \to \infty} f(k) = \lim_{z \to 1}(z - 1)F(z)$
7. Convolution: $\mathcal{Z}[f_{1(k)} \times f_{2(k)}] = F_{1(z)} F_{2(z)}$

A discrete-time dynamic system is often modeled with difference equations, which can be conveniently solved using $\mathcal{Z}$-transform (see references for details).

A table relating continuous-time signals to their Laplace and $\mathcal{Z}$-transform equivalents can be found at: http://lpsa.swarthmore.edu/LaplaceZTable/La placeZFuncTable.html

References

Nise, N. S. *Control Systems Engineering*, Seventh ed. Wiley, Hoboken, NJ, 2015.

Dorf, R. C. and Bishop, R. H. *Modern Control Systems*, 13th ed. Pearson, Hoboken, NJ, 2017.

Phillips, C. L. and Parr, J. M. *Feedback Control Systems*, Fifth ed. Prentice Hall, Upper Saddle River, NJ, 2014.

Fadali, A. S. and Visioli, A. *Digital Control Engineering*, Second ed. Associated Press, NY, 2012.

Friedland, B. *Control Systems Design, An Introduction to State-Space Methods*. McGraw Hill, NY, 1986. Reprinted by Dover, NY, 2005.

Index

159

About the Author

Kamran Iqbal obtained his B.E. in Aeronautical Engineering from NED University, his M.S. and Ph.D. in Electrical Engineering from the Ohio State University, and did his postdoctoral work at Northwestern University. He has held academic appointments at College or Aeronautical Engineering, GIK Institute of Engineering Science and Technology, University of California, Riverside, University of California, Irvine, California State University at Fullerton, and University of Arkansas at Little Rock where he currently serves as Professor of Systems Engineering in the College of Engineering and Information Technology. Control Systems and Digital Signal Processing are among his favorite subjects. His research interests include neuromechanics of human movement, computational intelligence, and biomedical engineering and signal processing. He is a member of IEEE (USA), IET (UK), ASEE, IASTED, and Sigma Xi, and has been a regular contributor to the IEEE Engineering in Medicine and Biology conference.

Autonomous Underwater Vehicles: Technology and Applications

Autonomous Underwater Vehicles: Technology and Applications

Edited by **Noah Carter**

LANRYE
INTERNATIONAL

New Jersey

Published by Clanrye International,
55 Van Reypen Street,
Jersey City, NJ 07306, USA
www.clanryeinternational.com

Autonomous Underwater Vehicles: Technology and Applications
Edited by Noah Carter

International Standard Book Number: 978-1-63240-074-1 (Hardback)

Printed in the United States of America.

Contents

Preface

The technology of autonomous underwater vehicles and their various applications are elucidated in a sophisticated and comprehensive way in this book. Autonomous Underwater Vehicles are robotic engines which travel underwater in order to study underwater activities. The successful developments of parallel research and technological studies that were underway; helped to conquer the challenges associated with autonomous operation in tough conditions. The ultimate aim behind all these advancements was to procure accurate data through economical means in lesser time using accurate geo locations. Some new models are already being employed to extract the best out of the present technology by making decisions according to the interpretation of the sensor data. This book is the comprehensive compilation of various aspects of AUV technology and its applications. Vehicle designing, navigation & control techniques and mission preparation & analysis are some of the topics that this book covers. It provides an overview of brighter prospects of vehicle technology and application.

This book is a comprehensive compilation of works of different researchers from varied parts of the world. It includes valuable experiences of the researchers with the sole objective of providing the readers (learners) with a proper knowledge of the concerned field. This book will be beneficial in evoking inspiration and enhancing the knowledge of the interested readers.

In the end, I would like to extend my heartiest thanks to the authors who worked with great determination on their chapters. I also appreciate the publisher's support in the course of the book. I would also like to deeply acknowledge my family who stood by me as a source of inspiration during the project.

Editor

Part 1

Vehicle Design

Development of a Hovering-Type Intelligent Autonomous Underwater Vehicle, P-SURO

Ji-Hong Li[1]*, Sung-Kook Park[1], Seung-Sub Oh[1], Jin-Ho Suh[1],
Gyeong-Hwan Yoon[2] and Myeong-Sook Baek[2]
1Pohang Institute of Intelligent Robotics
2Daeyang Electric Inc.
Republic of Korea

1. Introduction

P-SURO(PIRO-Smart Underwater RObot) is a hovering-type test-bed autonomous underwater vehicle (AUV) for developing various underwater core technologies (Li et al., 2010). Compared to the relatively mature torpedo-type AUV technologies (Prestero, 2001; Marthiniussen et al., 2004), few commercial hovering-type AUVs have been presented so far. This is partly because some of underwater missions of hovering-type AUV can be carried out through ROV (Remotely Operated Vehicle) system. But the most important reason is of less mature core technologies for hovering-type AUVs. To carry out its underwater task, hovering-type AUV may need capable of accurate underwater localization, obstacle avoidance, flexible manoeuvrability, and so on. On the other hand, because of limitation of present underwater communication bandwidth, high autonomy of an AUV has become one of basic function for hovering AUVs (Li et al., 2010).

As a test-bed AUV, P-SURO has been constructed to develop various underwater core technologies, such as underwater vision, SLAM, and vehicle guidance & control. There are four thrusters mounted to steer the vehicle's underwater motion: two vertical thrusters for up/down in the vertical plane, and 3DOF horizontal motion is controlled by two horizontal ones, see Fig. 1. Three communication channels are designed between the vehicle and the surface control unit. Ethernet cable is used in the early steps of development and program/file upload and download. On the surface, RF channel is used to exchange information and user commands, while acoustic channel (ATM: Acoustic Telemetry Modem) is used in the under water. A colour camera is mounted at the vehicle's nose. And three range sonar, each of forward, backward and downward, are designed to assist vehicle's navigation as well as obstacle avoidance and SLAM. An AHRS combined with 1-axis Gyro, 1-axis accelerometer, depth sensor consist of vehicle's navigation system.

In this chapter, we report the details of to date development of the vehicle, including SLAM, obstacle detection/path planning, and some of vehicle control algorithms. The remainder of this chapter is organized as follows. In Section II, we introduce the vehicle's general specifications and some of its features. Underwater vision for P-SURO AUV is discussed in Section III, and the SLAM algorithm in the basin environment is presented in Section IV. In Section V, we discuss some of control issues for P-SURO AUV. Finally in Section VI, we make a brief summary of the report and some future research issues are also discussed.

Fig. 1. P-SURO AUV and its open frame.

2. P-SURO AUV overview

As aforementioned, P-SURO AUV is a test-bed for developing underwater technologies. And most of its experimental tests will be carried out in an engineering basin in the PIRO with dimension of 12(L)×8(W)×6(D)m. Under these considerations, the vehicle is designed to be compact size with easiness of various algorithm tests (Li et al., 2010). The general specification of the vehicle is as Table. 1.

Item	Specifications
Depth rating	100m
Weight	53kg
Dimension	1.05(L)×0.5(W)×0.3(H)m
Max. speed	FW: 2.5knot; BW, UP/DW: 1.5knot
Battery system	400W·hr, Lithium Ion, Endurance: 2.5hrs
Payload	≤4kg

Table 1. General specification of P-SURO AUV.

2.1 Mechanical system

For the convenience of maintenance and also under the security consideration, we separate the battery system from other electronics systems, see Fig. 2. Main frame is made of AL-6061, fixing parts for camera and range sonar are made of POM. To increase the hydrodynamic mobility in the underwater horizontal plane, the open frame of vehicle is wrapped in a two-piece of FRP (Fibre-Reinforced Plastic) shell (Li et al., 2010).

Throughout its underwater missions, P-SURO is always keeping zero pitch angle using two vertical thrusters. With this kind of stability in its pitch dynamics, the vehicle's horizontal 3DOF motion is steered by two horizontal thrusters. From control point of view, this is a typical underactuated system. And how to design path tracking or following scheme for this kind of underactuated system has become one of most intense research area in the nonlinear control community (Jiang, 2002; Do et al., 2004; Li et al., 2008b).

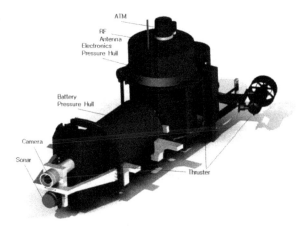

Fig. 2. Mechanical arrangement of P-SURO AUV.

Sensor Model (Maker)	Specifications
Super SeaSpy (Tritech)	- >480 TV lines - 1/3" Interline Transfer CCD - Composite output
Micron Echo Sounder (Tritech)	- Operating frequency: 500KHz - Beamwidth: 6° conical - Operating range: 50m
AHRS (Innalabs)	- Update rate: 1-100Hz - Heading accuracy: 1.0 deg - Attitude accuracy: <0.4 deg
CVG25 (Innalabs)	- Measurement range: ±200deg/sec - Bandwidth: 50Hz - Bias stability: 1.5deg/hr
AL-15M2.5 (Innalabs)	- Input range: ±2.5g - Bias: <4mg - Bandwidth: 300Hz
PR-36XW (Keller)	- Depth rating: 100m - Accuracy: 0.1% FS - Output rate: 100Hz
UWM2000H (LinkQuest)	- Payload data rate: 300-1200bits/sec - Working range: 1200m - Beam width: 210deg (omni-directional)
LinkWiser™-HP400 (Cellution)	- Half-duplex - Operating frequency: 400-470MHz - Air data rate: 4.8kbps
BTD-156 (SeaBotix)	- Continual thrust: 2.2kgf - Input voltage: 28V - Interface: RS485, 115200bps

Table 2. Sensor & thrust system of P-SURO AUV.

2.2 Sensor, thrust, and power system

For underwater vision, there is one colour camera mounted at the vehicle nose. And three range sonar (forward, backward and downward) are mounted on the vehicle. There sonar are designed for obstacle detection and also for assisting vehicle's underwater localization. For P-SURO AUV, we design a relatively simple but low grade of inertial navigation system which consists of AHRS, 1-axis Gyro, 1-axis accelerometer, one depth sensor.

SeaBotix BTD-156 thrusters are selected to steer the vehicle's underwater motion. This is a small size underwater thruster with 90W of average power consumption. For power system, the calculated total power consumption of vehicle system is about 450W. And correspondingly, we design the 1.2kW Lithium Ion battery system, which can support more than two hours of the vehicle's underwater continuous operation. The overall sensor & thrust system for P-SURO AUV is listed in Table. 2.

2.3 Embedded system

Three of PC104 type PCM3353 SBCs (Single Board Computers) are chosen as core modules, each of vision, navigation, and control. PCM3353 SBC provides 4 RS232 channels plus 4 USB channels. And using these USB channels, we can easily extend the necessary serial channels (RS232/422/485) using proper USB to serial converters. PCM3718HG analogue and digital I/O board is used for various peripheral interface. In addition, two peripheral boards, including DC/DC converter system, magnetic switch circuit, leakage detection circuit, are also designed. Fig. 3 shows the inner view of electronic pressure hull.

Fig. 3. Electronics system of P-SURO AUV.

3. Software architecture

As aforementioned, we choose Windows Embedded CE 6.0 as the near real-time OS for three of core modules; vision module, navigation module, and control module. For this, we design three different WinCE 6.0 BSPs (Board Support Package) for each of three core modules. Furthermore, these three core modules are connected to each other through Ethernet channel, and constructing a star topology of network structure.

Software frame for each core module consists of thread-based multi tasking structure. For each module, there are various sensors connected through serial and analogue channels. And these serial sensors, according to their accessing mechanism, can be classified into two types: active sensor (frequently output measurement) and passive sensor (trigger mode). For these passive sensors as well as analogue sensors, we read the measurements through *Timer( )* routine. And for each of active sensors, we design a corresponding thread. In most of time, this thread is in *Blocking* mode until there is measurement output. And this kind of real-time sensor interface also can be used to trigger other algorithm threads. For example, in the navigation module, there is a thread designed for interfacing with AHRS sensor (100kHz of output rate). After accessing each of attitudes, gyro, and accelerometer output measurement, the thread will trigger *Navigation( )* thread. Moreover, some of these threads are cautiously set with different priority values.

As with the most of other AUVs so far, the P-SURO AUV has the similar overall software frame, which can be divided into two parts: surface remote control system and the vehicle software system. For surface system, the main functions of it are to monitor the vehicle and deliver the user command. According to the user command (mission command in this case), the vehicle will plan a series of tasks to accomplish the mission. For P-SURO AUV, its most experimental field is in a small cuboid. In this kind of environment, it is well known that underwater acoustic channel is vulnerable. For this reason, the vehicle is required to possess relatively high level of autonomy, such as autonomous navigation, obstacle avoidance, path planning and so on.

From the control architecture point of view, the software architecture of P-SURO AUV can be classified into hybrid architecture (Simon et al., 1993; Healey et al., 1996; Quek & Wahab, 2000), which is a certain combinaiton of hierarchical architecture (Wang et al., 1993; Peuch et al., 1994; Li et al., 2005) and behavioral architecture (Brooks, 1986; Zheng, 1992; Bennett, 2000). As aforementioned, because of the limitation of underwater acoustic communication in the engineering basin in PIRO, it is strongly recommended for the vehicle to self-accomplish its mission without any of user interface in the water. For this consideration, the control architecture of P-SURO AUV is featured as a behavioral architecture based hybrid system (see Fig. 4).

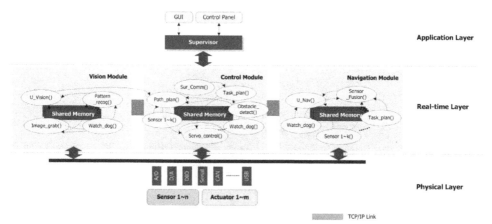

Fig. 4. Hybrid control architecture for P-SURO AUV.

If there is a pattern appeared in a certain area in front of the vehicle, the vision module will recognize the pattern and transmit the corresponding vehicle's pose information freqeuntly to the control module for aiding of path planning. According to the received mission command (user command is usually delivered to the vehicle on the surface through RF channel), the control module arranges a series of tasks to accomplish the mission. Also, this module carries out various thruster controls and other actuator controls. The main task of the navigation module is to carry out the real-time navigatin algorithm using acquired attitude, gyro, and accelerometer measurements. Other information including range sonar, depth sensor, underwater vision are served as aiding information for this inertial navigatin system.

4. Vision-based underwater localization

Visual localization methods usually can be classified into two types: one is based on natural feature points around the environment for recognition of robot pose, and the other one is using artificial landmarks which are usually known patterns pre-installed in the environment (Oh et al., 2010). PIRO engineering basin is surrounded by flat concrete walls, and it is difficult to extract specific feature points. For this reason, we use artificial landmark for visual localization of P-SURO.

4.1 Artificial landmarks

For P-SURO AUV, the main purpose of underwater vision is to provide localization information for the vehicle's path planning task. In the path decades, vision has become one of most intense research area in the robot community, and various artificial landmarks and corresponding recognition methods have been developed (Hartley & Zisserman, 2000). The pattern designed for P-SURO AUV is shown in Fig. 5, which consists of two rectangles and two sets of four dots with the same cross-ratio. The large number in the centre is used for distinguish the pattern (Li et al., 2010).

The eight dots in the pattern contain 3D pose information. For extracting these dots, we use the cross-ratio invariant, which is a basic projective invariant of perspective space (Hartley & Zisserman, 2000). The cross-ratio is defined as following

$$Cross(x_1, x_2, x_3, x_4) = \frac{|x_1 x_2||x_3 x_4|}{|x_1 x_3||x_2 x_4|}, \tag{1}$$

where

$$|x_i x_j| = det \begin{bmatrix} x_{i1} & x_{i2} \\ x_{j1} & x_{j2} \end{bmatrix}.$$

Fig. 5. Designed underwater pattern for P-SURO AUV.

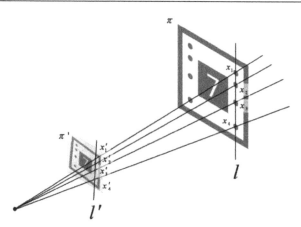

Fig. 6. Cross-ratio in the projective space.

The eight dots in the pattern contain 3D pose information. For extracting these dots, we use the cross-ratio invariant, which is a basic projective invariant of perspective space (Hartley & Zisserman, 2000). The cross-ratio is defined as following

$$Cross(x_1, x_2, x_3, x_4) = \frac{|x_1 x_2||x_3 x_4|}{|x_1 x_3||x_2 x_4|}, \tag{2}$$

where

$$|x_i x_j| = det \begin{bmatrix} x_{i1} & x_{i2} \\ x_{j1} & x_{j2} \end{bmatrix}.$$

The cross-ratio value defined in (2) is invariant under any projective transformation of the line. If $x' = H_{2 \times 2} x$, then $Cross(x'_1, x'_2, x'_3, x'_4) = Cross(x_1, x_2, x_3, x_4)$. As shown in Fig. 6, suppose the plane π is the pattern and the plane π' is the projected pattern image, then four dots on the line l have the same cross-ratio with the four dots on the line l'. Using this invariant, we can find the match between pattern dots and its image projection.

4.2 Auto-calibration of underwater camera
It is difficult to directly calibrate the camera in the underwater environment. For this reason, we apply a camera auto-calibration method using cross-ratio invariant (Zhang, 2000). If we denote the pattern point as $x = [u, v, 1]^T$ and the image point as $x' = [u', v', 1]^T$, then the relationship between them is given by

$$sx' = Hx, \tag{3}$$

where s is an arbitrary scale factor. In (3), homography H is defined as

$$H = [h_1 \ h_2 \ h_3]^T = \lambda A[r_1 \ r_2 \ t], \tag{4}$$

where λ is a scale factor, r_1, r_2 are part of rotation matrix $R = [r_1 \ r_2 \ r_3]$, t is a translation matrix and A is a camera intrinsic matrix. If camera moves, each matching point from the scene makes a homography. We can get the camera intrinsic matrix A from homography H (Zhang, 2000).

Given the camera intrinsic matrix A and the homography H in the image, we can get the three-dimensional relationship between the pattern and the camera (robot) using following equations

$$r_1 = \lambda A^{-1} h_1$$
$$r_2 = \lambda A^{-1} h_2,$$
$$r_3 = r_1 \times r_2,$$
$$t = \lambda A^{-1} h_3.$$
(5)

4.3 Lab-based evaluation

To evaluate the developed vision algorithm, we carry out a series of ground tests. Fig. 7 shows the test results. First, eight dots in the pattern are extracted from the image (Fig. 7-a). And Fig. 7-b shows the selected images for camera auto-calibration, and extracted camera pose is shown in Fig. 7-c,d.

While evaluating the performance of proposed visual localization method, we mainly consider two points: one is the pattern recognition rate, and the other one is the accuracy of pose estimation. For pattern recognition rate, we arbitrarily move the camera and take 306 images. 150 of them are correctly recognized, 85 are failed to recognize, 61 are blurred because of camera movement, and 10 do not include the full pattern. Except 61 blurred images and 10 of missed-pattern images, the recognition rate is about 64%. However, consider the fact that about half of the non-recognized images are rotated more than 40 degrees from the pattern, the recognition rate is about 81%.

To evaluate the accuracy of pose estimation, we fix the camera and locate the pattern at 12 known positions between 400mm to 700mm distance. Calculated average pose estimation error is 0.155mm and the standard deviation is 1.798mm.

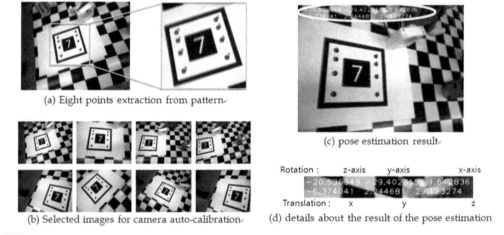

(a) Eight points extraction from pattern.

(c) pose estimation result.

(b) Selected images for camera auto-calibration.

Rotation :	z-axis	y-axis	x-axis
	-20.536349	-29.402345	1.642836
	-6.374041	2.24468	298.23274
Translation :	x	y	z

(d) details about the result of the pose estimation

Fig. 7. Process of pose estimation.

4.4 Camera underwater test

Given a pattern (landmark) and a camera, then the minimum and maximum pattern recognition range can be predetermined. And this information will be used for vehicle's underwater path planning. Under this consideration, we carry out a series of test measuring the minimum and maximum recognition range both in the air and in the water. Test results are shown in Fig. 8 and 9, from which we can see that the maximum recognition range in the water is approximately half of the one in the air. For the safety consideration, we force the vehicle to keep from the basin wall at least 1.5m throughout the various basin tests.

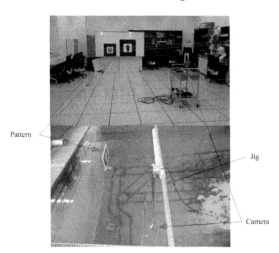

Fig. 8. Test environments.

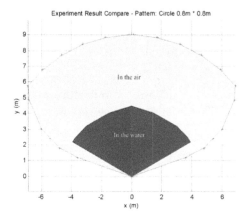

Fig. 9. Test results

5. SLAM using range sonar array

In the past decades, SLAM (Simultaneous Localization and Mapping) has been one of most hot issues in the robotics community (Leonard & Burrant-Whyte, 1992; Castellanos & Tardos, 1999; Thrun et al., 2005). SLAM problems arise when the robot does not have access to a map of the environment; nor does it have access to its own poses (Thrun et al., 2005).

For P-SURO AUV, as aforementioned, most of its underwater operations are carried out in the engineering basin in PIRO. For this reason, we have designed a relatively simple SLAM method with partially known environment.

5.1 Range sonar model

There are three Micron Echosounder@Tritech (6° conical beamwidth with 500kHz operating frequency) mounted on the vehicle; each of forward, backward, and downward. Because of their narrow beam-angle, we apply simple centreline model (Thrun et al., 2005) for all of these sonar behaviour.

Throughout its underwater operation, the vehicle is forced to keep away from the basin wall at least 2m. In this case, the maximum range error is about 2.2m (d_M in Fig. 10). Another significant error source is misalignment of range sonar with AHRS. For vehicle's dynamics, we observed that the yaw angular velocity of P-SURO was less than $10^o/s$. Consider the 1500m/s of acoustic velocity in the water, $10^o/s$ of yaw motion may cause less than 0.1^o of azimuth error. Therefore, the effect of vehicle dynamics on the range measurement can be neglected.

To investigate the range sonar behavior in a basin which is of a small cuboid, we performed a simple test where the vehicle is rotated (about $7.6^o/s$) several cycles with the center at the same point. Throughout the test, we force the vehicle to keep zero pitch angle. The resulted basin profile image is shown in Fig. 11, from which we can see that closer to the basin corner, more singular measurements are occurred.

5.2 Obstacle detection

At any point (x_t, y_t, ψ_t), through rotating the vehicle on the horizontal plane, we can easily get a 2D profile image of basin environment, see Fig. 12. And for 3D case, we simply extend the horizontal rotating motion with constant velocity of descent/ascent motion and get rough 3D profile image, see Fig. 13. According to this profile image, we detect the obstacle and further design corresponding vehicle path.

The obstacle block A in Fig. 12 is modelled as (a, b) with a obstacle start point and b the end point. Here $R_i = |P_c i|$ and ψ_i is yaw angle of $\overrightarrow{P_c i}$ with $i = a, b, c$. In the case of the vehicle facing point a and c, if $|ac| = |P_c c| - |P_c a| > d_c$ where $d_c = f(\psi_c)$ is a design parameter, then point a is taken as the start point of an obstacle. And in the case of b and d, if $|bd| = |P_c d| - |P_c b| > d_d$ with $d_d = f(\psi_d)$, then b is taken as the end point of the obstacle.

5.3 Path planning

Path planning is an important issue in the robotics as it allows a robot to get from point A to point B. Path planning can be defined as "determination of a path that a robot must take in order to pass over each point in an environment and path is a plan of geometric locus of the points in a given space where the robot has to pass through" (Buniyamin et al., 2011). If the robot environment is known, then the global path can be planned off line. And local path planning is usually constructed online when the robot face the obstacles. There are lots of path planning methodologies such as roadmap, probability roadmap, cell decomposition method, potential field have been presented so far (Choset et al., 2005; Khatib, 1986; Valavanis et al., 2000; Elfes, 1989; Amato & Wu, 1996; Li et al., 2008a).

For P-SURO AUV, there is only one range sonar mounted in front of vehicle. To get a profile image of environment, the vehicle has to take some specific motions such as rotating around it, and this usually takes quite an amount of time. In other word, it is not suitable for the vehicle to frequently take obstacle detecting process. For this reason, we design a relatively simple path planning method for autonomous navigation of P-SURO AUV. Consider the Fig. 12, to get to the point $P_{End} = (x_e, y_e, depth_e)$, the vehicle will turn around at start point P_c. According to detected obstacle A(a, b), we calculate d_a, d_b and $d_{max} = \max\{d_a, d_b\}$. If $d_{max} > d_{threshold}$ with $d_{threshold}$ design parameter, then we design the target point as $P_{Target} = (x_t, y_t, depth_e)$ with $y_t = R_a sin\psi_a + 0.5 d_a$ (see Fig. 12, in this case, we assume $d_a > d_b$ and $d_a > d_{max}$). In the case of $d_{max} \leq d_{threshold}$, the vehicle will take descent motion as shown in Fig. 13 until $depth = h_d$. Here d_R (see Fig. 13) is a design parameter. And in this case, the target point is set to $P_{Target} = (x_e, y_e, h_d)$.

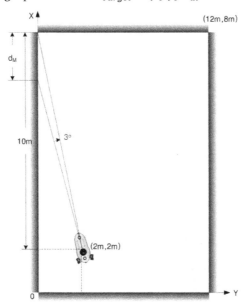

Fig. 10. Maximum range error.

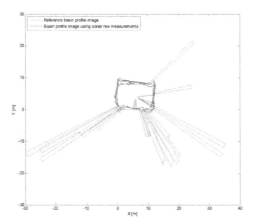

Fig. 11. Basin profile image using range sonar.

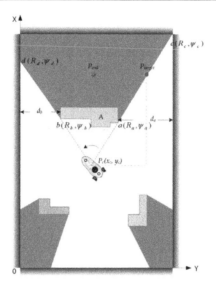

Fig. 12. Acquisition of 2D profile image.

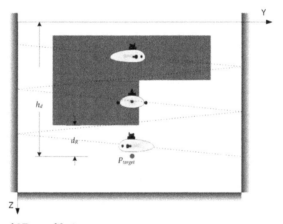

Fig. 13. Acquisition of 3D profile image.

6. Basin test

To demonstrate the proposed vision-based underwater localization and the SLAM methods, we carried out a series of field tests in the engineering basin in PIRO.

6.1 Preliminary of basin test

In its underwater mission, the vehicle is always forced to keep zero pitch angle. And in the horizontal plane, we design the vehicle's reference path to be always parallel to the axis X or axis Y, see Fig. 12. In this case, at any point, the vehicle's position can be easily got through simple rotation mode. However, considering the fact that the vehicle does not keep at the same point through its rotation, in other word, there is a drift for the vehicle's position in the rotating mode. So, though the accuracy of range sonar measurement is in the centimetres level, the total position error for this kind of rotation mode is significant. Through a number of basin tests, we observe that this kind of position error is up to 0.5m.

Consider this kind of forward/backward motion; the vehicle's forward/backward velocity can be calculated using range sonar measurements. For this purpose, the following filter is designed for acquisition of range sonar raw measurements

$$d_{FR}(k) = (1 - 2^{-n})d_{FR}(k-1) + 2^{-n}d_R(k), \tag{6}$$

where d_{FR} and d_R denote each of filtered and raw measurements of range sonar, and n is filtering order. The filtering results can be seen in Fig. 14.

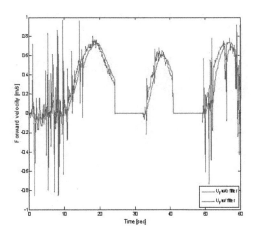

Fig. 14. Calculated forward speeds.

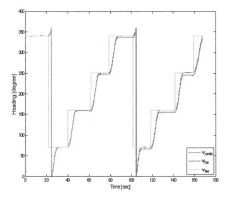

Fig. 15. Comparison of heading measurements

Another important issue for the basin test is about vehicle's AHRS sensor. The engineering basin in the PIRO is located in the basement of building, which is mainly constructed by steel materials. In this kind of environment, because of heavy distortion of earth magnetic field, AHRS cannot make proper initialization and Kalman filter compensation process. Therefore, there is significant drift in the AHRS heading output. However, fortunately, there is high accuracy 1-axis Gyro sensor horizontally mounted on the vehicle for the motion control purpose. And we estimate the vehicle's heading value using this Gyro output, whose bias value is also evaluated through lots of basin tests. Fig. 15 shows the comparison of these measurements.

6.2 P-SURO SLAM

To demonstrate the SLAM method proposed for P-SURO AUV, we perform the following three autonomous navigation tests: a) without obstacle, b) with one obstacle, c) with two obstacles, see Fig. 16. The autonomous navigation mission can be divided into following four phases

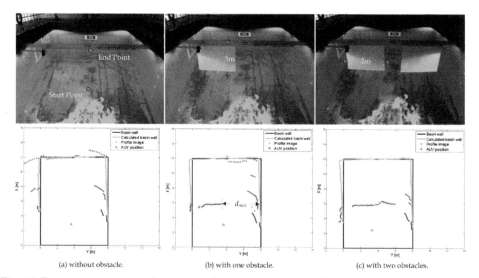

(a) without obstacle.   (b) with one obstacle.   (c) with two obstacles.

Fig. 16. Test environment and corresponding range sonar profile images.

Obstacle Detecting Phase: At start point $P_{Start} = (x, y, z, \psi)$=(3m,4m,1.5m, 90°), the vehicle turn a half cycle counter-clock wisely. In this period, the vehicle detects the obstacle using forward range sonar.

Path Planning Phase: According to the profile image got from the Obstacle Detecting Phase, the vehicle designs a target point $P_{Target} = (x_t, y_t, z_t, 0°)$.

Vision-based Underwater Localizing Phase: While approaching to P_{Target}, the vehicle recognizes the underwater pattern, from which defines the end point $P_{End} = (9, y_c - y_v, z_c - z_v, 0°)$. Here (x_c, y_c, z_c) denotes the vehicle's current position, and (x_v, y_v, z_v) is the vehicle's pose information acquired from pattern recognition.

Homing Phase: After approaching P_{End}, or failed to recognize the pattern, the vehicle returns to P_{Start} along with its previous tracking trajectory.

In Fig. 16, the blue line (calculated basin wall) is got through $f(x_c, y_c, \psi_c, R_{FSonar}, t)$ where R_{FSonar} is the forward range sonar measurement.

In the Path Planning Phase, the target points are set to different values according to three different cases. In the case of without obstacle, we set $P_{Target} = (8m, 4m, 1.5m, 0°)$; in the case of one obstacle, set $P_{Target} = (8m, y_c + 0.5d_{max}, 1.5m, 0°)$ with d_{max} is shown in Fig. 16(b); and in the case of two obstacles, $P_{Target} = (8m, 4m, h_d, 0°)$, where h_d is defined in Fig. 13 .

Autonomous navigation with obstacle avoidance and underwater pattern recognition test results are shown in Fig. 17. Through these field tests, we found that the proposed SLAM method for P-SURO AUV shown a satisfactory performance. Also, we found that the aforementioned drift in the vehicle's rotating motion is the main inaccuracy source of both the navigation and the path planning (specially, in the calculation of d_i with $i = a, b, c$ in Fig. 12).

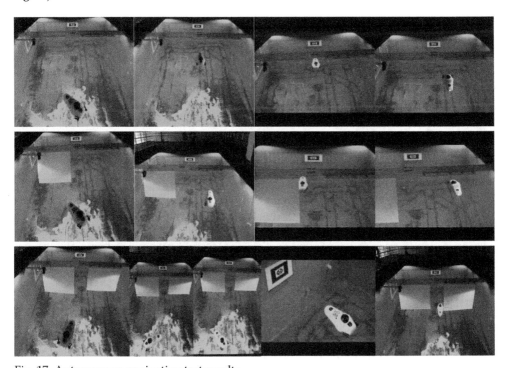

Fig. 17. Autonomous navigation test results.

7. Summary and future works

Recently, how to improve the vehicle's autonomy has been one of most hot issues in the underwater robotics community. In this chapter, we have discussed some of underwater intelligent technologies such as vision-based underwater localization and SLAM method only using video camera and range sonar, both of which are relatively cheap underwater equipments. Through a series of field tests in the engineering basin in PIRO using P-SURO

AUV, we observed that the proposed technologies provided satisfactory accuracy for the autonomous navigation of hovering-type AUV in the basin.

However, through the basin tests, we also observed that proposed vision algorithm was somewhat overly sensitive to the environmental conditions. How to improve the robustness of underwater vision is one of great interest in our future works. Besides, developing certain low-cost underwater navigation technology with partially known environmental conditions is also one of our future concerns.

8. Acknowledgment

This work was partly supported by the Industrial Foundation Technology Development project (No. 10035480) of MKE in Korea, and the authors also gratefully acknowledge the support from UTRC(Unmanned Technology Research Centre) at KAIST, originally funded by ADD, DAPA in Korea.

9. References

Amato, N. M. & Wu, Y. (1996). A randomized roadmap method for path and manipulation planning, *Proceedings of IEEE International Conference on Robotics and Automation*, pp. 113-120, Osaka, Japan, November 4-8, 1996

Bennet, A. A & Leonard, J. J. (2000). A behavior-based approach to adaptive feature detection and following with autonomous underwater vehicles. *IEEE Journal of Oceanic Engineering*, Vol.25, pp. 213-226, 2000

Brooks, R. A. (1986). A robust layered control system for a mobile robot. *IEEE Journal of Robotics and Automation*,Vol.2, pp. 14-23, 1986

Buniyamin, N., Sariff, N., Wan Ngah, W. A. J., Mohamad, Z. (2011). Robot global path planning overview and a variation of ant colony system algorithm. *International Journal of Mathematics and Computers in Simulation*, Vol.5, pp. 9-16, 2011

Castellanos, J. A. & Tardos, J. D. (1999). *Mobile Robot Localization and Map Building: A Multisensor Fusion Approach*. Boston, Mass.: Kluwer Academic Publishers, 1999

Choset, H., Lynch, K. M., Hutchinso, S., Kantor, G., Burgard, W., Kavraki, L. E., Thrun, S. (2005). *Principles of Robot Motion*. MIT Press, 2005

Do, K. D., Jiang, J. P., Pan, J. (2004). Robust adaptive path following of underactuated ships. *Automatica*, Vol.40, pp. 929-944, 2004

Elfes, A. (1989). Using occupancy grids for mobile robot perception and navigation. IEEE Transactions on Computer, Vol.22, pp. 46-57, 1989

Khatib, O. (1986). Real-time obstacle avoidance for manipulators and mobile robots. International Journal of Robotics Research, Vol.5, pp. 90-98, 1986

Hartley, R & Zisserman, A. (2000). *Multiple View Geometry in Computer Vision*. Cambridge University Press, June 2000

Healey, A. J., Marco, D. B., McGhee, R. B. (1996). Autonomous underwater vehicle control coordination using a tri-level hybrid software architecture, *Proceedings of the IEEE International Conferences on Robotics and Automation*, Minneapolis, Minnesota, pp. 2149-2159, 1996

Jiang, J. P. (2002). Global tracking control of underactuated ships by Lyapunov's direct method. *Automatica*, Vol.40, pp. 2249-2254, 2002

Marthiniussen, R., Vestgard, K., Klepaker, R., Storkersen, N. (2004). HUGIN-AUV concept and operational experience to date, *Proceedings of IEEE/MTS Oceans'04*, pp. 846-850, Kobe, Japan, November 9-12, 2004

Leonard, J. & Durrant-Whyte, H. (1992). *Directed Sonar Sensing for Mobile Robot Navigation*. London: Kluwer Academic Publishers, 1992

Li, J. H., Jun, B. H., Lee, P. M., Hong, S. K. (2005). A hierarchical real-time control architecture for a semi-autonomous underwater vehicle. *Ocean Engineering*, Vol.32, pp. 1631-1641, 2005

Li, J. H., Lee, P. M., Jun, B. H., Lim, Y. K. (2008a). *Underwater Vehicle*, InTech, ISBN 978-953-7619-49-7, Vienna, Austria

Li, J. H., Lee, P. M., Jun, B. H., Lim, Y. K. (2008b). Point-to-point navigation of underactuated ships. *Automatica*, Vol. 44, pp. 3201-3205, 2008

Li, J. H., Yoon, B. H., Oh, S. S., Cho, J. S., Kim, J. G., Lee, M. J., Lee, J. W. (2010). Development of an Intelligent Autonomous Underwater Vehicle, P-SURO, *Proceedings of Oceans'10 IEEE Sydney*, Sydney, Australia, May 24-27, 2010

Oh, S. S., Yoon, B. H., Li, J. H. (2010). Vision-based localization for an intelligent AUV, P-SURO, *Proceedings of KAOSTS Annual conference*, pp. 2602-2605, Jeju, Korea, 2010

Peuch, A., Coste, M. E., Baticle, D., Perrier, M., Rigaud, V., Simon, D. (1994). And advanced control architecture for underwater vehicles, *Proceedings of Oceans'94*, Brest, France, pp. I-590-595, 1994

Prestero, T. (2001). *Verification of a six-degree of freedom simulation model for the REMUS autonomous underwater vehicles*, Masters Thesis, MIT, USA

Quek, C & Wahab, A. (2000). Real-time integrated process supervision. *Engineering Applications of Artificial Intelligence*, Vol.13, pp. 645-658, 2000

Simon, D., Espiau, B., Castillo, E., Kapellos, K. (1993). Computer-aided design of a generic robot controller handling reactivity and real-time control issues. *IEEE Transactions on Control Systems Technology*, Vol.1, pp. 213-229, 1993

Thrun, S., Burgard, W., Fox, D. (2005). *Probabilistic Robotics*. The MIT Press, 2005

Valavanis, K. P., Hebert, T., Kolluru, R., Tsourveloudis, N. C. (2000). Mobile robot navigation in 2-D dynamic environments using electrostatic potential fields. *IEEE Transactions on Systems, Man and Cybernetics-part A*, Vol.30, pp. 187-197, 2000

Wang, H. H., Marks, R. L., Rock, S. M., Lee, M. J. (1993). Task-based control architecture for an untethered, unmanned submersible, *Proceedings of the 8th Symposium on Unmanned Untethered Submersible Technology*, Durham, New Hampshire, pp. 137-148, 1993

Zhang, Z. (2000). A flexible new technique for camera calibration. *IEEE Transactions on Pattern Analysis and Machine Intelligence*, Vol. 22, No. 11, pp. 1330-1334, 2000

Zheng, X. (1992). Layered control of a practical AUV, *Proceedings of the Symposium on Autonomous Underwater Vehicle Technology*, Washington DC, pp. 142-147, 1992

Development of a Vectored Water-Jet-Based Spherical Underwater Vehicle

Shuxiang Guo and Xichuan Lin
Kagawa University
Japan

1. Introduction

The applications of underwater vehicles have shown a dramatic increase in recent years, such as, mines clearing operation, feature tracking, cable or pipeline tracking and deep ocean exploration. According to different applications, the mechanical and electrical configuration and shape of an underwater vehicle are different. For instance, manipulators are necessary when doing mines clearing operation or some other tasks which need to deal with environment. If an underwater vehicle is used for underwater environment detection or observation, it is better to make this vehicle smaller and flexible in motion that it can go to smaller space easily. If the vehicle needs high speed moving in the water then a streamline body is required.

Different structures with different size of underwater vehicles are developed. Most of these underwater vehicles are torpedo-like with streamline bodies, like (Sangekar et al., 2009). And there are some small size AUVs like (Allen et al., 2002) and (Madhan et al., 2006). And also there are some other AUVs adopt different body shape, such as (Antonelli & Chiaverini, 2002). Meanwhile, the propulsion system is one of the critical facts for the performance of underwater vehicles, because it is the basis of control layers of the whole system. Propulsion devices have variable forms, for instance, paddle wheel, poles, magneto hydrodynamic drive, sails and oars.

Paddle wheel thrusters are the most common and traditional propulsion methods for underwater vehicles. Usually, there are at least two thrusters installed on one underwater vehicle, one for horizontal motion and the other for vertical motion. The disadvantages of paddle wheel thrusters are obvious, for example, it is easy to disturb the water around the underwater vehicles. Meanwhile, the more the paddle wheel thrusters are used, the weight, noise and energy consumption increases.

The steering strategies of traditional underwater vehicles are changing the angular of rudders or using differential propulsive forces of two or more than two thrusters. Of course, there are vectored propellers being used on underwater vehicles. Reference (Cavallo et al., 2004) and (Le Page & Holappa, 2002a) present underwater vehicles with vectored thrusters. Reference (Duchemin et al., 2007) proposes multi-channel hall-effect thrusters which involves vector propel and vector composition. Reference (Le Page & Holappa, 2002b) proposes an autonomous underwater vehicle equipped with a vectored thruster. At the same time, the design of vectoring thrusters used on aircrafts is also an example of vectored propulsion system (Kowal, 2002), (Beal, 2004) and (Lazic & Ristanovic, 2007).

The purpose of this research is to develop such a kind of underwater vehicle which can adjust its attitude freely by changing the direction of propulsive forces. Meanwhile, we would like to make the vehicle flexible when moving in the water. Inspired by jet aircraft, we adopt vectored water-jet propellers as the propulsion system. According to the design purpose, a symmetrical structure would be better for our underwater vehicle (Guo et al., 2009).

This spherical underwater vehicle has many implementation fields. Because of its flexibility, our vehicle can be used for underwater creatures observation. For example, we can install underwater cameras on the vehicle. It can track and take photos of fishes. Another example is that, due to its small size, we can use it to detect the inside situation of underwater oil pipes.

2. Mechanical and electrical design

2.1 Mechanical system design

Before the practical manufacture, we try to give a conceptual design of the whole structure for this spherical underwater vehicle. At this stage, we need to consider about the dimension, weight distribution, material, components installation, and so on. And we also need to consider about the configuration of the propulsion system, for example, how many water-jet propellers should we use for the purpose of optimizing power consumption without decreasing propulsion ability. Therefore, by all of that mentioned above, we give the conceptual designed structure of our spherical underwater vehicle as shown in Fig.1.

It adopts a spherical shape, all the components are installed inside the body. Its radius is $20cm$ which is smaller than that in (Antonelli et al., 2002). Its overall weight is about $6.5kg$. Its working depth is designed to 0 $10m$, with a max speed of about $1.5m/s$.

Inside the vehicle, there will be three water-jet propellers used as propulsion system, which is enough for surge, yaw and heave. One waterproof box is used for all the electronic components such as sensors, batteries and the control boards. And all of these are mounted on a triangle support which is fixed on the spherical hull. The whole structure is symmetrical in z-axis. Therefore, it can rotate along z-axis, and by doing this, the vehicle can change its orientation easily.

2.1.1 The spherical hull

As shown in Fig.2, the spherical hull of this underwater vehicle is made of acrylic which is light and easy to be cut. It is about $3mm$ thick and the diameter is $40cm$. Actually, we can see that this spherical hull is composed of two transparent hemisphere shells. There are three holes which can provide enough space for water-jet propellers to rotate for different motions. We will discuss the details about the principles of the water-jet propulsion system in the next section.

2.1.2 The waterproof box

Waterproof is essential for underwater vehicles. Fig.3 shows the design of the waterproof box. The whole size of this box is about $22cm$(hight) $\times$ $14cm$(inner diameter). An O-ring is used for seal, which has the ability to provide waterproof in our case. Inside the waterproof box, there will be two control boards, one or two lithium batteries, depending on tasks. Meanwhile, at the top part inside the box, there will be an digital rate gyro sensor for orientation feedback. The body of waterproof box is also transparent, therefore, we can easily observe the inside working status .

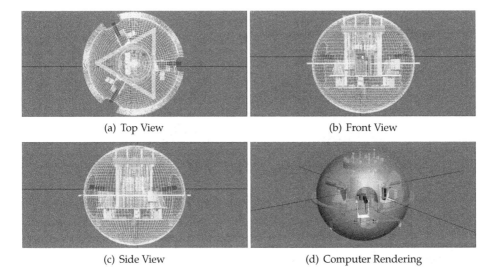

(a) Top View (b) Front View

(c) Side View (d) Computer Rendering

Fig. 1. Mechanical System Schematics of the Spherical Underwater Vehicle

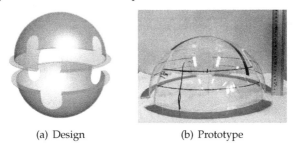

(a) Design (b) Prototype

Fig. 2. Spherical Hull

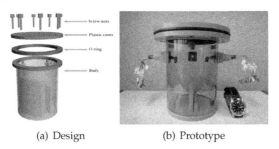

(a) Design (b) Prototype

Fig. 3. Design of Waterproof Box

2.1.3 Mechanism of the water-jet propulsion system

Fig.4 is the structure of one single water-jet propeller. It is composed of one water-jet thruster and two servo motors (above and side). The water-jet thruster is sealed inside a plastic box for waterproof. And we use waterproof glue on servo motors for waterproof. The thruster can be rotated by these two servo motors, therefore, the direction of jetted water can be changed in X-Y plane and X-Z plane, respectively.

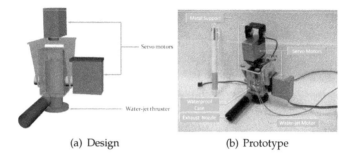

(a) Design   (b) Prototype

Fig. 4. Structure of a Water-jet Propeller

Three of the water-jet propellers are mounted on the metal support frame, as shown in Fig.5. Three of them are circumferentially $2\pi/3$ apart from each other.

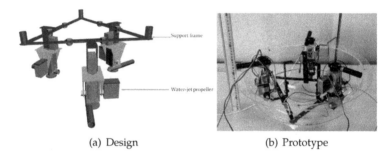

(a) Design   (b) Prototype

Fig. 5. Water-jet Propellers mounted on Support Frame

2.2 Electrical system design

We adopt a minimal hardware configuration for the experimental prototype vehicle. For a single spherical underwater vehicle, there are three major electrical groups, sensor group, control group and actuator group. Fig.6 gives the electrical schematics. At present, we only use one pressure sensor for depth control and one gyro sensor for surge control. One ARM7 based control board is used as central control, data acquisition, algorithm implement and making strategic decisions. One AVR based board is used as the coprocessor unit for motor control. It receives the commands from ARM and translates the commands into driving signals for the water-jet propellers.

Fig.7 gives the main hardware for this vehicle. Fig.7(a) is the ARM7 based board with S3C44B0X on it, which can fulfill our requirement at present. Fig.7(b) is the AVR based board with ATmega2560 on it. RS232 bus is used for the communication between ARM7 and AVR. In Fig.7(c) is the set of pressure sensor with the sensor body(right) and its coder (left). It use RS422 bus for data transmission. Digital gyro sensor CRS10 is shown in Fig.7(d), we use the build in AD converter of S3C44B0X for data acquisition.

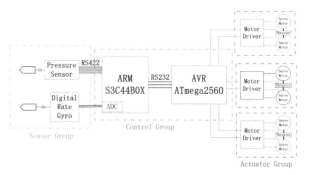

Fig. 6. Electrical Schematics for Prototype System

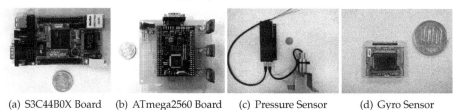

(a) S3C44B0X Board   (b) ATmega2560 Board   (c) Pressure Sensor   (d) Gyro Sensor

Fig. 7. Electrical Components for the Experimental Prototype Underwater Vehicle

2.3 Power supply

We adopt two power supply for the spherical underwater vehicle. The highest power consumption components in our vehicle are propellers. For each of them, the thruster has a working voltage of $7.2V$ and $3.5A$ current drain, servo motors can work under $5V$ with relatively small current. Therefore, we use two 2-cells LiPo batteries as the power supply for the propellers. The capacity of each battery is $5000mAh$ with parameter of $50c - 7.4V$. Besides, we use 4 AA rechargeable batteries for the control boards. We carried out the power consumption test for one LiPo battery, and Fig.8 gives the battery discharge graph of the power system.

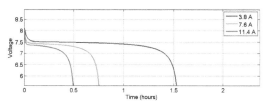

Fig. 8. Power Consumption of the Whole System. Blue line – one propeller working; green line – two propellers working; red line – three propellers working

3. Principles and modeling of the propulsion system

In this section, we will discuss about the working principles, modeling method and the identification experiment for the water-jet propeller. Many literatures have presented the computing formula for the torque and thrust exerted by a thruster. Most of them are base on the lift theory, and mainly focus on blades type propellers (Newman, 1977), (Fossen, 1995) and (Blanke et al., 2000). Our propellers are different with blades type propellers, therefore, we try to find another method for the modeling of water-jet propellers. In (Kim & Chung, 2006), the author presented a dynamic modeling method in which the flow velocity and incoming angle are taken into account. We will use this modeling method for our water-jet propellers.

3.1 Working principles
Before modeling of propulsion system, we want to give some basic working principles about the water-jet propellers. Fig.9(a) shows the top view of distribution of three propellers. They can work together to realize different motion, such as surge and yaw.

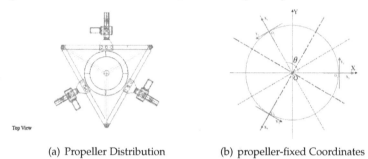

(a) Propeller Distribution   (b) propeller-fixed Coordinates

Fig. 9. Distribution and Coordination of Multiple Propellers

If we let θ be the interval angle of each water-jet propeller, as shown in Fig.9(b), then, for the purpose of kinematics transform, three propeller-fixed coordinates are introduced for propellers, which are fixed in the rotation center of the propellers. So we can see, these three propeller-fixed coordinates are actually transform results of vehicle-fixed coordinate reference frame. Meanwhile, it should be noted that, this transform only happens in X-Y plane. Let the matrix form of the coordinates transform be given as:

$$\begin{pmatrix} X_1 \\ Y_1 \\ Z_1 \end{pmatrix} = \begin{pmatrix} X \\ Y \\ Z \end{pmatrix} + \begin{pmatrix} -R \\ 0 \\ 0 \end{pmatrix} \tag{1}$$

$$\begin{pmatrix} X_2 \\ Y_2 \\ Z_2 \end{pmatrix} = \begin{pmatrix} c\theta & s\theta & 0 \\ -s\theta & c\theta & 0 \\ 0 & 0 & 1 \end{pmatrix} \begin{pmatrix} X \\ Y \\ Z \end{pmatrix} + \begin{pmatrix} \frac{1}{2}Rc\theta - Rs\theta c\frac{\pi}{6} \\ -\frac{1}{2}Rc\theta - Rc\theta c\frac{\pi}{6} \\ 0 \end{pmatrix} \tag{2}$$

$$\begin{pmatrix} X_3 \\ Y_3 \\ Z_3 \end{pmatrix} = \begin{pmatrix} c2\theta & s2\theta & 0 \\ -s2\theta & c2\theta & 0 \\ 0 & 0 & 1 \end{pmatrix} \begin{pmatrix} X \\ Y \\ Z \end{pmatrix} + \begin{pmatrix} \frac{1}{2}Rc2\theta + Rs2\theta c\frac{\pi}{6} \\ -\frac{1}{2}Rc2\theta + Rc2\theta c\frac{\pi}{6} \\ 0 \end{pmatrix} \tag{3}$$

where R is the radius of the vehicle, $s(\cdot) \equiv \sin(\cdot)$ and $c(\cdot) \equiv \cos(\cdot)$.

So, a general transform matrix can be obtained:

$$^pP_b = \Phi_p^b \cdot {}^pP_p + C \tag{4}$$

where ^pP_b is the position vector of propeller-fixed coordinate expressed in vehicle-fixed coordinate, $\Phi_p^b = (\Phi_{p1}^b, \Phi_{p2}^b, \Phi_{p3}^b)^T$ is the transform matrix from propeller-fixed coordinate to vehicle-fixed coordinate, ^pP_p is the position vector in propeller-fixed coordinate and the C is a constant vector.

Now, let us take a look at three motions, surge, heave and yaw. The definition of these three motions can be found in (Fossen, 1995). Before that, we define two angles which will be used for orientation of propellers. Fig.10 gives the definition of α and β. Fig.11 gives a demonstration of surge, heave and yaw.

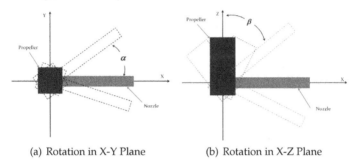

(a) Rotation in X-Y Plane   (b) Rotation in X-Z Plane

Fig. 10. Orientation of Propellers

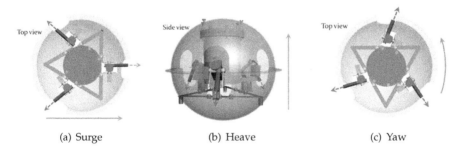

(a) Surge   (b) Heave   (c) Yaw

Fig. 11. Propulsion Forces for Surge, Heave and Yaw

The first case is surge. In this case, two of the water-jet propellers will work together, and the other one could be used for brake. So, from Fig.11(a), two water-jet propellers in the left will be used for propulsion, and if we want to stop the vehicle from moving, the third propeller can act as a braking propeller. From Equation 4, the resultant force for surge can be expressed in vehicle-fixed coordinate as:

$$\begin{cases} {}^pF_{xb} = \Phi_{p1}^{b^T} \sum_{i=1}^{3} ({}^pF_{ip} + e_1 C_i) \neq 0 \\ {}^pF_{yb} = 0 \\ {}^pF_{zb} = 0 \end{cases} \tag{5}$$

where, $\mathbf{e}_1 = (1, 0, 0)^T$.

Then, for the heave case, all the three water-jet propellers will work and the side servo motor will rotate to an angle that $\beta > \pi/2$. Therefore, in this case, the resultant force for heave can be expressed in vehicle-fixed coordinate as:

$$
\begin{cases}
{}^P F_{xb} = 0 \\
{}^P F_{yb} = 0 \\
{}^P F_{zb} = \mathbf{\Phi}_{p3}^{bT} \sum_{i=1}^{3} ({}^P \mathbf{F}_{ip} + \mathbf{e}_3 C_i) \neq 0
\end{cases}
\tag{6}
$$

where, $\mathbf{e}_3 = (0, 0, 1)^T$.

The third case is yaw which is rotating on z-axis. By denoting in propeller-fixed coordinates, α should have the same orientation, clockwise or counterclockwise, that means, $\alpha_i > 0$ or $\alpha_i < 0$. So in yaw, rotation moment will take effect. We can write the equation for yaw in vehicle-fixed coordinate as:

$$
\begin{cases}
{}^P F_{xb} = \mathbf{\Phi}_{p1}^{bT} \sum_{i=1}^{3} ({}^P \mathbf{F}_{ip} + \mathbf{e}_1 C_i) \neq 0 \\
{}^P F_{yb} = \mathbf{\Phi}_{p2}^{bT} \sum_{i=1}^{3} ({}^P \mathbf{F}_{ip} + \mathbf{e}_2 C_i) \neq 0 \\
{}^P F_{zb} = 0 \\
{}^P M_{xb} + {}^P M_{yb} + {}^P M_{zb} \neq 0
\end{cases}
\tag{7}
$$

where, $\mathbf{e}_2 = (0, 1, 0)^T$, $\mathbf{e}_3 = (0, 0, 1)^T$.

3.2 Modeling of single water-jet propeller

In the author's previous research (Guo et al., 2009), the modeling for orientation of water-jet propeller is presented. Therefore, in this part, we will only discuss about the hydrodynamics modeling of the water-jet thruster. The method we refer to is presented in (Kim & Chung, 2006). For the purpose of dynamic modeling of water-jet propeller, we give the flow model of the water-jet thruster, which is shown in Fig.12. The shaft is perpendicular to the nozzle, and there are two blades.

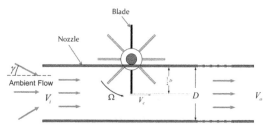

Fig. 12. Flow Model of the Water-jet Thruster (top view)

where,

Ω is angular velocity of the thruster

V_i is velocity of incoming flow

V_c is central flow velocity in the nozzle

D is diameter of the nozzle

V_o is velocity of outlet flow

γ is incoming angle of ambient flow

Because the diameter of the nozzle is small, the velocity difference in the nozzle can be ignored, so we consider the axis flow velocity V_a as a linear combine of incoming flow velocity and the central flow velocity,

$$V_a = k_1 V_i + k_2 V_c$$
$$V_c = \frac{1}{2} D\Omega \qquad (8)$$
$$V_i = V_f cos\gamma$$

By assuming that the flow is incompressible, therefore, from equation of continuity, we know that the volume of incoming flow must equal to the outlet flow, then we get:

$$\rho_a V_a A_a = \rho_o V_o A_o \qquad (9)$$

where, $\rho_a = \rho_o$ is density of flow, $A_a = A_o$ is cross-section of the nozzle. Therefore, we can also get:

$$V_a = V_o \qquad (10)$$

Meanwhile, we know that, the propulsive force of the water-jet thruster is:

$$F_t = \rho A V_a^2 \qquad (11)$$

By substituting Equation 8 in Equation 11, we can get:

$$F_t = \frac{\pi}{4} \rho D^2 (k_1^2 V_i^2 + 2k_1 k_2 V_i D\Omega + k_2^2 D^2 \Omega^2) \qquad (12)$$

By rewriting Equation 12, we get:

$$\frac{F_t}{\rho D^4 \Omega^2} = \frac{\pi}{4} (k_1^2 (\frac{V_i}{D\Omega})^2 + 2k_1 k_2 \frac{V_i}{D\Omega} + k_2^2) \qquad (13)$$

Then, we can let the non-dimensional parameter be:

$$K_T(J_0) = \frac{\pi}{4} (k_1^2 (\frac{V_i}{D\Omega})^2 + 2k_1 k_2 \frac{V_i}{D\Omega} + k_2^2) \qquad (14)$$

where

$$J_0 = \frac{V_i}{D\Omega} = \frac{V_f cos\gamma}{D\Omega} \qquad (15)$$

J_0 is the advance ratio.

Now, the modeling becomes measuring of three parameters, flow velocity, incoming angle and angular velocity of thruster. For this purpose, we designed an experiment to measure these parameters and find out their relationship.

Flow velocities	Depth	Incoming angles	Control voltages
0.1m/s 0.2m/s	80cm	$0 - \pi$	$3V - 7V$ (DC)

Table 1. Experiment Condition

3.3 Experiments for the dynamics modeling
In this part, we try to identify the dynamics model of the water-jet propeller by experiment. What we are interested in is the relation of flow incoming angles, flow velocities and propulsive forces. Experiment condition is listed in Table 1.

3.3.1 Experiment design
Fig.13 gives the experiment principle. We use one NEC 2301 stain gage as the force sensor and use NEC AC AMPLIFIER AS 1302 to amplify the output signal from strain gage, which are shown in Fig.14.

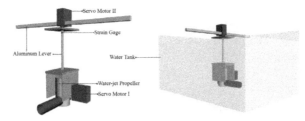

Fig. 13. Experiment Design for Identification

Fig. 14. Strain Gage and Amplifier

Firstly, we give a brief illustration for the strain gage measurement. Let F_d be deformation force, and ε be the deformation of aluminium lever used in the experiment. Therefore, from the theory of mechanics of materials, we can get the relation of F_d and ε as:

$$F_d = \frac{ZE}{X}\varepsilon \tag{16}$$

where, Z is second moment of area, E is the Young's modulus, X is the distance from acting point of force to stain gage.

3.3.2 Experiments and analysis
In the experiment, there are four variables we need to consider, the equivalent cross-section of propeller, flow velocity, incoming angle and control voltage. What we are interested in is the variation of propulsive force in different incoming angles and different control voltage.

3.3.2.1 Equivalent cross-section variation of propellers

As a vectored water-jet-based propulsion system, it should be noted that both the propulsive force and its direction can be changed. Therefore, when the propeller changes its direction, actually, the incoming angle of flow is also changing, and the equivalent cross-section of the propeller is changing. From Equation 11 we know the propulsive force will change if cross-section A changes. Fig.15 gives a demonstration of this case. When the propeller rotate from position I to II, the equivalent cross-section will change from cross-section I to cross-section II. So we try to find an equation to describe this variation.

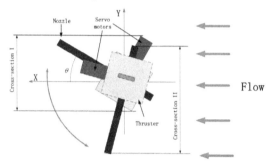

Fig. 15. Variation of Equivalent Cross-section of Propellers

Considering that the measured force from stain gage is actually a resultant force of propulsive force and fluid force. And we also know that the fluid force acted on the propeller depends on the equivalent cross-section.

So the first experiment is measurement of the equivalent cross-section variation. The propeller is submerged in the flow which has a speed of $0.2m/s$, propeller is powered off. And we only change its orientation in $X - Y$ plane. Because of experiment limits, we can not change the flow direction, so in the experiment, the incoming angle equals to the orientation angle of the propeller. Fig.15 gives a demonstration of the equivalent cross-section. We give some special angles, $0, \pi/6, \pi/3, 2\pi/3, 5\pi/6, \pi$, for this experiment. Fig.16 gives the experiment data of equivalent cross-section. You may notice that, we did not adopt the orientation angle of $\pi/2$. Because, when the propeller rotate to $\pi/2$, which means that the measure surface of the strain gage is parallel to the flow direction, the strain gage can not measure the flow force.

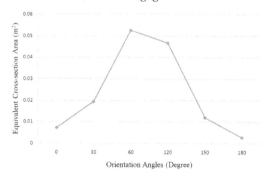

Fig. 16. Variation of Equivalent Cross-section

From Fig.16 we can see, the data curve is similar with a sinusoid, so we use a sine function to fit this experiment data:

$$A_e(\phi) = \lambda_1 sin(\lambda_2\phi + \lambda_3) \tag{17}$$

where ϕ is incoming angle, $\lambda_1, \lambda_2, \lambda_3$ are coefficients.

3.3.2.2 Incoming angle and deformation force

In this case, the flow velocity is seen as constant. Two groups of experiment are carried out at flow velocity of $0.1m/s$ and $0.2m/s$. The control voltage to thruster is from $3V$ to $7V$ every $1V$. From the data shown in Fig.17, we can see that the deformation force does not simply increase

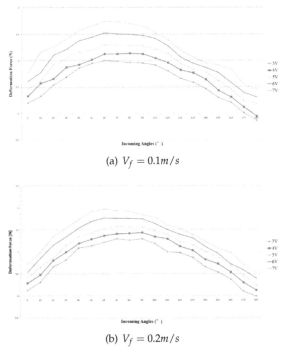

(a) $V_f = 0.1m/s$

(b) $V_f = 0.2m/s$

Fig. 17. Deformation Forces with Different Incoming Angles

with the increasing of incoming angle, the maximum deformation of the lever happens at about 60 degree of the incoming angles. They are not a linear relation. Then, how about the real propulsive force?

3.3.2.3 Incoming angle and propulsive force

As we mentioned, what we measured by stain gage is actually a resultant force of propulsive force and fluid force. Therefore, the real propulsive force F_t should be calculated using deformation force F_d and the equivalent cross-section A_e.

$$F_d = F_t - F_f cos\gamma \tag{18}$$

If we consider about the fluid force of the flow, we can refer to Equation 11 and 17, then write the fluid force as:

$$F_f = \rho V_c^2 A_e(\phi) \tag{19}$$

We substitute Equation 19 into 18 we get:

$$F_t = F_d + \rho V_c^2 A_e(\phi)cos\gamma \tag{20}$$

So now, we can calculate the real propulsive force by using equivalent cross-section, deformation force and incoming angles. The results is shown in Fig.18. Fig.18(a) and Fig.18(b) are results at the flow velocity of $V_f = 0.1m/s$ and $V_f = 0.2m/s$, respectively.

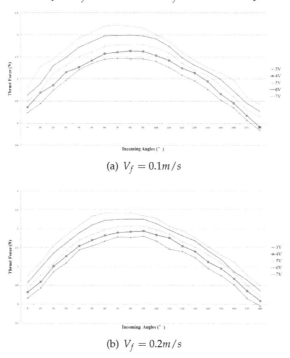

(a) $V_f = 0.1m/s$

(b) $V_f = 0.2m/s$

Fig. 18. Propulsive forces with Different Incoming Angles

3.3.2.4 Control voltage and propulsive force

From the results of 3.3.2.3, we can obtain the relation of control voltage and propulsive force. First, we give the experiment data, in Fig.19. From the diagram, we can see that the relation of control voltage and propulsive force can be described using linear equation.

4. Underwater experiments for basic motions

Because of the symmetrical shape of the hull, it is obvious that motion characteristics of surge, sway and heave should be similar. However, from another point of view, surge and sway are motions in $X - Y$ plane while heave is a motion that its motion surface perpendicular to $X - Y$ plane. Therefore, we carry out experiments for horizontal motion surface and vertical motion surface respectively. Besides, for the experimental prototype vehicle, we only consider one rotational DOF in Z axis, so the third experiment is yaw motion.

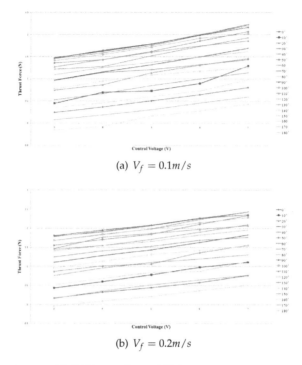

(a) $V_f = 0.1m/s$

(b) $V_f = 0.2m/s$

Fig. 19. Propulsive forces with Different Control Voltages

4.1 Experiment of horizontal motion
This experiment combines surge and sway together to verify the motion characteristics of the vehicle in horizontal plane. We carried out three experiments:

Case 1:

step 1. Surge (Move forward in X axis);

step 2. Right steering (Turn right about 90°);

step 3. Sway (Move forward along Y axis.)

Case 2:

step 1. Surge (Move forward in X axis);

step 2. Left steering (Turn left about 90°);

step 3. Sway (Move forward along Y axis.)

Case 3:

step 1. Surge (Propeller I and II work together, propeller III powered off);

step 2. Brake (Propeller I and II powered off, Propeller III works to produce brake force).

In case 1, timing of step 1 is about 10s, and step 2 takes about 12s. And timing of case 2 is relatively the same with case 1, because of the same hydrodynamics characteristics of turning right and left. In case 3, it takes 15s reaching a stable speed, and the brake effect happens in about 3s which is effective for low speed underwater vehicles. From Fig.20, we can see,

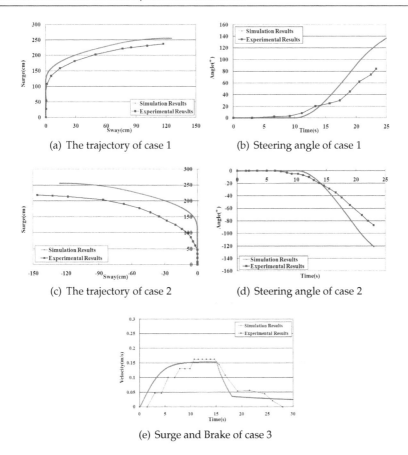

(a) The trajectory of case 1

(b) Steering angle of case 1

(c) The trajectory of case 2

(d) Steering angle of case 2

(e) Surge and Brake of case 3

Fig. 20. Experimental Results of Horizontal Motion

the experimental results fit well with simulation results in surge stage, but when the vehicle rotating, errors become large. The reason of this is because the simulation experiment only considered linear damping force and quadratic damping force, but in reality, there are other hydrodynamic forces act on the vehicle.

4.2 Experiment of vertical motion
Even though we design the working depth of the vehicle to $10m$, because the depth of experimental pool is only $1.2m$, we can only make the experiments in shallow water. So we set the vertical motion time in a relatively small range. We also carried out two experiments:
Case 1:

step 1. Set the top point of the spherical hull as the start point;

step 2. Move downward in Z axis for about $7s$;

step 3. Float up to the surface.

Case 2:

step 1. Set the top point of the spherical hull as the start point;

step 2. Move downward in Z axis for about $7s$;

step 3. Stop the vehicle.

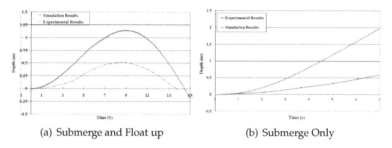

(a) Submerge and Float up	(b) Submerge Only

Fig. 21. Experimental Results of Vertical Motion

From Fig.21(a) and Fig.21(b), we can see, the experimental results does not fit well with simulation results very well, errors exceed 100%. When we analyze the reasons, we find that, the simulation experiment does not consider the variation of water pressure. The control voltage to the thrusters is $7V$ as a constant. That means, the propulsive force will not change. But with the increasing of depth, water pressure increases. As a result, the effective propulsive force are weaken by water pressure.

4.3 Experiment of yaw

We let the vehicle rotate about 90^0 then stop. From Fig.22(a) and Fig.22(b), the maximum error between simulation results and experimental results happens at about $2.8s$ where is nearly the maximum angular velocity. The reason of this result is that, we simplified the model of our vehicle, especially the hydrodynamic damping forces. Only linear damping force and quadratic damping force are taken into account in our case. But in the real experiment, there are many other velocity related hydrodynamic damping forces, therefore, when the angular velocity increasing, the damping effect of ignored forces become obvious.

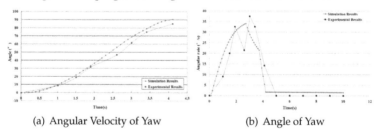

(a) Angular Velocity of Yaw	(b) Angle of Yaw

Fig. 22. Experimental Results of Yaw

5. Conclusions

In this paper, we proposed a spherical underwater vehicle which uses three water-jet propellers as its propulsion system. We introduced the design details of mechanical and electrical system.

Based on the design of the vehicle, we introduced the principles of the water-jet propulsion system including the force distribution of three water-jet propellers, the working principles of different motions. And then we discussed about the modeling of one single propeller by identification experiments. For the modeling, the flow velocity and equivalent cross-section of the propeller are taken into account for dynamics model.

One experimental prototype of this spherical underwater vehicle is developed for the purpose of evaluation. Underwater experiments are carried out to evaluate the motion characteristics of this spherical underwater vehicle. Experimental results are given for each experiment, and the analysis are also given.

From the underwater experiments of the prototype vehicle, the availability of the design is proved, and the water-jet propulsion system can work well for different motions. But there are also some problems needed to be resolved. Firstly, the propulsive force of the water-jet propellers needed to be increased; secondly, the variation of water pressure on the propulsive force should be considered when building the dynamics model of propellers; thirdly, the gravity distribution should be re-regulated to improve stability; finally, from experiments, it is necessary to improve the accuracy of the dynamics model of the vehicle for precise control.

6. References

Allen, B., Stokey, R., Austin, T., Forrester, N., Goldsborough, R., Purcell, M. & von Alt, C. (2002). REMUS: a small, low cost AUV; system description, field trials and performance results, *OCEANS'97. MTS/IEEE Conference Proceedings*, Vol. 2, IEEE, pp. 994–1000.

Antonelli, G. & Chiaverini, S. (2002). Adaptive tracking control of underwater vehicle-manipulator systems, *Proceedings of the 1998 IEEE International Conference on Control Applications, 1998*, Vol. 2, IEEE, pp. 1089–1093.

Antonelli, G., Chiaverini, S., Sarkar, N. & West, M. (2002). Adaptive control of an autonomous underwater vehicle: experimental results on ODIN, *IEEE Transactions on Control Systems Technology* 9(5): 756–765.

Beal, B. (2004). *Clustering of Hall effect thrusters for high-power electric propulsion applications*, PhD thesis, The University of Michigan.

Blanke, M., Lindegaard, K. & Fossen, T. (2000). Dynamic model for thrust generation of marine propellers, *Proceedings of the IFAC Conference on Maneuvering and Control of Marine Craft (MCMC 2000)*, Citeseer.

Cavallo, E., Michelini, R. & Filaretov, V. (2004). Conceptual design of an AUV Equipped with a three degrees of freedom vectored thruster, *Journal of Intelligent & Robotic Systems* 39(4): 365–391.

Duchemin, O., Lorand, A., Notarianni, M., Valentian, D. & Chesta, E. (2007). Multi-Channel Hall-Effect Thrusters: Mission Applications and Architecture Trade-Offs, *30th International Electric Propulsion Conference*, Florence, Italy.

Fossen, T. I. (1995). *Guidance and Control of Ocean Vehicles*, John Wiley & Sons Ltd., USA.

Guo, S., Lin, X. & Hata, S. (2009). A conceptual design of vectored water-jet propulsion system, *International Conferenceon Mechatronics and Automation, 2009. ICMA 2009*, IEEE, pp. 1190–1195.

Kim, J. & Chung, W. (2006). Accurate and practical thruster modeling for underwater vehicles, *Ocean Engineering* 33(5-6): 566–586.

Kowal, H. (2002). Advances in thrust vectoring and the application of flow-control technology, *Canadian aeronautics and space journal* 48(2): 145–151.

Lazic, D. & Ristanovic, M. (2007).   Electrohydraulic thrust vector control of twin rocket engines with position feedback via angular transducers, *Control Engineering Practice* 15(5): 583–594.

Le Page, Y. & Holappa, K. (2002a).   Hydrodynamics of an autonomous underwater vehicle equipped with a vectored thruster, *OCEANS 2000 MTS/IEEE Conference and Exhibition*, Vol. 3, IEEE, pp. 2135–2140.

Le Page, Y. & Holappa, K. (2002b). Simulation and control of an autonomous underwater vehicle equipped with a vectored thruster, *OCEANS 2000 MTS/IEEE Conference and Exhibition*, Vol. 3, IEEE, pp. 2129–2134.

Madhan, R., Desa, E., Prabhudesai, S., Sebastião, L., Pascoal, A., Desa, E., Mascarenhas, A., Maurya, P., Navelkar, G., Afzulpurkar, S. et al. (2006). Mechanical design and development aspects of a small AUV–Maya, *7th IFAC Conference MCMC2006*.

Newman, J. (1977). *Marine hydrodynamics*, The MIT press.

Sangekar, M., Chitre, M. & Koay, T. (2009). Hardware architecture for a modular autonomous underwater vehicle STARFISH, *OCEANS 2008*, IEEE, pp. 1–8.

3

Hydrodynamic Characteristics of the Main Parts of a Hybrid-Driven Underwater Glider PETREL

Wu Jianguo[1], Zhang Minge[2] and Sun Xiujun[2]
[1]Shenyang Institute of Automation Chinese Academy of Sciences
[2]Tianjin University
China

1. Introduction

Autonomous Underwater Vehicle (AUV), Remotely Operated Vehicle (ROV) and Autonomous Underwater glider (AUG) are the main autonomous underwater platforms available currently, which play important role in the marine environmental monitering. The relationships between those three types of vehicles were shown in Figure 1.

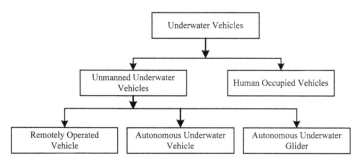

Fig. 1. Underwater Vehicles

As a special type of AUV, underwater gliders have many advantages, such as long endurance, low noise and low energy cost. A glider can periodically change its net buoyancy by a hydraulic pump, and utilize the lift from its wings to generate forward motion. The inherent characteristics of a glider can be summarized as buoyancy-driven propulsion, sawtooth pathway, high endurance and slow speed. There exist three legacy gliders named respectively Seaglider, Spray and Slocum [1~6]. In spite that underwater gliders features low level of self noise and high endurance, they also have weaknesses like the lack of maneuverability and the inability to perform a fixed depth or level flight[7].

Driven by a propeller with carried energy source, autonomous underwater vehicles is preprogrammed to carry out an underwater mission without assistance from an operator on the surface. However, they can only cover a relatively short range after each recharge due to the high power consumed for propulsion and generate much more noise than the AUGs because of its propeller and motors [8~10]. The range of AUV's is restricted by the amount of energy carried on board, can was not more than several hundreds kilometers in general [11]. The performances of the underwater vehicle are compared in Figure 2.

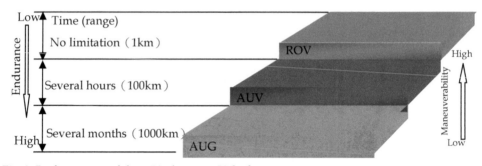

Fig. 2. Performances of three Underwater Vehicles

By combining the advantages of the glider and the propeller-driven AUVs, A hybrid-driven underwater glider PETREL with both buoyancy-driven and propeller-driven systems is developed. Operated in buoyancy-driven mode, the PETREL carries out its mission to collect data in a wide area like a legacy glider. When more exact measurements of a smaller area or level flight are needed, the PETREL will be operated by using the propeller-driven system [5, 7]. This flexible driven glider contributes to have a long range while operated in the buoyancy driven mode like a glider, as well as improve the robust performance to deal with some wicked circumstances by the propeller driven system [7].

Proper hydrodynamic design is important for the improvement of the performance of an underwater vehicle. A bad shape can cause excessive drag, noise, and instability even at low speed. At the initial stage of design, there are two ways to obtain the hydrodynamic data of the underwater vehicle, one is to make model experiment and the other is to use the computational fluid dynamics (CFD). With the development of the computer technology, some accurate simulation analysis of hydrodynamic coefficients have been implemented by using the computational fluid dynamic (CFD) software, instead of by experiments at a much higher cost over the past few years [12-13] . In consideration of the reduced time, lower cost, more flexible and easier optimumal design, the CFD method was used in this article. The fluent Inc.'s (Lebanon,New Hampshire) CFD software FLUENT 6.2 was adopted by this article.

This chapter focuses on the hydrodynamic effects of the main parts of a hybrid-driven underwater glider especially in the glide mode. By analyzing the results of the three main hydrodynamic parts, the wings, the rudders and the propeller, the characteristics of drag, glide efficiency and stability will be discussed, and suggestions for altering the HUG's design to improve its hydrodynamic performance are proposed.

2. Computational details

2.1 Mathematical model

A criterion for determining of the flow regime of the water when the vehicles moving in it is proposed by Reynolds number [14-15]:

$$R_e = \rho v L / \mu \tag{1}$$

Here ρ is the density of water, v is the velocity of vehicle, L is the characteristic length, μ is the dynamic coefficient of viscosity. The transition point occurred when the Reynolds number is near 10^6 for the external flow field, which is called critical Reynolds numbers. It was laminar boundary layer when the $Re < 5 \times 10^5$, it was seem as turbulent flow while $Re > 2 \times 10^6$. The Reynolds number of the hybrid underwater glider PETREL at two different steering modes is shown in table 1.

steering mode	velocity v / (m/s)	Reynolds number
Glider	0.5	1.25×10[6]
AUV	2	5×10[6]

Table 1. The Reynolds number at different steering modes

The turbulence model will be adopted because the Reynolds numbers of the PETREL in two steering modes are all above the critical Reynolds numbers. Computations of drag, lift and moment and flow field are performed for both the model over a range of angles of attack by using the commercially available CFD solver FLUENT6.2. The Reynolds averaged Navier–Stokes equation based on SIMPLAC algorithm and the finite volume method were used by our study. In our study RNG k-ε model was adopted and the second-order modified scheme was applied to discrete the control equations to algebra equations. Assuming that the fluids were continuous and incompressible Newtonian fluids. For the incompressible fluid, the RNG $k - \varepsilon$ transport equations are [12, 16]:

$$\rho \frac{\partial}{\partial t}(k) + \rho \frac{\partial}{\partial x_i}(ku_i) = \frac{\partial}{\partial x_j}\left(\alpha_k \mu_{eff} \frac{\partial k}{\partial x_j}\right) + G_k - \rho\varepsilon + S_k \tag{2}$$

$$\rho \frac{\partial}{\partial t}(\varepsilon) + \rho \frac{\partial}{\partial x_i}(\varepsilon u_i) = \frac{\partial}{\partial x_j}\left(\alpha_\varepsilon \mu_{eff} \frac{\partial \varepsilon}{\partial x_j}\right) + C_{1\varepsilon}\frac{\varepsilon}{k}G_k - C_{2\varepsilon}\rho\frac{\varepsilon^2}{k} - R_\varepsilon + S_\varepsilon \tag{3}$$

Here S_k and S_ε are source items, μ_{eff} is effective viscosity, G_k is turbulence kinetic energy induced by mean velocity gradient.

$$G_k = -\rho \overline{u_i' u_j'} \frac{\partial u_j}{\partial x_i} \tag{4}$$

σ_k and σ_ε is respectively the reversible effect Prandtl number for k and ε. $C_{1\varepsilon} = 1.42$, $C_{2\varepsilon} = 1.68$
In the RNG model,a turbulence viscosity differential equation was generated in the non-dimensional treatment.

$$d\left(\frac{\rho^2 k}{\sqrt{\varepsilon\mu}}\right) = 1.72 \frac{\hat{v}}{\sqrt{\hat{v}^3 - 1 + C_v}} d\hat{v} \tag{5}$$

here, $\hat{v} = {\mu_{eff}}/{\mu}$, $C_v \approx 100$. Taking the integral of the(5),the exact description of active turbulence transport variation with the effective Reynolds number can be acquired, which makes the mode having a better ability to deal with low Reynola number and flow near the wall. For the large Reynola number, the equation(5)can be changed into (3-6).

$$\mu_t = \rho C_\mu \frac{k^2}{\varepsilon}$$

(6)

Here , $C_\mu = 0.0845$ The RNG $k - \varepsilon$ model was adopted due to the initial smaller Reynola number of boundary layer, and the more exact results can be gained by substituting the differential model into the RNG $k - \varepsilon$ model.

2.2 Meshing and boundary conditions
The size function and unstructured meshes were adopted to keep the meshes distributing reasonably and make the meshes generating expediently. The examples of meshing are shown as figure 3, figure 4 and figure 5.

Fig. 3. Stern meshes

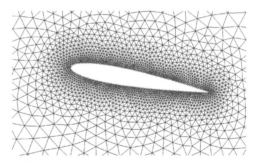

Fig. 4. Two-dimension rudder meshes

Fig. 5. Whole meshes of the vehicle

Boundary conditions:
1. inlet boundary condition: setting the velocity inlet in front of the head section with a distance of one and a half times of the length .
2. outlet boundary condition: setting the free outflet behind the foot section with a distance of double length of the vehicle.
3. wall boundary condition: setting the vehicle surface as static non-slip wall.
4. pool wall boundary condition: non-slip wall.

2.3 Results verification

To verify the precision of the calculation, we computed the drag coefficients of Slocum underwater glider [17] at different angle of attack as shown in table 2. The table 3 shows the verification of numerical simulation results of drag of AUV shell of Tianjin University. The error percentage of our calculation is less than 9.35%.

Angle of attack α(degree)	Reynolds number R_e	C_D (experiment)	C_D (CFD)	Error percentage
-2.9	7.5×10^5	0.31	0.281	9.35%
2.3	6.3×10^5	0.25	0.268	7.20%
2.7	5.8×10^5	0.27	0.274	1.46%

Table 2. Verification of numerical simulation results of C_D

Velocity /(m/s)	Reynolds number	Drag Experiment(N)	Drag CFD(N)	Error percentage
0.81	2.5×10^6	7.4	6.903	6.72%
1.4	4.4×10^6	20.3	19.92	1.87%
2.0	6.2×10^6	37.5	37.34	0.427%

Table 3. Verification of numerical simulation results of drag of AUV shell

3. Wing hydrodynamic design [18]

3.1 Orthogonal experimental design and results analysis
3.1.1 Orthogonal experimental design
An orthogonal experimental with four factors and three levels was conducted by keeping the main body size of the vehicle as constant. The four factors are wing chord, aspect ratio, backswept and distance between the center of wing root and the center of body. The simulation experiments were done at the situation of angle of attack is $\alpha = 6°$ and the velocity is $v = 0.5 m / s$. The airfoil of the wings was NACA0010. The orthogonal experimental table was shown as table 4.

level	chord(mm)	aspect ratio	backswept (°)	distance(mm)
1	100	6	20	100
2	150	8	40	0
3	200	10	60	-100

Table 4. Orthogonal experimental table

3.1.2 Analysis indexes

The design of the wing will generate important impacts on glide efficiency and glide stability of the vehicle. The lift to drag ratio L/D is chosen for measurement of the glide efficiency, the bigger values correspond to the more efficient gliding. The inverse of L/D expresses the glide slope [7, 19]. Existing oceanographic gliders are designed for static stability in steady glides, and the static stability can be measured by the non-dimensional hydrodynamic lever l'_α, the equations are[20~21]:

$$l'_\alpha = l_\alpha / l \tag{7}$$

$$l_\alpha = -M_\alpha / L_\alpha \tag{8}$$

Here, l is the vehicle length, M_α is the hydrodynamic moment induced by angle of attack α, L_α and is the Lift induced by the angle of attack α. It is static instability while $l'_\alpha > 0$, the moment induced by incremental angle of attack makes the angle of attack become bigger; It is neutral stability while $l'_\alpha = 0$; It is called static stability while $l'_\alpha < 0$, the moment induced by incremental angle of attack makes the vehicle to turn to the original state.

3.1.3 Influencing factors analysis

The orthogonal experimental table L18(37) and the simulation results are shown in table5.

The trend charts were shown as Figure 6 and Figure 7. The L/D increase with the growth of chord and aspect ratio, and decrease with the growth of backswept,it has little relationship with the location of the wings. The l'_α increase as chord and backswept increase when the wings is located after the hydrodynamic center, which means the stability increase as chord and backswept raise. The stability gets higher as the wing location becomes father away from behind the center of the body.

The ranges of chord, aspect ratio, backswept and distance of the wings is separately 2.448, 1.077, 1.303 and 0.312 for the L/D, which was gained by the range method. It is shows that the effects significance series for glide efficiency is chord, backswept, aspect ratio and the location of wings. The chord was dramatic for the index L/D at the significance level 0.10 and 0.05 adopted by the range method.

In like manner, the range of chord, aspect ratio, backswept and distance of the wings is separately 0.051, 0.037, 0.095 and 0.031 for the l'_α, which was gained by the range method. It is shows that the effects significance series for glide stability is backswept, chord, aspect ratio and the location of wings. The backswept was dramatic for the index l'_α at the significance level 0.10 and 0.05 adopted by the range method.

Simulation times	factors				results	
	chord(mm)	aspect ratio	backswept (°)	distance(mm)	L/D	l'_α
1	100	6	20	100	2.86	0.0274
2	100	8	40	0	3.11	-0.0323
3	100	10	60	-100	2.74	-0.109
4	150	6	20	0	4.07	-0.00059
5	150	8	40	-100	4.37	-0.0854
6	150	10	60	100	3.82	-0.127
7	200	6	40	100	4.88	-0.0158
8	200	8	60	0	4.47	-0.172
9	200	10	20	-100	6.81	-0.0713
10	100	6	60	-100	2.33	-0.0630
11	100	8	20	100	3.31	0.0212
12	100	10	40	0	3.44	-0.0444
13	150	6	40	-100	3.78	-0.0594
14	150	8	60	100	3.40	-0.0858
15	150	10	20	0	5.30	-0.0237
16	200	6	60	0	3.94	-0.115
17	200	8	20	-100	6.17	-0.0562
18	200	10	40	100	6.21	-0.0756

Table 5. Orthogonal experimental table and the results

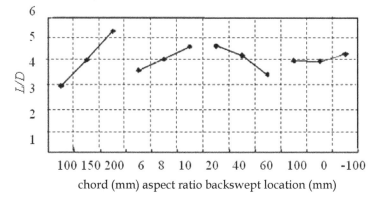

Fig. 6. L/D tendency chart

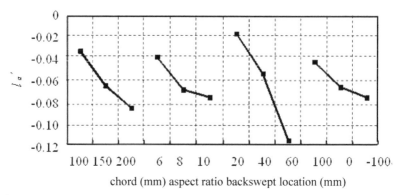

Fig. 7. l'_α Tendency chart

It is well known that the chord and aspect ratio of the wings should be increased, and the backswept decreased for the higher glide efficiency when PETREL is operated in the gilde mode. Simutaneously, the backswept of the wings should be increased and the wings should be moved backward father behind the center of the vehicle for the higher stability. It indicates that the effects from the increment of the backswept of the wings are inversed in increasing the glide efficiency and the stability. The backswept of the wings should be determined in terms of other capability indexes of the underwater vehicle.

3.2 Concrete models analysis

Four concrete models with varied wing parameters listed in table 6 were choosen for some further investigation. We carried out this new series of experiments in the hope of providing the effects of wings, rudders and propeller on the L / D and l'_α at different glide angle of attack when the velocity is 0.5m/s and the angle of attack on the rang of $0°\sim20°$.The model one has the highest glide efficiency and glide stability in the table6; The models 2~4 were proposed in order to evaluate the affects of location, aspect ratio and chord of wings on the analysis index as shown in the table 6. The Figure 8 gives the pressure distribution chart of model 3. The calculation results of different models are shown as Fig. 9~ Fig. 12.

models	chord(mm)	aspect ratio	backswept(°)	location(mm)
1	200	10	40	0
2	200	10	40	100
3	200	8	40	0
4	150	10	40	0

Table 6. The parameter of the concrete model

The location of the wings has little influence on the L/D, which means it has little influence on the glide efficiency illustrated in Figure 9, but it has dramatic effects on the glide stability which can be seen in the Figure 10. From the figure 9 and 10, it can be seen that the L/D decreased and the l'_α increased obviously when the aspect ratio and chord reduced, but the effects is more dramatically to decrease the chord of the wings for the L/D. It has the biggest lift to drag ratio when the angle of attack at 6 degree shown in Figure 9, that is means the maximum glide efficiency can be gain when the angle of attack at 6 degree.

Fig. 8. The pressure distribution chart of model 3

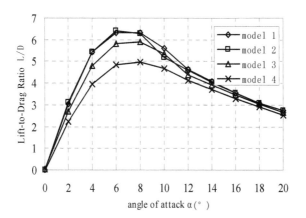

Fig. 9. The relationship between L/D and angle of attack

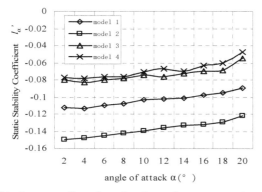

Fig. 10. The relationship between l'_α and angle of attack

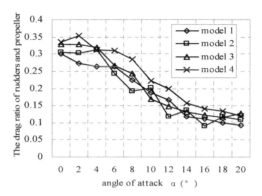

Fig. 11. The drag ratio of rudders and propeller

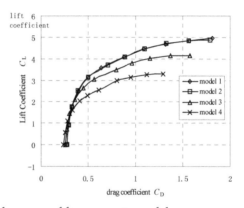

Fig. 12. The lift-drag polar curve of four concrete models

The drag of the hybrid glider will be increased because of the drag generated by the rudders and propeller compared with the legacy gliders in the glide mode. The range in the glide mode will be decreased because of the drag of these parts . The ratio of drag on the propeller and rudders to whole drag is illustrated in Fig. 11, where we find that the ratio changed as the angle of attack increases, and the values is within the range of 10%~35%. Compared with the legacy gliders, the range of the vehicles with the same configuration as PETREL will be decrease 10%~35%. The Lift to Drag polar curves of the four concrete models are shown as figure 12. The model 3 and model 4 have the bigger lift than the model 1 and model 2 when the drag coefficients from the figure 3-9 is less than 0.5, but the lift of model 1 and model 2 increases greatly when drag coefficients gets bigger than 0.5. Due to the drag of the vehicle need overcome by the variable buoyancy B in the end and there is equation (9), so the net buoyancy supplied by the buoyancy driven system and glide angle should be taken into consideration.

$$B \sin \theta = D \qquad (9)$$

Here B is the net buoyancy, θ is the glide angle, D is the drag of the glider.

3.3 Results and discussion

The orthogonal experiment shows that glide efficiency is most significantly influenced by the chord length while stability of the vehicle is most remarkably affected by the sweep angle.

Further numerical calculations based on four specific models with the attack angle in the range of 0°-20° indicate that location of the wings mainly affects glide stability but has little influence on glide efficiency.

When the vehicle glides at about 6° attack angle it has the maximum ratio of lift to drag. The range of the hybrid glider with the same configuration as PETREL will be decrease 10%~35% compared with the legacy gliders.

4. Rudder hydrodynamic design [22]

4.1 Rudder parameters

The rudders parameters include root chord, half span, aerofoil and backswept, which are shown in Figure 13. As defined in the [23], the chord is denoted by C , the distance from the leading edge to trailing edge in a given two-dimensional section. The chord is measured in parallel with the section at the root of the rudder. In general, the chord can vary along the span, in which case the geometric mean chord, $\bar{C}$, is used in computations unless noted[21]. The $\bar{C}$ is defined based on Figure 14 as

$$\bar{C} = \frac{C_t + C_r}{2} \tag{10}$$

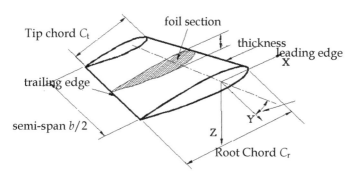

Fig. 13. Rudders parameters

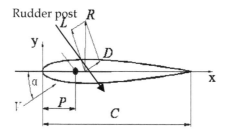

Fig. 14. Foil section and hydrodynamic force

The semi-span, denoted by $b\,/\,2$, measures the distance from the rudder root to tip along the line perpendicular to the root section. The span, in this work, is twice as long as the root-to-tip distance for an isolated plan. The hydrodynamic forces including lift and drag acted on the aerofoil is shown in Figure 14 and can be expressed as

$$L = 1/2\,\rho C_L A V^2 \tag{11}$$

$$D = 1/2\,\rho C_D A V^2 \tag{12}$$

Here, ρ is the density of the water; C_L is the lift coefficient; C_D is the drag coefficient; A is the area of rudder; V is the velocity of water; α is the angle of attack. The rudderpost location is expressed by P, which is shown in Figure 14.

4.2 Foil section

The geometry of a rudder is mainly defined by the two-dimensional foil section. The symmetrical foil sections are generally used by the underwater vehicles. Many types of the foil sections are proposed by many countries to improve the hydrodynamic performance. The famous foil sections series include NACA series, НЕЖ series, ЦАГИ series, and JFS series [21], among which the four-digit NACA sections are most widely used for underwater vehicle rudders in that it provides the higher lift and the lower drag. The four-digit NACA section series is a low velocity foil sections series, and have a bigger radius of leading edge and a plumpy head section, which is suitable for the rudder of underwater vehicles at low velocity. In this work, the four digit NACA00×× section was used, where the ×× denote the thickness-to-chord ratio. The lift coefficient and drag coefficient of the foil sections can be calculated as

$$C_L = \frac{L}{1/2\,\rho V^2 C} \tag{13}$$

$$C_D = \frac{D}{1/2\,\rho V^2 C} \tag{14}$$

Here, L is the profile lift, D is the profile drag, C is the chord. The NACA0008, NACA0012, NACA0016, NACA0020 and NACA0025 are usually used for the rudders of miniature underwater vehicles, their hydrodynamic characteristics were calculated by using computational fluid dynamics. According to the most often adopted velocity of the autonomous underwater vehicles and the velocity of PETREL in AUV mode, the calculation velocity was determined as 2m/s. An example of CFD meshing result is shown in figure 15, where the unstructured mesh was adopted and the wall of section was made dense. The calculating results were shown in the Figure 16~ Figure 18

The relationship of lift coefficient and angle of attack is illustrated in Figure 16, where we can see that there was a bigger angle of stalling and bigger maximal lift coefficient when the section becomes much thicker. From the figure 17 we can see that the thinner wing section has a lower drag cofficient when the angle of attack is small, but the thicker wing section has a lower drag cofficient when the angle of attack is bigger than a certain critical angle of attack. The NACA0008 section has the maximal L/D and NACA0025 has the minimal L/D than other sections which is shown in the Figure 18. The NACA0012 section with angle of stall about 20° and a higher L/D was adopted by the Hybrid glider PETREL.

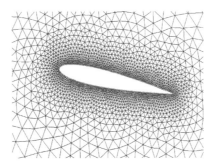

Fig. 15. CFD meshing results

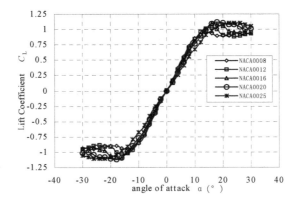

Fig. 16. The relationship of profile lift coefficient and angle of attack

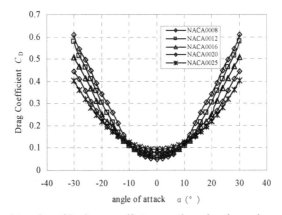

Fig. 17. The relationship of profile drag coefficient and angle of attack

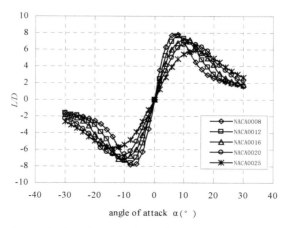

Fig. 18. The relationship of L/D and angle of attack

4.3 Area of rudder calculation

The area of rudder as an important parameter for maneuverability of the underwater vehicle is related to the size and shape of the body. The area of rudder can be design by cut and try method, master model method and empirical formula design method. For the high maneuverable ship, the control surfaces can be designed according to Det Norske Veritas, (DNV) rudder sizing rules [24].

$$Area = \frac{DL}{100}[1 + 25(\frac{B}{L})^2]$$

(15)

Here, D is the diameter of the vehicle, L is the length of the vehicle, B is the width of the vehicle, and $B = D$ for revolution body. It suggested 30% increase in area if rudders in front of the propeller, and then increased by an additional 50% to match empirical data from other underwater vehicles by the DNV rules. The turn diameter induced by single rudder is about triple-length of the vehicle in terms of the design by DNV rules. The rudder design for the hybrid glider PETREL is shown in Figure 19 and the parameters of the rudder shown in table 7.

Fig. 19. The photo of the rudder of PETREL

parameters	Tip chord C_t	Root chord C_r	Semi-span $b/2$	section
Value	125mm	200mm	120mm	NACA0012

Table 7. The parameters of the rudder

4.4 Hinge moment analysis

The hinge moment is produced by a hydrodynamic force about the hinge line of a control surface. It makes an impact on maneuverability of the underwater vehicle in that the hinge moment must be overcome during steering. The bigger hinge moment will make the turning velocity of rudders become slowly and make the control action slow-witted.

The hydrodynamic performance of three dimension rudders at different angles of attack was simulated by using CFD methods. The inlet velocity was set to be 2m/s.

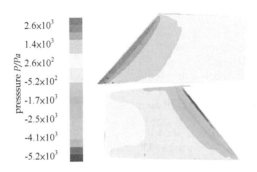

Fig. 20. Pressure distribution chart when angle of attack is 20°

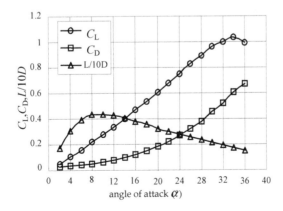

Fig. 21. C_L, C_D and $L/10D$ variation curve with different angles of attack

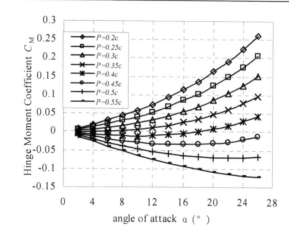

Fig. 22. Hinge moment with different angles of attack

The pressure distribution of the rudder is illustrated in Figure 20, where we can find that there is higher pressure on the front flow face and was local higher pressure area on the back flow face of the tail, that means there exist roundabout flow at the tail of the rudder. Figure 21 shows the relationship between lift, drag and angle of attack. The relationship between L/D and angle of attack is also illustrated in the figure 21, the L/D value reduces ten times for the same scale with other two curves. It can be known that the maximal lift to drag ratio was about 8° and the angle of stall about 34°, so the angles of stall of three dimensional rudders are greater than two-dimension section. The hinge moment of rudders with different axis of rudder position is shown in Figure 22, where we can seen that the hinge moment varied with the angle of attack. The hinge moments are little while $P = 0.4c$ for the rudder we design no matter how the angle of attack changed.

4.5 Results and discussion

Aiming at the key problems of the rudder design for autonomous underwater vehicle, the hydrodynamic characteristic of the NACA00xx series section at different angles of attack were simulated when velocity was 2m/s by using the two-dimensional computational fluid dynamics (CFD). For the rudder we design, the stall angle is about 34° for the three dimensional rudders and about 20° for the two-dimensional foil section, so the angle of stall of three dimensional rudders are greater than two-dimension foil section. The area of the rudder of PETREL was calculated using the DNV rules ; The hinge moments are little when $P = 0.4c$ for the rudder we design no matter how the angle of attack changed.

5. Shroud hydrodynamic effects analysis[25]

For the PETREL, the propeller plays a significant role in the vehicle's hydrodynamic performance, so analysis of the hydrodynamic effect of a propeller with a shroud on a winged HUG was performed with Fluent Inc.'s (Lebanon,New Hampshire) CFD software FLUENT6.2.

5.1 Models description

To analyze the effects of the shroud, two simulations were performed, where one model is with the shroud and the other without. Two models are shown in Figure 23.

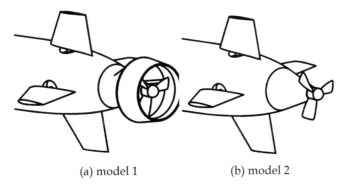

(a) model 1   (b) model 2

Fig. 23. The models studied in the paper

5.2 Effect of shroud on the glide drag

The drag on the vehicle can be expresses as equation (16).

$$D = \frac{1}{2}\rho V^2 C_D A \qquad (16)$$

Where, D is the force of drag in Newton, ρ is the density of water in kg/m³, V is the velocity of the vehicle in m/s, A is the reference area in m², C_D is the drag coefficient (dimensionless). The reference area A of the PETREL is 0.096m².

Figure 24 shows the overall drag of the two models in the glide mode. The propeller in this mode doesn't rotate. The overall drags of two models are calculated by CFD firstly and then are fitted by the semi-empirical formulae (16). The drag coefficients of two models are respectively 0.32 and 0.26. The average relative error of overall drag between CFD and semi-empirical formulae is 4.7%. The overall drag increase 21%-26% with the propeller shroud compared with the model two according to the CFD computation results, so the shroud greatly increased the drag of the hybrid in glide mode. The drag components of the model at the speed of 0.5m/s without angle of attack are shown in Fig. 25. The drag on the body, rudders and wings is mainly viscous forces, while the drags on the propeller, shroud and GPS antenna pole are primarily the pressure forces,. As shown in Figure 26, the propeller and its shroud make up over 30% of total resistance and the percentage will increase with the increment of the velocity. The reason for the high percentage is because of the great pressure drags on the shroud in the glide mode. The local velocity streamline diagram near the shroud of model one shown in the Figure 27. In the Figure, we can see that v_{in} and P_{in} are the velocity and pressure inside the shroud of water, v_{out} and P_{out} are the velocity and pressure outside the shroud of water. Because the propeller doesn't rotate in the glide mode, the velocity of water inside the shroud is slower than that outside the shroud, so there exits $v_{out} > v_{in}$. According to the Bernoulli equation there was $P_{in} > P_{out}$, so a pressure force f is produced by the pressure difference. The percentage of the shroud drag to total resistance is 26%-35% at the different speed due to the pressure force in the glide mode.

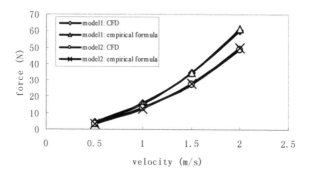

Fig. 24. The overall drags of the two models at difference velocities

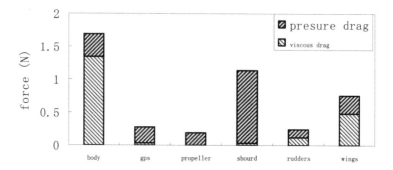

Fig. 25. The drag components of the mode 1 at the speed 0.5m/s

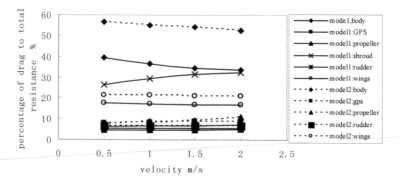

Fig. 26. The drag distribution of vehicle at the different velocities

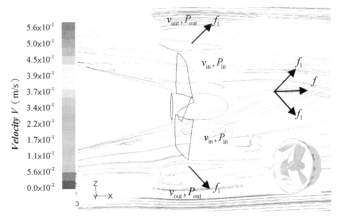

Fig. 27. The local velocity streamline diagram of model1 ($v = 0.5m/s$)

5.3 Effect of shroud on the glide efficiency

The specific energy consumption can be defined using classical aerodynamics [7] as

$$E_e = \frac{DU}{Bu} = \frac{Bw}{Bu} = \frac{w}{u} = \frac{D}{L} = \frac{C_D}{C_L} \tag{17}$$

Underwater gliders will have a higher glide efficiency when E_e is lower. So the lift to drag ratio L/D is a measure of glide efficiency, where bigger values represent higher glide efficiency [7].

The Lift-to-drag ratio versus angle of attack is plotted in Fig. 28, the relations of model one is indicated by the solid lines. The Lift-to-drag ratio of model one is lower than the model two at different angles of attack, that means the vehicle with the shroud will have a lower glide efficient than that without. The Lift-to-drag ratio of model one is less than model two by 20% to 5% for the varied angles of attack within the range from 2°to 20°. The maximum lift-to-drag ratio occurred at the angle of attack 6°-8°for both the models at different speed.

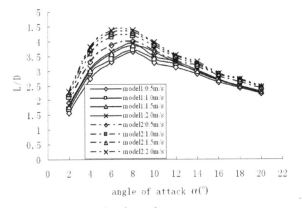

Fig. 28. Lift-to-drag ratio versus angle of attack

5.4 Effect of shroud on the glide stability

The underwater gliders usually are designed for static stability [17], the dimensionless hydrodynamic moment arm $l_\alpha^{'}$ often used to represent the static stability of the underwater vehicles motion. The equations of the $l_\alpha^{'}$ are shown in equations(7)and (8).

Existing oceanographic gliders are designed to be static stable in steady glides for the easy control and high energy economy. The hybrid-driven underwater glider PETREL was designed as static stability for the high energy economy in the glide mode.

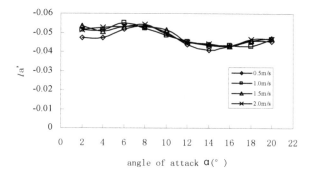

Fig. 29. The static stability coefficient $l_\alpha^{'}$ versus angle of attack of model one

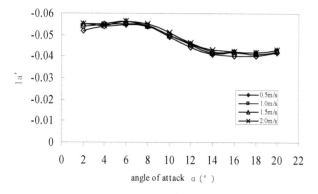

Fig. 30. The static stability coefficient $l_\alpha^{'}$ versus angle of attack of model two

Figure 29 show the static stability coefficient $l_\alpha^{'}$ versus angle of attack of model one and model two. It is static stability for both of the two models in terms of our design intention. The stability decreases when the angle of attack gets bigger than 8°, but the stability slightly increases for model one when the angle of attack is more than 12°. The glide speed has little effect on the stability as shown in the Figure 29 and Figure 30. Figure 31 shows the moment of the shroud versus angle of attack of model one. The values of the moment were positive when the angle of attack is lower than 8° for the $v = 0.5$ m/s and $v = 1$ m/s, and the angle of attack is less than 10° for the $v = 1.5$ m/s and $v = 2.0$ m/s. The values of the moment were negative when the angle of attack gets higher than those critical angles. So the effect of shroud on the static stability of model one is that, when the angle of attack is lower than the critical angle the shroud will makes the stability decreasing but makes the stability increasing when the angle of attack is higher than the critical angle. as shown in the Figure 31, the action of the shroud makes the stability slightly increased when the attack angle is higher than 12° .

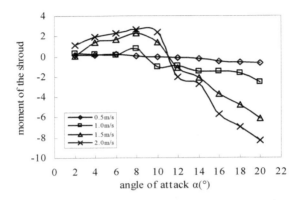

Fig. 31. The moment of the shroud versus angle of attack of model one

5.5 Conclusions

It was found that overall drag increased by 21 to 26 percent for the model with the propeller shroud compared with the one without a shroud, but with the same structure and size, the shroud's resistance is mainly pressure force.

The shroud made the lift-to-drag of the vehicle in glide mode decrease by as much as 20 percent when the angle of attack was 2º. As the angle of attack increased, the shroud's effect was minimized, and the decrease in lift-to-drag ratio ranged down to five percent at an angle of attack of 20º, meaning glide efficiency decreased due to the propeller shroud.

Finally, the shroud decreases the stability of the HUG when the angle of attack is lower than the critical angle, but increases it when the angle of attack is higher than the critical angle. The critical angle is between 8º and 10º for velocities lower than one meter per second, and between 10º and 12º for velocities in the range of one to two meters per second.

These findings indicate that for an underwater glider, the shroud will increase drag and decrease the glide efficiency, but it is good for stability when the angle of attack is larger than 8º. Therefore, the shroud is not a successful design element for the HUG in glide mode, but in propeller mode the shroud can increase the thrust of the vehicle.

Using CFD to analyze the shroud's hydrodynamic effects shows that the vehicle should only be equipped with this feature for activities requiring operation in propeller mode.

6. Flow field analysis

6.1 Velocity field

The direct route flow field with the velocity of the hybrid underwater glider PETREL at 0.5m/s、1m/s、1.5m/s and 2m/s was simulated by using CFD ways. The simulation results are shown in Figure 32.

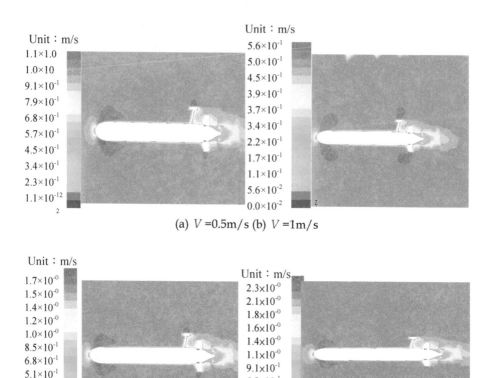

(a) V =0.5m/s (b) V =1m/s

(c) V =1.5m/s (d) V =2.0m/s

Fig. 32. The flow field at different velocity

It is seen that the flow field patterns in the figures are nearly the same. There was high flow rate region near the abrupt curve surfaces of the vehicle head, ballast of the GPS, rudders, while there was also the low flow field domain on the front of those parts and near the tail of the vehicle. The high flow rate region area decreases as the velocity increases. The existence of the mast of GPS makes the flow field behind it disturbed, and makes the flow field asymmetrical. These changes will increase the drag and hydrodynamic moment on the vehicle.

The steady turning flow field in longitudinal vertical and horizontal plane with the velocity of vehicle at 0.5m/s, is shown in figure 33.

It is noted from Figure 33 that the pattern of the steady turning flow field in longitudinal vertical plane and in horizontal plane has notability difference. Due to the rotational speed, the flow field is obviously asymmetric and appears large scale high flow rate region and low flow rate region in the back of the field. An extra hydrodynamic moment is induced because of the asymmetry of the flow field.

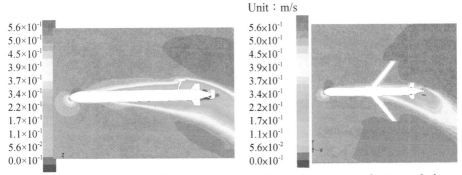

Unit：m/s

(a) Steady turning in longitudinal vertical plane (b) Steady turning in horizontal plane

Fig. 33. The steady turning flow field

6.2 Pressure distribution

The pressure distributions on the vehicle at the speeds 0.5m/s and 2m/s when the angle of attack α is zero are shown in Figure 34 and Figure 35. There has a tendency that the pressure on the vehicle gradual reduction from head to tail of the vehicle, a high pressure region on the head and a low pressure region on the tail, which induced the pressure drag on the vehicle. The pressure of the high pressure region become higher and the low pressure region become lower with the speed of the vehicle increasing, it means that the pressure drag on the vehicle increase with the speed increasing. It can be known from the pressure distribution on the propeller shroud that pressures drag act on the shroud because of there has higher pressure inside the shroud and lower pressure outside the shroud. The reason for thus pressure distribution is that the propeller doesn't rotating in the glide mode which makes the velocity of flow inside the shroud slower than the outside. So the shroud should be removed or the profile changed to reduce the drag on the vehicle in glide mode.

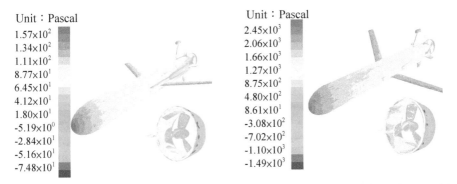

Fig. 34. Pressure distribution (V =0.5m/s)   Fig. 35. Pressure distribution (V =2m/s)

The pressure distributions on the vehicle at the speed 0.5m/s when the angle of attack α isn't zero are shown in Figure 36. The pressure distribution on the vehicle isn't symmetry, the pressure of front flow surface higher than back flow surface, when glide with an angle of attack. The wing has the biggest degree of asymmetry of the pressure distribution which makes the wings the main lift generating parts. The asymmetry of the pressure distribution on the vehicle also induces the hydrodynamic moment on the vehicle.

Unit : Pascal

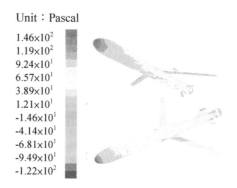

Fig. 36. Pressure distribution (V =0.5m/s, $\alpha = 6$)

7. Conclusions

This chapter focuses on the hydrodynamic effects of the main parts of a hybrid-driven underwater glider especially in the glide mode, and conducts analysis of the simulation results of the three main hydrodynamic parts by using the computational fluid dynamics (CFD) ways. The fluent Inc.'s (Lebanon, New Hampshire) CFD software FLUENT 6.2 was adopted by this article. The main conclusions are:

It is found that the glide efficiency is most significantly influenced by the chord length while stability of the vehicle is most remarkably affected by the sweep angle, and the location of the wings mainly affects glide stability but has little influence on glide efficiency. When the vehicle glides at about 6°attack angle it has the maximum ratio of lift to drag. The endurance of the hybrid glider with the same configuration as PETREL will decrease by 10%~35% compared with the legacy gliders.

For the rudder we design, the angle of stall is about 34° for the three dimensional rudders and about 20° for the two-dimensional foil section, so the angle of stall of three dimensional rudder is greater than two-dimension foil section. The area of the rudder of PETREL was calculated using the DNV rules ; The hinge moments are little when $P = 0.4c$ for the rudder we design no matter how the angle of attack changes.

It was found that overall drag increased by 21 to 26 percent for the model with the propeller shroud compared with the one without a shroud, but with the same structure and size, the shroud's resistance is mainly pressure force. The shroud made the lift-to-drag of the vehicle in glide mode decrease by as much as 20 percent when the angle of attack was 2°. As the angle of attack increased, the shroud's effect was minimized, and the decrease in lift-to-drag ratio ranged down to five percent at an angle of attack of 20°, meaning glide efficiency decreased due to the propeller shroud. Finally, the shroud decreases the stability of the HUG when the angle of attack is lower than the critical angle, but increases it when the angle of attack is higher than the critical angle. The critical angle is between 8° and 10° for velocities lower than one meter per second, and between 10°and 12°for velocities in the range of one to two meters per second.

These findings indicate that the shroud of the underwater glider will increase drag, decrease the glide efficiency, but it improves the stability when the angle of attack is larger than 8°. Therefore, the shroud is not a successful design element for the HUG in glide mode, but it can increase the thrust of the vehicle in propeller mode.

Using CFD to analyze the shroud's hydrodynamic effects shows that the vehicle should only be equipped with this feature for activities requiring operation in propeller mode.

Finally, the velocity field, pressure distribution of the hybrid glider PETREL were analyzed, which make us understand how those main parts effect on the hydrodynamic characteristic of the vehicle.

8. References

[1] C. C. Eriksen, T. J. Osse, R. D. Light, T, et al, (2001)"Sea glider: A long range autonomous underwater ve hicle for oceanographic research," *IEEE Journal of Oceanic Engineering*, Vol. 26, 2001, pp. 424–436

[2] J. Sherman, R. E. Davis, W. B. Owens, et al , "The autonomous underwater glider "Spray"," *IEEE Journal of Oceanic Engineering*, vol.26, 2001, pp. 437–446

[3] D. C. Webb, P. J. Simonetti, C. P. Jones, "SLOCUM, an underwater glider propelled by environmental energy," *IEEE Journal of Oceanic Engineering*, vol.26, 2001, pp. 447–452

[4] R. E. Davis, C. C. Eriksen and C. P. Jones, "Autonomous Buoyancy-driven Underwater Gliders," The Technology and Applications of Autonomous Underwater Vehicles. G.. Griffiths, ed., Taylor and Francis, London, 2002, pp. 37-58

[5] R. Bachmayer, N. E. Leonard, J. Graver, E. Fiorelli, P. Bhatta, D. Paley, "Underwater gliders: Recent development and future applications," Proc. *IEEE International Symposium on Underwater Technology (UT'04)*, Tapei, Taiwan, 2004, pp.195- 200

[6] D. L. Rudnick, R. E. Davis, C. C. Eriksen, D. M. Fratantoni, and M. J. Perry, "Underwater gliders for ocean research, " *Marine Technology Society Journal*, vol. 38, Spring 2004, pp. 48–59

[7] S. A. Jenkins, D. E. Humphreys, J. Sherman, et al. "Underwater glider system study." *Technical Report*, Office of Naval Research, 2003

[8] Marthiniussen, Roar ; Vestgård, Karstein ; Klepaker, Rolf Arne ; Strkersen, Nils ,Fuel cell for long-range AUV ,*Sea Technology*, vol. 38, pp. 69-73

[9] Blidberg, D.Richard ,Solar-powered autonomous undersea vehicles. *Sea Technology*, vol. 38, pp. 45-51

[10] Ferguson, J , Pope, A.,Butler, B.1, Verrall, R.I.1 ,Theseus AUV-Two record breaking missions, *Sea Technology*, vol. 40, pp. 65-70

[11] Albert M. Bradley, Michael D. Feezor, Member, IEEE, Hanumant Singh, and F. Yates Sorrell, Power Systems for Autonomous Underwater Vehicles, *IEEE JOURNAL OF OCEANIC ENGINEERING*, VOL. 26, NO. 4, OCTOBER 2001,pp.526-538

[12] Amit Tyagi*, Debabrata Sen, Calculation of transverse hydrodynamic coefficients using computational fluid dynamic approach, *Ocean Engineering 2006 (33)* , pp. 798–809

[13] Douglas E.H, Correlation and validation of a CFD based hydrodynamic &dynamic model for a towed undersea vehicle, *OCEANS*, 2001,Vol3, pp.1755-1760

[14] Li Wanping. (2004).*Computational fluid dynamics*. Huazhong University of Science and Technology Press, 978-7-5609-3214-9. Wuha (in Chinese)

[15] Wu   Wangyi.   (1983).*Fluid   mechanics   (volume   I)*.   Higher   Education   Press, 7301001991,Beijin. (in Chinease).

[16] Wu Ziniu. (2001) .The basic principles of computational fluid dynamics , science press , 7-03-008128-5, Beijin. (in Chinese)

[17] Graver J G. *Underwater gliders: dynamics, control and design*. The USA: Princeton University, 2005

[18] Wu Jianguo, Chen Chaoyin, Wang Shuxin, et al, Hydrodynamic Characteristics of the Wings of Hybrid-Driven Underwater Glider in Glide Mode, *Journal of Tianjin University*, 2010,Vol43,No.1,pp.84-89, (in Chinease)

[19] Jenkins S A, Humphreys D E, Sherman J, et al, Alternatives for enhancement of transport economy in underwater gliders, *IEEE Proceedings of OCEANS, 2003*. 948-950

[20] Wu Baoshan, Xing Fu, Kuang Xiaofeng, et al, Investigation of hydrodynamic characteristics of submarine moving close to the sea bottom with CFD methods, *Journal of Ship Mechanics*, 2005, Vol.9,No.3,pp. 19-28

[21] Shi Shengda. 1995. *Submarine Maneuverability*. National Defense Industry Press , 9787118013498,Beijing(in Chinese).

[22] Wu Jianguo, Zhang Hongwei, Design and research on the rudder of Mini-type AUV, *ocean technology*, 2009 , Vol3,   No.2, pp.5-8.

[23] P.M.Ostafichuk, *AUV hydrodynamics and modeling for improve control*, Canada: University of British Columbia, 2004

[24] Timothy Curtis, B.Eng, The design, conatruction, outfitting, and preliminary testing of the C-SCOUT autonomous underwater vehicle (AUV), Canada, *Faculty of Engineering and Applied Science Memorial University of Newfoundland*, 2001

[25] Jianguo Wu, Chaoying Chen and Shunxin Wang, Hydrodynamic Effects of a shroud Design For a Hybrid-Driven Underwater Glider,*Sea Technology*,2010, Vol.51, No.6,pp.45-47

Part 2

Navigation and Control

Modeling and Motion Control Strategy for AUV

Lei Wan and Fang Wang
Harbin Engineering University
China

1. Introduction

Autonomous Underwater Vehicles (AUV) speed and position control systems are subjected to an increased focus with respect to performance and safety due to their increased number of commercial and military application as well as research challenges in past decades, including underwater resources exploration, oceanographic mapping, undersea wreckage salvage, cable laying, geographical survey, coastal and offshore structure inspection, harbor security inspection, mining and mining countermeasures (Fossen, 2002). It is obvious that all kinds of ocean activities will be greatly enhanced by the development of an intelligent underwater work system, which imposes stricter requirements on the control system of underwater vehicles. The control needs to be intelligent enough to gather information from the environment and to develop its own control strategies without human intervention (Yuh, 1990; Venugopal and Sudhakar, 1992).

However, underwater vehicle dynamics is strongly coupled and highly nonlinear due to added hydrodynamic mass, lift and drag forces acting on the vehicle. And engineering problems associated with the high density, non-uniform and unstructured seawater environment, and the nonlinear response of vehicles make a high degree of autonomy difficult to achieve. Hence six degree of freedom vehicle modeling and simulation are quite important and useful in the development of undersea vehicle control systems (Yuh, 1990; Fossen 1991, Li et al., 2005). Used in a highly hazardous and unknown environment, the autonomy of AUV is the key to work assignments. As one of the most important subsystems of underwater vehicles, motion control architecture is a framework that manages both the sensorial and actuator systems (Gan et al., 2006), thus enabling the robot to undertake a user-specified mission.

In this chapter, a general form of mathematical model for describing the nonlinear vehicle systems is derived, which is powerful enough to be applied to a large number of underwater vehicles according to the physical properties of vehicle itself to simplify the model. Based on this model, a simulation platform "AUV-XX" is established to test motion characteristics of the vehicle. The motion control system including position, speed and depth control was investigated for different task assignments of vehicles. An improved S-surface control based on capacitor model was developed, which can provide flexible gain selections with clear physical meaning. Results of motion control on simulation platform "AUV-XX" are described.

2. Mathematical modeling and simulation

Six degree of freedom vehicle simulations are quite important and useful in the development of undersea vehicle control systems. There are several processes to be modeled in the simulation including the vehicle hydrodynamics, rigid body dynamics, and actuator dynamics, etc.

2.1 AUV kinematics and dynamics

The mathematical models of marine vehicles consist of kinematic and dynamic part, where the kinematic model gives the relationship between speeds in a body-fixed frame and derivatives of positions and angles in an Earth-fixed frame, see Fig.1. The vector of positions and angles of an underwater vehicle $\eta = [x, y, z, \varphi, \theta, \psi]^T$ is defined in the Earth-fixed coordinate system(E) and vector of linear and angular $v v = [u, v, w, p, q, r]^T$ elocities is defined in a body-fixed(B) coordinate system, representing surge, sway, heave, roll, pitch and yaw velocity, respectively.

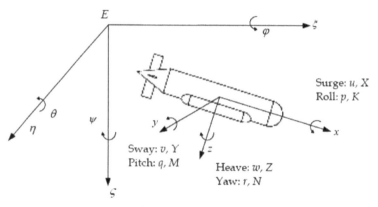

Fig. 1. Earth-fixed and body-fixed reference frames

According to the Newton-Euler formulation, the 6 DOF rigid-body equations of motion in the body-fixed coordinate frame can be expressed as:

$$
\begin{cases}
m\left[(\dot{u}_r - v_r r + w_r q) - x_G\left(q^2 + r^2\right) + y_G(pq - \dot{r}) + z_G(pr + \dot{q})\right] = X \\
m\left[(\dot{v}_r - w_r p + u_r r) - y_G(r^2 + p^2) + z_G(qr - \dot{p}) + x_G(qp + \dot{r})\right] = Y \\
m\left[(\dot{w}_r - u_r q + v_r p) - z_G(p^2 + q^2) + x_G(rp - \dot{q}) + y_G(rq + \dot{p})\right] = Z \\
I_x \dot{p} + (I_z - I_y)qr + m\left[y_G(\dot{w}_r + pv_r - qu_r) - z_G(\dot{v}_r + ru_r - pw_r)\right] = K \\
I_y \dot{q} + (I_x - I_z)rp + m\left[z_G(\dot{u}_r + w_r q - v_r r) - x_G(\dot{w}_r + pv_r - u_r q)\right] = M \\
I_z \dot{r} + (I_y - I_z)pq + m\left[x_G(\dot{v}_r + u_r r - pw_r) - y_G(\dot{u}_r + qw_r - v_r r)\right] = N
\end{cases}
\tag{1}
$$

where m is the mass of the vehicle, I_x, I_y and I_z are the moments of inertia about the x_b, y_b and z_b-axes, x_g, y_g and z_g are the location of center of gravity, u_r, v_r, w_r are relative translational velocities associated with surge, sway and heave to ocean current in the body-fixed frame, here assuming the sea current to be constant with orientation in yaw only, which can be described by the vector $U_c = [u_c, v_c, w_c, 0, 0, \alpha_c]^T$. The resultant forces X, Y, Z, K, M, N includes positive buoyant $B - W = \Delta P$ (since it is convenient to design underwater vehicles with positive buoyant such that the vehicle will surface automatically in the case of an emergency), hydrodynamic forces $X_H, Y_H, Z_H, K_H, M_H, N_H$ and thruster forces.

2.2 Thrust hydrodynamics modeling

The modeling of thruster is usually done in terms of advance ratio J_0, thrust coefficients K_T and torque coefficient K_Q. By carrying out an open water test or a towing tank test, a unique curve where J_0 is plotted against K_T and K_Q can be obtained for each propeller to depict its performance. And the relationship of the measured thrust force versus propeller revolutions for different speeds of advance is usually least-squares fitting to a quadratic model.

Here we introduce a second experimental method to modeling thruster dynamics. Fig.2 shows experimental results of thrusters from an open water test in the towing tank of the Key Lab of Autonomous Underwater Vehicles in Harbin Engineering University. The results are not presented in the conventional way with the thrust coefficient K_T plotted versus the open water advance coefficient J_0, for which the measured thrust is plotted as a function of different speeds of vehicle and voltages of the propellers.

The thrust force of the specified speed of vehicle under a certain voltage can be finally approximated by Atiken interpolation twice. In the first interpolation, for a certain voltage, the thrust forces with different speeds of the vehicle (e.g. 0m/s, 0.5m/s, 1.0m/s, 1.5m/s) can be interpolated from Fig.1, and plot it versus different speeds under a certain voltage. Then based on the results of the first interpolation, for the second Atiken interpolation we can find the thrust force for the specified speed of the vehicle.

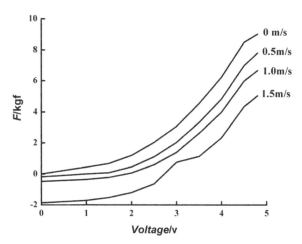

Fig. 2. Measured thrust force as a function of propeller driving voltage for different speeds of vehicle

Compared with conventional procedure to obtain thrust that is usually done firstly by linear approximating or least-squares fitting to K_T - J_0 plot (open water results), then using formulation $F_t = K_t n^2 D^4$ to compute the thrust F_t. The experimental results of open water can be directly used to calculate thrust force without using the formulation, which also can be applied to control surface of rudders or wings, *etc.*

2.3 General dynamic model

To provide a form that will be suitable for simulation and control purposes, some rearrangements of terms in Eq.(1) are required. First, all the non-inertial terms which have velocity components were combined with the fluid motion forces and moments into a fluid vector denoted by the subscript vis (viscous). Next, the mass matrix consisting all the coefficient of rigid body's inertial and added inertial terms with vehicle acceleration components $\dot{u}, \dot{v}, \dot{w}, \dot{p}, \dot{q}, \dot{r}$ was defined by matrix E, and all the remaining terms were combined into a vector denoted by the subscript else, to produce the final form of the model:

$$E\dot{X} = F_{\text{vis}} + F_{\text{else}} + F_t \tag{2}$$

where $X = [u, v, w, p, q, r]^{\text{T}}$ is the velocity vector of vehicle with respect to the body-fixed frame.

Hence, the 6 DOF equations of motion for underwater vehicles yield the following general representation:

$$\begin{cases} \dot{X} = E^{-1}(F_{\text{vis}} + F_{\text{else}} + F_t) \\ \dot{\eta} = J(\eta)X \end{cases} \tag{3}$$

with

$$E = \begin{bmatrix} m-X_{\dot{u}} & 0 & 0 & 0 & mz_G & -my_G \\ 0 & m-Y_{\dot{v}} & 0 & -mz_G-Y_{\dot{p}} & 0 & mx_G-Y_{\dot{r}} \\ 0 & 0 & m-Z_{\dot{w}} & my_G & -mx_G-Z_{\dot{q}} & 0 \\ 0 & -mz_G-K_{\dot{v}} & 0 & I_x-K_{\dot{p}} & 0 & -K_{\dot{r}} \\ mz_G & 0 & -mx_G-M_{\dot{w}} & 0 & I_y-M_{\dot{q}} & 0 \\ 0 & mx_G-N_{\dot{v}} & 0 & -N_{\dot{p}} & 0 & I_z-N_{\dot{r}} \end{bmatrix} \tag{4}$$

where $J(\eta)$ is the transform matrix from body-fixed frame to earth-fixed frame, η is the vector of positions and attitudes of the vehicle in earth-fixed frame.

The general dynamic model is powerful enough to apply it to different kinds of underwater vehicles according to its own physical properties, such as planes of symmetry of body, available degrees of freedom to control, and actuator configuration, which can provide an effective test tool for the control design of vehicles.

3. Motion control strategy

In this section, the design of motion control system of AUV-XX is described. The control system can be cast as two separate designs, which include both position and speed control in horizontal plane and the combined heave and pitch control for dive in vertical plane. And an improved S-surface control algorithm based on capacitor plate model is developed.

3.1 Control algorithm

As a nonlinear function method to construct the controller, S-surface control has been proven quite effective in sea trial for motion control of AUV in Harbin Engineering University (Li et al., 2002). The nonlinear function of S-surface is given as:

$$u = 2.0/(1.0 + \exp(-k_1 e - k_2 \dot{e})) - 1.0 \tag{5}$$

where $e, \dot{e}$ are control inputs, and they represent the normalized error and change rate of the error, respectively; u is the normalized output in each degree of freedom; k_1, k_2 are control parameters corresponding to control inputs e and $\dot{e}$ respectively, and we only need to adjust them to meet different control requirements.

Based on the experiences of sea trials, the control parameters k_1, k_2 can be manually adjusted to meet the fundamental control requirements, however, whichever combination of k_1, k_2 we can adjust, it merely functions a global tuning which dose not change control structure. Here the improved S-surface control algorithm is developed based on the capacitor with each couple of plates putting restrictions on the control variables $e, \dot{e}$ respectively, which can provide flexible gain selection with proper physical meaning.

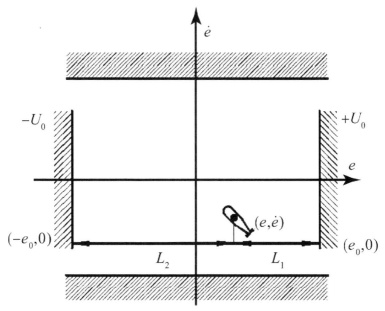

Fig. 3. Capacitor plate model

The capacitor plate model as shown in Fig.3 demonstrates the motion of a charged particle driven by electrical field in capacitor is coincident with the motion of a controlled vehicle from current point $(e, \dot{e})$ to the desired point, for which the capacitor plate with voltage serves as the controller, and the equilibrium point of electrical field is the desired position that the vehicle is supposed to reach.

Due to the restriction of two couples of capacitor plates put on control variables e and $\dot{e}$, the output of model can be obtained as

$$y = u^{+U_0} + u^{-U_0} = F(L_1, L_2)(+U_0) + F(L_2, L_1)(-U_0) \tag{6}$$

where L_1, L_2 are horizontal distances from the current position of the vehicle to each capacitor plate, respectively, and the restriction function $F(*, *)$ is defined to be hyperbolic function of L_1, L_2 by Ren and Li (2005):

$$\begin{cases} F(L_1, L_2) = \dfrac{L_1^{-k}}{L_1^{-k} + L_2^{-k}} \\ F(L_2, L_1) = \dfrac{L_2^{-k}}{L_1^{-k} + L_2^{-k}} \end{cases} \tag{7}$$

The restriction function $F(*, *)$ reflects the closer the current position $(e, \dot{e})$ of vehicle moving to capacitor plate, the stronger the electrical field is. Choosing $U_0 = 1$, the output of capacitor plate model yields:

$$u = \frac{L_1^{-k} - L_2^{-k}}{L_1^{-k} + L_2^{-k}} U_0 = \frac{(e_0 + e)^k - (e_0 - e)^k}{(e_0 + e)^k + (e_0 - e)^k} \tag{8}$$

where e_0 is the distance between the plate and field equilibrium point of capacitor.

An improved S-surface controller based on the capacitor plate model is proposed, that is

$$\begin{cases} u_{ei} = [2.0 / \left(1.0 + (\frac{e_0 - e_i}{e_0 + e_i})^{ki1} \right) - 1.0] \\ u_{\dot{e}i} = [2.0 / \left(1.0 + (\frac{e_0 - \dot{e}_i}{e_0 + \dot{e}_i})^{ki2} \right) - 1.0] \\ f_i = K_{ei} \cdot u_{ei} + K_{\dot{e}i} \cdot u_{\dot{e}i} \end{cases} \tag{9}$$

where f_i is the outputted thrust force of controller for each DOF, and $K_{ei} = K_{\dot{e}i} = K_i$ is the maximal thrust force in i th DOF, therefore the control output can be reduced to

$$\begin{cases} u_i = u_{ei} + u_{\dot{e}i} = [2.0 / \left(1.0 + (\frac{e_0 - e_i}{e_0 + e_i})^{ki1} \right) - 1.0] + [2.0 / \left(1.0 + (\frac{e_0 - \dot{e}_i}{e_0 + \dot{e}_i})^{ki2} \right) - 1.0] \\ f_i = K_i \times u_i \end{cases} \tag{10}$$

The capacitor model's S-surface control can provide flexible gain selections with different forms of restriction function to L_1, L_2 to meet different control requirements for different phases of control procedure.

3.2 Speed and position control in horizontal plane

Since AUV-XX is equipped with two transverse tunnel thrusters in the vehicle fore and aft respectively and two main thrusters (starboard and port) aft in horizontal plane, which can produce a force in the x-direction needed for transit and a force in the y-direction for maneuvering, respectively. So both speed and position controllers are designed in horizontal plane.

Speed control is to track the desired surge velocity with fixed yaw angle and depth, which is usually used in long distance transfer of underwater vehicles. Before completing certain kind of undersea tasks, the vehicle needs to experience long traveling to achieve the destination. In this chapter, speed control is referred to a forward speed controller in surge based on the control algorithm we introduced in above section, its objective is to make the vehicle transmit at a desired velocity with good and stable attitudes such as fixed yaw and depth.

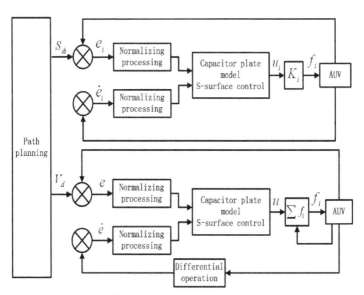

Fig. 4. Position and speed control loop

Position control enables the vehicle to perform various position-keeping functions, such as maintaining a steady position to perform a particular task, following a prescribed trajectory to search for missing or seek after objects. Accurate position control is highly desirable when the vehicle is performing underwater tasks such as cable laying, dam security inspection and mine clearing. To ensure AUV-XX to complete work assignments of obstacles avoiding, target recognition, and mine countermeasures, we design position controllers for surge, sway, yaw and depth respectively for equipping the vehicle with abilities of diving at fixed deepness, navigating at desired direction, sailing to given points and following the given track, etc.

As for the desired or target position or speed in the control system, it is the path planning system who decides when to adopt and switch control scheme between position and speed, the desired position that the vehicle is supposed to reach, and the velocity at which the vehicle should navigate with respect to the present tasks and motion states of the vehicle as well as operation environment.

Fig.4 shows both position and speed control procedures. It can be seen that it is easier to realize the control algorithm. For position control of i th DOF, the control inputs are the position error and the change rate of position error, that is the velocity obtained from motion sensors; while for speed control, the velocity error and acceleration are control inputs, since AUV-XX is not equipped with IGS to acquire the acceleration of the vehicle, acceleration is calculated differential the velocity in each control step.

3.3 Combined control of pitch and heave in vertical plane

Since when the vehicle is moving at a high speed, the thrust that tunnel thrusters can provide will strongly degrade, it is difficult to control the depth merely using tunnel thrusters. Considering that once the depth or height of the vehicle changes, the pitch will change with it, and vice versa, so we combine pitch with heave control for diving when the vehicle is moving at some high speed to compensate for the thrust reduction of tunnel thrusters. In that case, the desired pitch angle can be designed as a function of the surge velocity of the vehicle:

$$\theta_T = \begin{cases} -\dfrac{k_\theta \cdot \Delta z}{\sqrt{u}} - \theta_0 & u \geq 0.8 m/s \\ 0 & u < 0.8 m/s \end{cases} \tag{11}$$

where θ_T is the target pitch angle, k_θ is a positive parameter to be adjusted, Δz is the depth deviation, u is the surge velocity of vehicle, θ_0 is the pitch angle when the vehicle is in static equilibrium. Since the change of pitch angle is usually associated with depth change and they affect each other, the target pitch is the output of the proportional control with respect to depth change with the proportional parameters k_θ, the velocity threshold of 0.8m/s is chosen based on the engineering experience of the sea trial and the capability of the thruster system of the vehicle.

When the vehicle is moving at a low speed($u < 0.8 m/s$), tunnel thrusters can normally provide the needed force, so the command of target pitch will not be sent to motion control. According to control law (9), we can get the output of the heave controller. Since the vehicle is usually designed with positive buoyancy, the final output of control of heave can be obtained by

$$f_3 = K_3 \cdot u_3 + \Delta P \tag{12}$$

where ΔP is the positive buoyancy.

And as the surge velocity grows, the thrust of vertical tunnel will experience worse reduction and degradation, so the role it plays in the depth control will be greatly abridged, as a result, the output of pitch control of the main vertical thrusters aft of the vehicle should compensate for that, so the output of control will finally yields as

$$f_5 = \begin{cases} K_5 u_5 & u \leq 0.8 \\ K_5(\varepsilon_1 \cdot u_3 + \varepsilon_2 \cdot u_5) & u > 0.8 \end{cases} \tag{13}$$

with

$$\begin{cases} \varepsilon_1 = \alpha_1 / \sqrt{\beta u} \\ \varepsilon_2 = \alpha_2 \cdot \exp(u \cdot u / 10) \end{cases} \tag{14}$$

where $\alpha_1, \alpha_2, \beta$ can be manually adjusted based on experiences. The block diagram of combined control of pitch and depth in vertical plane can be in Fig.5.

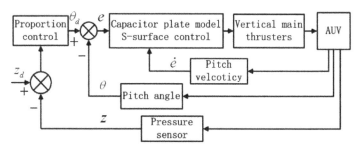

Fig. 5. Combined control of pitch and depth in vertical plane

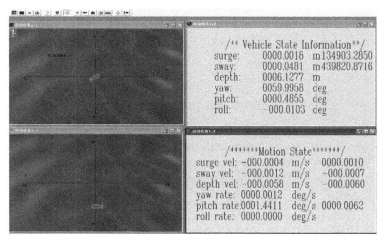

Fig. 6. AUV-XX simulation platform

4. Simulation results and analysis

To verify the feasibility and effectiveness of the motion control system for the vehicle AUV-XX, simulations are carried out in the AUV-XX simulation platform. The vehicle researched in this chapter named by AUV-XX, AUV-XX's configuration is basically a long cylinder of 0.5m in diameter and 5m in length with crossed type wings near its rear end. On each edge of the wings, a thruster is mounted, which is used for both turn and dive. AUV-XX is also equipped with a couple of lateral tunnel thrusters for sway and a couple of vertical tunnel thrusters fore and aft of the vehicle, respectively. Based on the modeling method described in above section, we established the AUV-XX simulation platform to carry out fundamental tests on its motion characteristics, stability and controllability. The states of the vehicle including positions, attitudes and velocities are obtained at each instant by solving the mathematical model equation by integration using a time step of 0.5s. Fig.6 shows the interface of AUV-XX simulation platform. Fig.7 shows the data flow of the simulation platform connecting with motion control system.

Figs.8–10 show the simulation results of capacitor plate model's S-surface control for separate position and speed control in surge, sway and yaw respectively, and also combined control of heave and pitch for diving of the vehicle. The roll is left uncontrolled. And all the vehicle states, including speed, are initialized to zero at the beginning of the each of the simulation. The solid lines denote the actual responses of the vehicle while the dashed denote the desired position or speed that the vehicle is commanded to achieve.

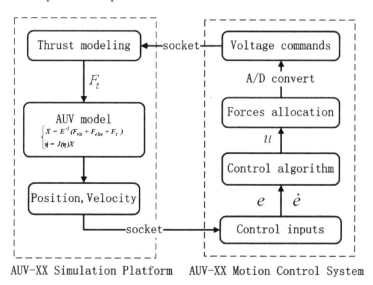

AUV–XX Simulation Platform   AUV–XX Motion Control System

Fig. 7. Data flow of AUV-XX motion control in simulation platform

It can be seen from Fig.8 that, the vehicle is commanded to move to some specified positions in surge, sway and yaw, respectively. For the case of surge as shown in Fig.8(a) and Fig.8(b), the larger position deviation from the target position produces faster response of surge velocity. Compared with responses of surge and yaw, the sway is slower with a rise time of 150s for the desired position is 16m, which may result from that the lateral resistance is much larger longitudinally and the thrust of transverse tunnel thrusters can provide is smaller than the main aft thrusters.

Fig.9 shows the speed control results of surge with constant yaw and depth keeping. The desired velocity is 1m/s and once the vehicle is moving stably with such speed then the vehicle is commanded to track the specified depth (5m) and yaw (45°) commands. It can be seen that there is no overshoot in surge speed and system response is fast with the yaw experiencing an accepted overshoot of ±2°, and the depth is stably maintained, hence the vehicle is able to move at a desired speed with a desired fixed heading and depth. The speed control simulation results prove the feasibility of the proposed speed control strategy.

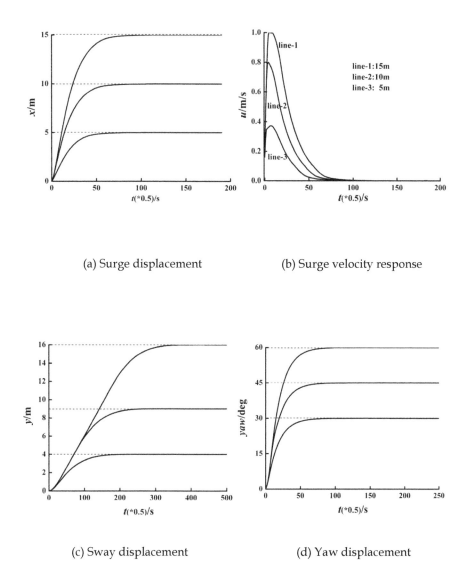

(a) Surge displacement   (b) Surge velocity response

(c) Sway displacement   (d) Yaw displacement

Fig. 8. Position control results of surge, sway and yaw

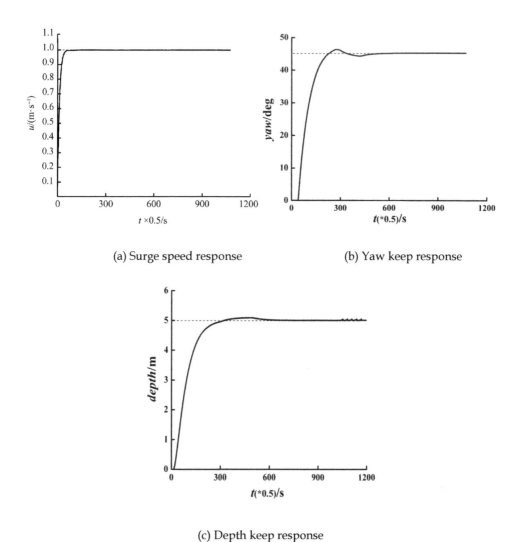

(a) Surge speed response   (b) Yaw keep response

(c) Depth keep response

Fig. 9. Speed control results of surge with yaw and depth keeping

Fig.10 shows the combined control of heave and pitch for diving in vertical plane. For this case, the velocity of surge that the vehicle is commanded to track is 1.5m/s, which is not very large so that the vertical tunnel thrusters will suffer thrust reduction to some extent but still can work to provide a portion of vertical thrust for diving. Hence when the vehicle is commanded to dive, the pitch will not experience a large change, which is reasonable design consideration in the case of large inertial vehicles.

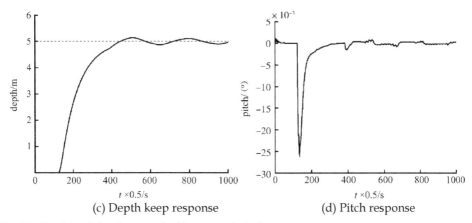

(c) Depth keep response   (d) Pitch response

Fig. 10. Combined depth control of heave and pitch

5. Conclusions

In this chapter, the design of motion control system for Autonomous Underwater Vehicles is described, which includes both position and speed control in horizontal plane and combined control of heave and pitch in vertical plane. To construct the control system, a 6 DOF general mathematical model of underwater vehicles was derived, which is powerful enough to apply it to different kinds of underwater vehicles according to its own physical properties. Based on the general mathematical model, a simulation platform was established to test motion characteristics, stability and controllability of the vehicle. To demonstrate the performance of the designed controller, simulations have been carried out on AUV-XX simulation platform and the capacitor plate model S-surface control shows a good performance.

6. References

Fossen T. I. (1991). *Nonlinear modeling and control of underwater vehicles*. Ph.D. thesis, Norwegian Institute of Technology-NTH, Trondheim, Norway

Fossen T. I. (2002). *Marine control system: guidance, navigation and control of ships, rigs and underwater vehicles*. Marine Cybernetics, Trondheim, 254-260

Gan Yong, Sun Yushan, Wan Lei, Pang Yongjie (2006). Motion control system architecture of underwater robot. *Proceedings of the 6th World Congress on Intelligent Control and Automation*, Dalian, China, 8876-8880

Gan Yong, Sun Yushan, Wan Lei, Pang Yongjie (2006). Motion control system of underwater robot without rudder and wing. *Proceeding of the 2006 IEEE International Conference on Intelligent Robotics and Systems*, Beijing, China, 3006-3011

Li Xuemin, Xu Yuru (2002). S-control of automatic underwater vehicles. *The Ocean Engineering*, 19(3), 81-84

Li Ye, Liu Jianchen, Shen Minxue (2005). Dynamics model of underwater robot motion control in 6 degrees of freedom. *Journey of Harbin Institute of Technology*, 12(4), 456-459

Ren Yongping, Li Shengyi (2005). Design method of a kind of new controller. *Control and Decision*, 20(4), 471-474

Venugopal KP, Sudhakar R (1992). On-line learning control of autonomous underwater vehicles using feedforward neural networks. *IEEE Journal of Oceanic Engineering*, 17(4), 308-319

Yuh J (1990). Modeling and control of underwater robotic vehicles. *IEEE Transaction on Systems, Man, and Cybernetics*, 20(6), 1475-1483

Formation Guidance of AUVs Using Decentralized Control Functions

Matko Barisic, Zoran Vukic and Nikola Miskovic
University of Zagreb, Faculty of Electrical Engineering and Computing
Croatia

1. Introduction

Autonomous Underwater Vehicles (AUVs) are the most complex type of unattended marine systems, being *mobile*, with challenging dynamics and non-holonomic kinematics. They are increasingly being recognized as a keystone technology for projecting human scientific and economical interests into the deep Ocean (Papoulias et al., 1989). A recent report by Bildberg (2009) delivers the verdict of several key researchers that the AUVs are rapidly moving towards maturity.

The *autonomy* of AUVs is their key capability. They autonomously explore Ocean phenomena relevant to human scientific and economic interests. Well engineered autonomous control allows them to act robustly and predictably with regards to waves, currents, wind, sea-state and numerous other disturbances and operational conditions in nature. As a consequence, they are today being cast in the leading role in projecting human presence and human interests in the Ocean, in an increasingly diverse gamut of topics:

- Physical oceanography (Plueddemann et al., 2008; Tuohy, 1994),

- Marine biology, conservationist biology, marine ecology management, biological oceanography (Farrell et al., 2005; Pang, 2006; Pang et al., 2003),

- Geology, petrology, seismology, hydrography (for the benefit of e.g. the oil and gas industry, maritime civil engineering etc.),

- Maritime and naval archaeology, submerged cultural heritage protection and management,

- Marine traffic management, search and rescue, hazardous material and waste management, emergencies and catastrophes management and first responding (Carder et al., 2001; Pang, 2006)

- Maritime security, customs enforcement, border protection and defense (Allen et al., 1997; 2004; Clegg & Peterson, 2003; Curtin et al., 1993; Eisman, 2003; US Navy, 2004).

To increase the effectiveness, safety, availability, economics and applicability of AUVs to these and other topics of interest, this chapter proposes a *decentralized cooperative cross-layer formation-control* paradigm for entire groups of AUVs collaborating in exploration tasks. The AUVs are assumed to navigate on a common "flight ceiling" by using robust altitude controllers, based on altimeter echosounder measurements. The proposed virtual potential framework allows for the 2D organization of individual trajectories on such a "flight ceiling".

The goal is to provide decentralized consensus-building resulting in synoptical situational awareness of, and coordinated manoeuvring in the navigated waterspace. The paradigm is formally developed and tested in a hardware-in-the-loop simulation (HILS) setting, utilizing a full-state hydrodynamical rigid-body dynamic model of a large, sea-capable, long-endurance Ocean-going vehicle. Existence of realistic, technically feasible sensors measuring proxy variables or directly the individual kinematic or dynamic states is also simulated, as is the presence of realistic, non-stationary plant and measurement noise.

1.1 The cooperative paradigm

Since 1970s, robotics and control engineers have studied the cooperative paradigm. Cooperative control is a set of complete, halting algorithms and machine-realized strategies allowing multiple *individual agents* to complete a given task in a certain optimal way. This optimality results from the agents' leveraging each other's resources (e.g. manoeuvring abilities) to more effectively minimize some cost function that measures a "budget" of the entire task, in comparison to what each agent would would be capable of on their own (without the benefit of the group).

In the marine environment, such "social" leveraging is beneficial in several ways. Firstly, deployment of more AUVs significantly reduces the time needed to survey a given theater of operations. This has enormous economic repercussion in terms of conserved hours or days of usually prohibitively expensive ship-time (for the vessel that is rendering operational support to the AUV fleet). Secondly, deployment of a larger number of AUVs *diversifies the risk to operations*. In a group scenario, loss of a (small) number of AUVs doesn't necessarily preclude the achievement of mission goals. Lastly, if each of the group AUVs are furnished with adaptive-sampling algorithms, such as in the chemical plume-tracing applications (Farrell et al., 2005; Pang, 2006; Pang et al., 2003), deployment of multiple vehicles guarantees much faster convergence to the points of interest.

Cooperative control frameworks are split into *centralized* and *decentralized* strategies. A centralized cooperative control system's task is to determine the actions of each agent based on a perfectly (or as near perfectly as possible) known full data-set of the problem, which consists of the state vectors of *every agent* for which the problem is stated. The centralized system instantiates a globally optimal solution based on the assessment of momentary resource-disposition of the entire ensemble, as well as based on the total, if possibly non-ideal knowledge of the environment. The state data are usually collected by polling all agents through a communication network. After the polling cycle, the centralized system communicates the low-level guidance commands back to individual agents. This approach allows for the emergence of a global optimum in decision-making on grounds of *all obtainable information*, but heavily depends on *fault-intolerant, quality-assured, high-bandwidth communication*.

In a *decentralized* approach, such as we have chosen to present in this chapter, each agent possesses imperfect state and perception data of every other agent and of the observable portion of the environment, and *locally* decides its own course of action. The greatest issue in decentralized cooperative control is the achievement of a *consensus* between separately reasoning autonomous agents.

2. The virtual potentials framework

To address the issue of reactive formation guidance of a number of AUVs navigating in a waterspace, a method based on *virtual or artificial potentials* is hereby proposed. The virtual potentials alleviate some of the most distinct problems encountered by competing reactive formation guidance strategies, which are prone to the following problems:

- Reliance on the *perfect knowledge* of a map of the waterspace,

- Lack of *reaction* to the decentralized, agent-local process of accumulating or perfecting knowledge of the environment on top of the initially imperfect situational awareness of each individual agent,

- Trajectory planning that is sub-optimal, or optimal based on a *hard-coded* criterion, without possibility of adjusting or restating that criterion at run-time, because the cost function is implicit in the choice of mathematical tools (such as a distinct set of curve formulations used for trajectories etc.).

Stemming from these considerations, we propose a scheme where each AUV in a 2D formation imbedded in the "flight ceiling" plane as previously discussed maintains a local imperfect map of the environment. Every possible map only ever consists of a finite number of instantiations of any of the three types of features:

1. A way-point that is commanded for the entire formation, $w \in \mathbb{R}^2$

2. Obstacles which need to be circumnavigated in a safe and efficient manner, (O_i), $\forall i = 1 \ldots n_{obs} O_i \subset \mathbb{R}^2$,

3. Vertices of the *characteristic cell* of the chosen formation geometry, covered in more detail in sec. 2.3.3 and 3.

With this in mind, let the *virtual potential* be a real, single valued function $P : \mathbb{R}^2 \to \mathbb{R}$, mapping almost every attainable position of an AUV on the "flight ceiling" to a real. Let P-s total differential exists almost wherever the function itself is defined. P can be said to live on the subspace of the full-rank state-space of the AUVs, $C = \mathbb{R}^6 \times SE^3$. The state-space of the AUV is composed of the Euclidean 6-space $\mathbb{R}^6$ spanned by the angular and linear velocities, $\{[\mathbf{v}^T \ \boldsymbol{\omega}^T]^T\} = \{[u \ v \ w \mid p \ q \ r]^T\} \equiv \mathbb{R}^6$ and a full 3D, 6DOF configuration-space $\{[\mathbf{x}^T \ \boldsymbol{\Theta}^T]^T\} = \{[x \ y \ z \mid \varphi \ \vartheta \ \psi]^T\}$ which possesses the topology of the *Special Euclidean group of rank 3*, SE^3. Function P therefore maps to a real scalar field over that same C.

Furthermore, this framework will be restricted to only those P that can be expressed in terms of a sum of finitely many terms:

$$\exists n \in \mathbb{N} \mid P_\Sigma = \sum_{i=1}^{n} P_i \tag{1}$$

Where P_i is of one of a small variety of considered function forms. Precisely, we restrict our attention to three function forms with each one characteristic of each of the three mentioned *types of features* (way-point, obstacle, vertices of formation cells).

The critical issue in the guidance problem at hand is Euclidean 2D distance (within the "flight ceiling") between pairs of AUVs in the formation, and each AUV and all obstacles. Therefore, our attention is further restricted to only such $\{P_i\} \subset \mathcal{L}(C \to \mathbb{R})$ with $\mathcal{L}$ being the space of all functions mapping C to $\mathbb{R}$ whose total differential exists almost wherever each of the functions is defined on C, which can be represented as the composition $P_i \equiv p_i \circ d_i$, $p_i : \mathbb{R}_0^+ \to \mathbb{R}$, and $d_i : C \to \mathbb{R}_0^+$ a Euclidean 2D metric across the "flight ceiling". Consequently, P_i is completely defined by the choice of $p_i(d)$, the *isotropic potential contour generator*. Choices and design of $p_i(d)$-s will be discussed in sec. 2.3.

With all of the above stated, a *decentralized total control function* $f : \mathbb{Z} \to \mathbb{R}^2$ is then defined as a sampling, repeated at sample times $k \in \mathbb{Z}_0$, of the 2D vector field $E : \mathbb{W}_i \to \mathbb{R}^2$ over a subspace $\mathbb{W}_i \subseteq \mathbb{R}^2 \subset \mathcal{C}$, the *navigable waterspace*:

$$\forall x \in \mathbb{W}_i \subseteq \mathbb{R}^2 \subset \mathcal{C}, \; \mathbf{E}(x) = -\nabla P_{\Sigma}(x) \tag{2}$$

Where $\mathbb{W}_i = \mathbb{R}^2 \setminus \left(\bigcup_i O_i \cup \bigcup_j O_j^{(ag)} \right)$ contains all of $\mathbb{R}^2$ to the exclusion of closed connected subsets of $\mathbb{R}^2$ that represent interiors of obstacles, $\{O_i\}$ and those that represent safety areas around all the j-th AUVs ($j \neq i$) other than the i-th one considered. $\mathbb{W}_i$ is an open, connected subset of $\mathbb{R}^2$, the "flight ceiling", inheriting its Euclidean-metric-generated topology and always containing the way-point w. Sampling $\mathbf{E}$ at the specific $x_i(k) \in \mathbb{W}_i$, the location of the i-th AUV, results in $f_i(k)$, the *total decentralized control function* for the i-th AUV at time k and location x_i.

2.1 Passivity
The decentralized total control function f is used as the forcing signal of an idealized dimensionless charged particle of unit mass, modeled by a *holonomic 2D double integrator*. If any AUV were able to behave in this manner, the AUV would follow an *ideal conservative trajectory* given by:

$$x_i(t) = \iint_{\tau=0}^{t} \mathbf{E}\left[x_i(\tau)\right] d\tau^2 + x_i^{(0)} \tag{3}$$

This ideal conservative trajectory, while stable in the BIBO sense, is in general not asymptotically stable, nor convergent by construction. The simplest case when this doesn't hold is when $\mathbf{E}(x)$ is an *irrotational*[1] vector field whose norm is affine in the $\|x - w\|$ 2D Euclidean distance:

$$\|\mathbf{E}(x)\| = e\|x - w\| + E_0; \; e \in [0, \infty) \tag{4}$$

And whose direction is always towards w:

$$\forall x, \; \mathbf{E}(x) \cdot (x - w) \overset{\text{id}}{=} e\|x - w\|^2 + E_0\|x - w\| \tag{5}$$

In that case (3) can be regarded as a linear second or third order system with two of the poles in $\pm i$. Such a system exhibits borderline-stable oscillation – a hallmark of its *conservativeness*. An example of such BIBO-stable non-convergent oscillation is given in figure 1.
Note that this analysis is irrespective of the initial condition $\dot{x}_0$ as long as (4, 5) approximate $\mathbf{E}(x)$ sufficiently well in some open ε-ball centered on w. However, AUVs are in general not able to actuate as ideal holonomic 2D double integrators. The introduction of *any* finite non-zero lag in the above discussion, which is sure to exist from first physical principles in a real AUV, is sufficient to cause *dissipation* and as a consequence passivity and convergence to w.

2.2 Local minima
In addition to the problem of passivity, the virtual potential approach suffers from the *existence of local minima*. Without further constraints, the nature of $\mathbf{E}(\mathbb{W}_i)$ so far discussed doesn't preclude a dense, connected, closed state-subspace $C_{i0}^{(j)} \subseteq C$, containing uncountably many initial vectors $\{x_0^{(l)} | l \in \mathbb{R}\}$ of "related" trajectories (with the indexing by AUV denoted by

---

[1] Whose *rotor* or *curl* operator is identically zero.

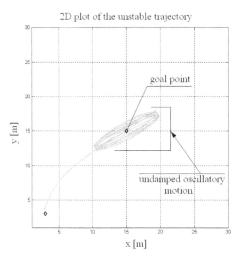

Fig. 1. An example of an oscillatory trajectory due to the conservativeness of the virtual potential system.

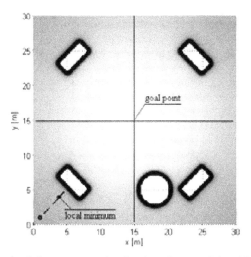

Fig. 2. Example of a local minimum occurring in virtual potential guidance in a 2D waterspace.

i, and the enumeration of the distinct points of convergence other than the way-point by j omitted for clarity) that *do not* converge to the way-point w or a finitely large orbit around it, but rather to another point $x_\infty^{(j)}$ (or a finitely large orbit around *it*). Therefore, for each of these uncountably many "nearby" trajectories (to be visualized as a "sheaf" of trajectories emanating from a distinct, well defined neigbourhood in $\mathbb{W}_i$ for some range of initial linear and angular velocities) there exists a lower bound t_l after which $\|x_i(t > t_l) - x_\infty^{(l)}\| \leq \|x_i(t > t_l) - w\|$ almost always. The set $\{t_l\}$ is also dense and connected.

Furthermore, there is no prejudice as to the *number* of such $C_{i0}^{(j)}$-s, i.e. there exists $C_{i0}^{\Sigma} \subseteq C$, $C_{i0}^{\Sigma} = \bigcup_j C_{i0}^{(j)}$. There may be multiple disjoint dense, connected, closed sets of initial conditions of the trajectory of the i-th AUV which all terminate in the same, or distinct local minima. The enumerator j may even come from $\mathbb{R}$ (i.e. there may be *uncountably many distinct local minima*, perhaps arranged in dense, connected sets – like curves or areas in $\mathbb{R}^2$).

An example of an occurrence of a local minimum is depicted in figure 2. In order to resolve local minima, an intervention is required that will ensure that either one of the following conditions is fulfilled:

1. The set C_{i0}^{Σ} is empty by construction.

2. A halting P-complete algorithm is introduced that for every $x_0 \in C_{i0}^{\Sigma}$, triggering at t_0 with $\varepsilon(t_0) = \sup \|x(t > t_0) - x_\infty | x_0\|$ characterizing a ε-ball centered on the particular x_∞ and containing all $x(t > t_0)$, to intervene in $\mathbf{E}(\mathbf{W}_i)$ *guaranteeing* that this entire ball is outside (a possibly existing) new $C_{i0}^{\Sigma'}$ (with $x_0' \leftarrow x(t_0)$).

Out of the two listed strategies for dealing with local minima, the authors have published extensively on strategy 2 (Barisic et al., 2007a), (Barisic et al., 2007b). However, strategy 1 represents a much more robust and general approach. A method guaranteeing $C_{i0}^{\Sigma} \overset{id}{=} \varnothing$ by designing in *rotors* will be described in sec. 2.4.

2.3 Potential contour generators and decentralized control functions

As for the potential contour generators $p_i(d) : \mathbb{R}_0^+ \to \mathbb{R}$, their definition follows from the global goals of guidance for the formation of AUVs. Bearing those in mind, the potential contour generators of each *feature type*, $p_{(o,w,c)}$ (for obstacle, way-point and formation cell vertex, accordingly) are specified below.

2.3.1 Obstacles

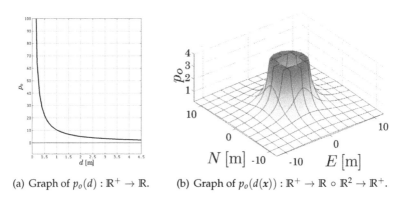

(a) Graph of $p_o(d) : \mathbb{R}^+ \to \mathbb{R}$.   (b) Graph of $p_o(d(x)) : \mathbb{R}^+ \to \mathbb{R} \circ \mathbb{R}^2 \to \mathbb{R}^+$.

Fig. 3. The potential contour generator of an obstacle, $p_o(d(x))$.

$$p_o(d) = \exp\left(\frac{A^+}{d}\right) - 1; \quad \lim_{d \to \infty} p_o(d) = 0; \quad \lim_{d \to 0^+} p_o(d) = \infty \tag{6}$$

$$\frac{\partial}{\partial d} p_o(d) = -\frac{A^+}{d^2} \exp\left(\frac{A^+}{d}\right); \quad \lim_{d \to \infty} \frac{\partial}{\partial d} p_o(d) = 0; \quad \lim_{d \to 0^+} \frac{\partial}{\partial d} p_o(d) = \infty \tag{7}$$

Where:
- $p_o(d) : \mathbb{R}^+ \to \mathbb{R}$ is the potential contour generator of obstacles, a strictly monotonously decreasing smooth single-valued Lebesgue-integrable function mapping a non-negative real to a real,
- $A^+ \in \mathbb{R}^+ \setminus \{0\}$ is a positive real independent parameter dictating the scale of the acceleration away from the obstacle.

2.3.2 Way-points

$$p_w(d) = \begin{cases} d \leq d_0 : & \frac{A_p^-}{2} d^2 \\ d > d_0 : & A_0^-(d - d_0) + p_0 \end{cases} ; \quad d_0 \overset{id}{=} \frac{A_0^-}{A_p^-}; \quad p_0 \overset{if}{=} \frac{A_0^{-2}}{2A_p^-} \tag{8}$$

$$\therefore p_w(d, d > d_0) = A_0^- d - \frac{A_c^{-2}}{2A_p^-}; \quad \lim_{d \to \infty} p_w(d) = \infty; \quad \lim_{d \to 0^+} p_w(d) = 0 \tag{9}$$

$$\frac{\partial}{\partial d} p_w(d) = \max(A_p^- d, A_0^-); \quad \lim_{d \to \infty} \frac{\partial}{\partial d} p_w(d) = A_0^-; \quad \lim_{d \to 0^+} \frac{\partial}{\partial d} p_w(d) = 0 \tag{10}$$

Where:
- $p_w(d) : \mathbb{R}^+ \to \mathbb{R}$ is the potential contour generator of way-points, a strictly monotonously increasing smooth single-valued Lebesgue-integrable function mapping a non-negative real to a real,
- $A_p^- \in \mathbb{R}^+ \setminus \{0\}$ is a positive real independent parameter dictating the scale of acceleration towards the way-point in the area of proportional attraction,
- $A_0^- \in \mathbb{R}^+ \setminus \{0\}$ is a positive real independent parameter dictating the constant acceleration towards the way-point outside the area of proportional attraction,
- $d_p \in \mathbb{R}^+ \setminus \{0\}$ is a positive real independent parameter dictating the radius of the open ball ion.

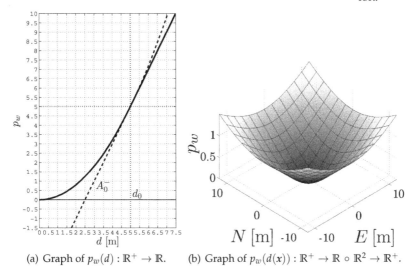

(a) Graph of $p_w(d) : \mathbb{R}^+ \to \mathbb{R}$.   (b) Graph of $p_w(d(x)) : \mathbb{R}^+ \to \mathbb{R} \circ \mathbb{R}^2 \to \mathbb{R}^+$.

Fig. 4. The potential contour generator of a way-point, $p_w(d(x))$.

2.3.3 Cell vertices

A good candidate potential contour generator of formation cell vertices, which behaves similar to a function *with local support*, is the *normal distribution curve*, adjusted for attractiveness (i.e. of inverted sign).

$$p_c(d) = -A_c^- d_c \exp\left(1 - \frac{d^2}{2d_c^2}\right); \quad \lim_{d\to\infty} p_c(d) = 0; \quad \lim_{d\to 0^+} p_w(d) = -A_c^- \tag{11}$$

$$\frac{\partial}{\partial d} p_c(d) = \frac{A_c^-}{d_c} d \exp\left(1 - \frac{d^2}{2d_c^2}\right); \quad \frac{\partial}{\partial d} p_c(d)|_0 = 0; \quad \lim_{d\to\infty} \frac{\partial}{\partial d} p_c(d) = 0 \tag{12}$$

$$\frac{\partial^2}{\partial d^2} p_c(d) = \frac{A_c^-}{d_c}\left(1 - \frac{d^2}{d_c^2}\right)\exp\left(1 - \frac{d^2}{2d_c^2}\right) \tag{13}$$

$$\therefore \ d_{max} = \arg\left\{\frac{\partial^2}{\partial d^2} p_c(d) \overset{!}{=} 0\right\} \tag{14}$$

$$d_{max} \overset{id}{=} \pm d_c \tag{15}$$

$$\frac{\partial}{\partial d} p_c(d)\Big|_{d_{max}\overset{id}{=}d_c} = A_c^- \tag{16}$$

Where:
- $p_c(d) : \mathbb{R}^+ \to \mathbb{R}$ is the potential contour generator of cell vertices of *characteristic cells* of a formation, a strictly monotonously increasing smooth single-valued Lebesgue-integrable function mapping a non-negative real to a real,
- $A_c^- \in \mathbb{R}^+ \setminus \{0\}$ is a positive real independent parameter dictating the scale of acceleration towards the cell vertex at the distance of maximum acceleration towards the vertex (equivalent to the valuation of $A_c^- \cdot \mathcal{N}(\pm\sigma)$ on a Gaussian normal distribution curve),
- $d_c \in \mathbb{R}^+ \setminus \{0\}$ is a positive real independent parameter dictating the radius of a sphere at which the inflection in the potential contour generator occurs, i.e. the distance at which maximum acceleration towards the vertex occurs (taking the place of σ in (12), which is analogous to a Gaussian normal distribution curve).
The potential of a square formation cell surrounding an agent that figures as an obstacle is represented in figure 5.

2.3.4 Reformulation in terms of decentralized control functions

The monotonicity of (6, 8, 11) ensures that the direction of the gradient of the potential, $\nabla P(x) / \|\nabla P(x)\| \in SO^2$, is always $\pm n_i = (x - x_i)/\|x - x_i\|$. Therefore, since (1, 2) are linear, (2) can be solved *analytically* for any finite sum of terms of the form specified by (6, 8, 11) up to the values of the independent parameters $(A^+, A_p^-, A_0^-, A_c^-, d_c)$. The procedure follows:

$$-\nabla P_\Sigma(x) = -\nabla \sum_i P_i(x)$$

$$= \sum_i (-\nabla p_i(d_i(x))) \tag{17}$$

$$= -\sum_i \frac{\partial}{\partial d_i(x)} p_i[d_i(x)] \cdot n_i(x) \tag{18}$$

Equation (18) can be summarized by designating the terms in (7, 10, 12) as $a^{(o,w,c)}$ respectively. The terms $a_i^{(o,w,c)} \cdot n_i$, can likewise be denoted $a_i^{(o,w,c)}$, respectively, and represent the

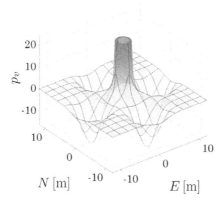

Fig. 5. The potential contour generator of a formation agent.

potential-based decentralized control functions due to the i-th feature.

$$-\nabla P_\Sigma = \overbrace{\sum_i^{\text{obstacles}} \underbrace{a_i^{(o)}(x) \cdot n_i(x)}_{a_i^{(o)}(x)} + \underbrace{a_i^{(w)}(x) \cdot n_w(x)}_{a^{(w)}(x)} + \sum_i^{\text{vertices}} \underbrace{a_i^{(c)}(x) \cdot n_i(x)}_{a_i^{(c)}(x)}} \tag{19}$$

$$= \sum_i^{\text{obstacles}} a_i^{(o)}(x) + a^{(w)}(x) + \sum_i^{\text{vertices}} a_i^{(c)}(x) \tag{20}$$

$$= \sum_i^{\text{obstacles}} \frac{A^+}{d_i(x)^2} \exp\left[\frac{A^+}{d_i(x)}\right] n_i(x) + \min[A_p^- d(x), A_c^-]\frac{w-x}{\|w-x\|}$$
$$+ \sum_i^{\text{vertices}} \frac{A_c^-}{d_c} d_i(x) \exp\left(1 - \frac{d_i(x)^2}{2d_c^2}\right) \tag{21}$$

2.4 Rotor modification

As mentioned in sec. 2.2, the virtual potential approach to guidance is extremely susceptible to the appearance of local minima. A robust and simple approach is needed to assure local minima avoidance.

In terms of the vector field introduced by (2), the analytical solution of which is presented in (20 – 21), stable local minima occur due to the *irrotationality* of the field, rot $\mathbf{E}(x) \overset{\text{id}}{=} 0$.

In order to avoid irrotationality, and thereby local minima, decentralized control functions proposed in (7, 10, 12) are redesigned, adding a *rotor* component:

$$a_i^{(s)} \leftarrow a_i \qquad a_i' \xleftarrow{\text{redef}} a_i^{(s)} + a_i^{(r)} \tag{22}$$

Where:
- a_i' is the redefined *total decentralized control function* due to the i-th feature (the dash will hereafter be omitted),
- $a_i^{(s)}$ is the *stator* decentralized control function as introduced in the preceding section,

denoted with the superscript (s) to contrast it with the newly introduced $a_i^{(r)}$,
$-a_i^{(r)}$ is the *rotor decentralized control function*, all of which are continuous real 2D vector fields over the Euclidean 2-space (mapping $\mathbb{R}^2$ to itself) such that they Jacobians exist wherever each of them is defined.

The introduction of $a_i^{(r)}$ establishes a non-zero rot($\mathbf{E}$) by design, as follows:

$$\text{rot } \mathbf{E}(x) = \sum_i a_i(x) \neq 0$$

$$= \text{rot} \underbrace{\sum_i a_i^{(s)}(x)}_{\overset{\text{id}}{=}0} + \text{rot} \sum_i a_i^{(r)}(x)$$

$$= \text{rot} \sum_i a_i^{(r)}(x) \tag{23}$$

With respect to the way-point, its potential influence on an AUV in this framework must not be prejudiced in terms of the direction of approach. If a decentralized control function of a way-point were augmented with a rotor part, the direction of a_w would deviate from line-of-sight. The same is true of formation cell vertices. Therefore, the only non-zero rotor decentralized control functions are those of *obstacles*. As a result, (23) can be further simplified to:

$$\text{rot } \mathbf{E}(x) = \text{rot} \overset{\substack{\text{obstacles,}\\ \text{w.p.,}\\ \text{vertices}}}{\sum_i} a_i^{(r)}(x) = \text{rot} \overset{\text{obstacles}}{\sum} a_i^{(r)}(x) \tag{24}$$

An individual *obstacle rotor* decentralized control function is defined below:

$$\forall i = \text{enum}(\text{obstacles})$$

$$a(x) = a_r(x)\hat{a}_r(x) \tag{25}$$

$$a_r(x) = \frac{A_i^{(r)}}{d_i(x)^2} \exp\left(\frac{A_i^{(r)}}{d_i(x)}\right) \tag{26}$$

$$\hat{a}_r(x) = \begin{bmatrix} 1 & 0 & 0 \\ 0 & 1 & 0 \\ (0 & 0 & 1) \end{bmatrix} \cdot (r_i(x) \times [n_i(x)|(0)]^{\mathrm{T}}) \tag{27}$$

$$r_i(x) = \left[\frac{w - x_i}{\|w - x_i\|}\middle|(0)\right] \cdot [n_i(x)|(0)]^{\mathrm{T}} \tag{28}$$

$$r_i(x) = \begin{cases} r_i = 1: & n_i(x) \times \left[\frac{v}{\|v\|} - (\frac{v}{\|v\|} \cdot n_i)n_i\middle|(0)\right]^{\mathrm{T}} \\ 0 \leq r_i < 1: & \left[\frac{w - x_i}{\|w - x_i\|}\middle|(0)\right]^{\mathrm{T}} \times [n_i(x)|(0)]^{\mathrm{T}} \\ \text{otherwise}: & \vec{0} \end{cases} \tag{29}$$

Where:
- $A_i^{(r)} \in \mathbb{R} \setminus \{0\}$ is a positive real independent parameter dictating the scale of acceleration perpendicular to the direction of fastest flight from the obstacle,
- $a_i^{(r)} \in \mathbb{R}^+$ is the magnitude of the rotor decentralized control function,

- $\hat{a}_i^{(r)} \in SO^2$ is the direction of the rotor decentralized control function,
- $x_i \in \mathbb{R}^2$ is the center of the i-th obstacle,
- $n_i(x) \in SO^2$ is the unit vector in the direction of fastest flight from the i-th obstacle,
- $r_i(x)$ is the *unit rotor direction generator*, such that $\hat{a}_i^{(r)}(x) \overset{\text{id}}{=} r_i \times n_i(x)$,
- $v \in \mathbb{R}^2$ is the current true over-ground velocity of the AUV (including possible sideslip) projected onto the "flight ceiling".

The rotor decentralized control function and the *total* decentralized control function consisting of the superposition of the rotor and stator parts, are displayed in figure 6.

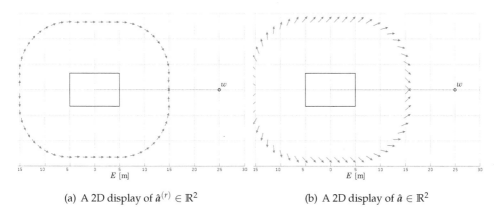

(a) A 2D display of $\hat{a}^{(r)} \in \mathbb{R}^2$ (b) A 2D display of $\hat{a} \in \mathbb{R}^2$

Fig. 6. Direction of the rotor decentralized control function $a_i^{(r)}$ and the two-term $a_i = a_i^{(s)} + a_i^{(r)}$ decentralized control function.

3. Potential framework of formations

The formation introduced by the proposed framework is the *line graph occurring at the tile interfaces of the square tessellation* of $\mathbb{R}^2$, represented in figure 7. Due to a non-collocated nature of AUV motion planning, an important feature of candidate tessellations is that they be *periodic and regular*, which the square tessellation is.

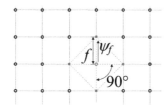

Fig. 7. The square tiling of the plane.

Each AUV whose states are being estimated by the current, i-th AUV, meaning j-th AUV, $j \neq i$) is considered to be a center of a formation cell. The function of the presented framework for potential-based formation keeping is depicted in figure 8. In an unstructured motion of the cooperative group, only a small number of cell vertices attached to j-th AUVs $\forall j \neq i$, if any,

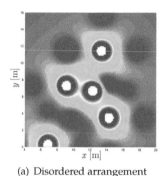

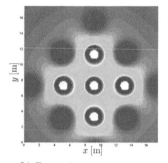

(a) Disordered arrangement   (b) Formation arrangement

Fig. 8. The potential masking of agents out of and in formation.

are *partially* masked by nodes. The i-th AUV is attracted strongest to the closest cell vertex, in line with how attractiveness of a node varies with distance expressed in (11). In the structured case in 8.b), presenting an ideal, undisturbed, non-agitated and stationary formation, all the j-th AUVs in formation are masking each the attractiveness (w.r.t. the i-th AUV) of the vertex they already occupy. At the same time they reinforce the attractiveness of certain unoccupied vertices at the perimeter of the formation. The vertices that attract the i-th AUV the strongest thus become those that result in the most compact formation. Notice in figure 8.b) how certain vertices are colored a deeper shade of blue than others, signifying the lowest potential.

The square formation cell is a cross figure appearing at the interstice of four squares in the tessellation, comprised of the j-th AUV and the four cell vertices attached to it, in the sense that their position is completely determined based on the i-th AUV's local estimation of j-th AUV's position, $(\hat{x}_j^{(i)})$, as in figure 7. The cell vertices are uniquely determined by $\hat{x}_j^{(i)}$ and an independent positive real scaling parameter f.

4. The platform – A large *Aries*-precursor AUV

The vehicle whose dynamic model will be used to demonstrate the developed virtual potential framework is an early design of the NPS[2] *Aries* autonomous underwater vehicle which was resized during deliberations preceding actual fabrication and outfitting. The resulting, smaller *Aries* vehicle has been used in multiple venues of research, most notably (An et al., 1997; Marco & Healey, 2000; 2001). As Marco & Healey (2001) describe, the vehicle whose model dynamics are used has the general body plan of the *Aries*, displayed in figure 9, albeit scaled up. The body plan is that of a chamfered cuboid-shaped fuselage with the bow fined using a nose-cone. The modeled *Aries*-precursor vehicle, the same as the *Aries* itself, as demonstrated in the figure, combines the use of two stern-mounted main horizontal thrusters with a pair of bow- and stern-mounted rudders (four hydrofoil surfaces in total, with dorsal and ventral pairs mechanically coupled), and bow- and stern-mounted elevators.

Healey & Lienard (1993) have designed sliding mode controllers for the *Aries*-precursor vehicle, considering it as a full-rank system with states $\mathbf{x} = [\mathbf{v}^T\ \boldsymbol{\omega}^T\ \mathbf{x}^T\ \Theta] = [u\,v\,w|p\,q\,r|x\,y\,z|\varphi\,\vartheta\,\psi]^T$, relying on the actuators:

$$\mathbf{u}(t) = [\delta_r(t)\ \delta_s(t)\ n(t)]^T \tag{30}$$

[2] Naval Postgraduate School, 700 Dyer Rd., Monterey, CA, USA.

Fig. 9. The *Aries*, demonstrating the body-plan and general type of the model dynamics. Image from the public domain.

Where:
- $\delta_r(t)$ is the stern rudder deflection command in radians,
- $\delta_s(t)$ is the stern elevator planes' command in radians,
- $n(t)$ is the main propellers' revolution rate in rad/s.

4.1 Model dynamics of the vehicle

The dynamics published by Healey and Lienard are used in the HILS[3] in the ensuing sections, and were developed on the grounds of hydrodynamic modelling theory (Abkowitz, 1969; Gertler & Hagen, 1967), exploited to great effect by Boncal (1987). The equations for the six degrees of freedom of full-state rigid-body dynamics for a cuboid-shaped object immersed in a viscous fluid follow, with the parameters expressed in Table 1.

4.1.1 Surge

$$m\left[\dot{u} - vr + wq - x_G(q^2 + r^2) + y_G(pq - \dot{r}) + z_G(pr + \dot{q})\right] = \frac{\rho}{2}L^4\left[X_{pp}p^2 + X_{qq}q^2 + X_{rr}r^2\right.$$

$$\left.+X_{pr}pr\right] + \frac{\rho}{2}L^2\left[X_{\dot{u}}\dot{u} + X_{wq}wq + X_{vp}vp + X_{vr}vr + uq\left(X_{q\delta_s}\delta_s + X_{q\delta_b}/2\delta_{bp} + X_{q\delta_b}/2\delta_{bs}\right)\right.$$

$$\left.+X_{r\delta_r}ur\delta_r\right] + \frac{\rho}{2}L^2\left[X_{vv}v^2 + X_{ww}w^2 + X_{v\delta_r}uv\delta_r + uw\left(X_{w\delta_s}\delta_s + X_{w\delta_b}/2\delta_{bs} + X_{w\delta_b}/2\delta_{bp}\right)\right.$$

$$\left.+u^2\left(X_{\delta_s\delta_s}\delta_s^2 + X_{\delta_b\delta_b}/2\delta_b^2 + X_{\delta_r\delta_r}\delta_r^2\right)\right] - (W - B)\sin\vartheta + \frac{\rho}{2}L^3X_{\delta_s n}uq\delta_s\epsilon(n) + \frac{\rho}{2}L^2\left[X_{w\delta_s n}uw\delta_s\right.$$

$$\left.+X_{\delta_s\delta_s n}u^2\delta_s^2\right]\epsilon(n) + \frac{\rho}{2}L^2u^2X_{prop} \tag{31}$$

---

[3] Hardware-in-the-loop simulation.

4.1.2 Sway

$$m\left[\dot{v} + ur - wp + x_G(pq + \dot{r}) - y_G(p^2 + r^2) + z_G(qr - \dot{p})\right] = \frac{\rho}{2}L^4\left[Y_{\dot{p}}\dot{p} + Y_{\dot{r}}\dot{r} + Y_{pq}pq\right.$$

$$+Y_{qr}qr\right] + \frac{\rho}{2}L^3\left[Y_{\dot{v}}\dot{v} + Y_p up + Y_r ur + Y_{vq}vq + Y_{wp}wp + Y_{wr}wr\right] + \frac{\rho}{2}L^2\left[Y_v uv + Y_{vw}vw\right.$$

$$+Y_{\delta_r}u^2\delta_r\right] - \frac{\rho}{2}\int_{x_{nose}}^{x_{tail}}\left[C_{dy}h(x)(v + xr)^2 + C_{dz}b(x)(w - xq)^2\right]\frac{v + xr}{U_{cf}(x)}dx + (W - B)\cos\vartheta\sin\varphi$$

$$\tag{32}$$

4.1.3 Heave

$$m\left[\dot{w} - uq + vp + x_G(pr - \dot{q}) + y_G(qr + \dot{p}) - z_G(p^2 + q^2)\right] = \frac{\rho}{2}L^4\left[Z_{\dot{q}}\dot{q} + Z_{pp}p^2 + Z_{pr}pr\right.$$

$$+Z_{rr}r^2\right] + \frac{\rho}{2}L^3\left[Z_{\dot{w}}\dot{w} + Z_q uq + Z_{vp}vp + Z_{vr}vr\right] + \frac{\rho}{2}L^2\left[Z_w uw + Z_{vv}v^2 + u^2\left(Z_{\delta_s}\delta_s + Z_{\delta_b/2}\delta_{bs}\right.\right.$$

$$\left.\left.+Z_{\delta_b/2}\delta_{bp}\right)\right] + \frac{\rho}{2}\int_{x_{tail}}^{x_{nose}}\left[C_{dy}h(x)(v + xr)^2 + C_{dx}b(x)(w - xq)^2\right]\frac{w - xq}{U_{cf}(x)}dx$$

$$+(W - B)\cos\vartheta\cos\varphi + \frac{\rho}{2}L^3 Z_{qn}uq\epsilon(n) + \frac{\rho}{2}L^2\left[Z_{wn}uw + Z_{\delta_s n}u\delta_s\right]\epsilon(n)$$

$$\tag{33}$$

4.1.4 Roll

$$I_y\dot{q} + (I_x - I_z)pr - I_{xy}(qr + \dot{p}) + I_{yz}(pq - \dot{r}) + I_{xz}(p^2 - r^2) + m\left[x_G(\dot{w} - uq + vp)\right.$$

$$\left.-z_G(\dot{v} + ur - wp)\right] + \frac{\rho}{2}L^5\left[K_{\dot{p}}\dot{p} + K_{\dot{r}}\dot{r} + K_{pq}pq + K_{qr}qr\right] + \frac{\rho}{2}L^4\left[K_{\dot{v}}\dot{v} + K_p up + K_r ur\right.$$

$$\left.+K_{vq}vq + K_{wp}wp + K_{wr}wr\right] + \frac{\rho}{2}L^3\left[K_v uv + K_{vw}vw + u^2\left(K_{\delta_b/2}\delta_{bp} + K_{\delta_b/2}\delta_{bs}\right)\right]$$

$$+(y_G W - y_B B)\cos\vartheta\cos\varphi - (z_G W - z_B B)\cos\vartheta\sin\varphi + \frac{\rho}{2}L^4 K_{pn}up\epsilon(n)$$

$$+\frac{\rho}{2}L^3 u^2 K_{prop}$$

$$\tag{34}$$

4.1.5 Pitch

$$I_x\dot{p} + (I_z - I_y)qr + I_{xy}(pr - \dot{q}) - I_{yz}(q^2 - r^2) - I_{xz}(pq + \dot{r}) + m\left[y_G(\dot{w} - uq + vp)\right.$$

$$\left.-z_G(\dot{u} - vr + wq)\right] + \frac{\rho}{2}L^5\left[M_{\dot{q}}\dot{q} + M_{pp}p^2 + M_{pr}pr + M_{rr}r^2\right] + \frac{\rho}{2}L^4\left[M_{\dot{w}}\dot{w} + M_q uq\right.$$

$$\left.+M_{vp}vp + M_{vr}vr\right] + \frac{\rho}{2}L^3\left[M_{uw}uw + M_{vv}v^2 + u^2\left(M_{\delta_s}\delta_s + M_{\delta_b/2}\delta_{bs} + M_{\delta_b/2}\delta_{bp}\right)\right]$$

$$-\frac{\rho}{2}\int_{x_{tail}}^{x_{nose}}\left[C_{dy}h(x)(v + xr)^2 + C_{dz}b(x)(w - xq)^2\right]\frac{w + xq}{U_{cf}(x)}x\,dx - (x_G W - x_B B)\cdot$$

$$\cdot\cos\vartheta\cos\varphi - (z_G W - z_B B)\sin\vartheta + \frac{\rho}{2}L^4 M_{qn}qn\epsilon(n) + \frac{\rho}{2}L^3\left[M_{wn}uw\right.$$

$$\left.+M_{\delta_s n}u^2\delta_s\right]\epsilon(n)$$

$$\tag{35}$$

4.1.6 Yaw

$$I_z\dot{r} + (I_y - I_x)pq - I_{xy}(p^2 - q^2) - I_{yz}(pr + \dot{q}) + I_{xz}(qr - \dot{p}) + m[x_G(\dot{v} + ur - wp)$$
$$-y_G(\dot{u} - vr + wq)] + \frac{\rho}{2}L^5[N_{\dot{p}}\dot{p} + N_{\dot{r}}\dot{r} + N_{pq}pq + N_{qr}qr] + \frac{\rho}{2}L^4[N_{\dot{v}}\dot{c} + N_p up$$
$$+N_r ur + N_{vq}vq + N_{wp}wp + N_{wr}wr] + \frac{\rho}{2}L^3[N_v uv + N_{vw}vw + N_{\delta_r}u^2\delta_r] - \frac{\rho}{2}\int_{x_{tail}}^{x_{nose}}[C_{dy} \cdot$$
$$\cdot h(x)(v + xr)^2 + C_{dz}b(x)(w - xq)^2]\frac{w + xq}{U_{cf}(x)}x\,dx + (x_G W - x_B B)\cos\vartheta\sin\varphi + (y_G W - y_B B) \cdot$$
$$\cdot \sin\vartheta + \frac{\rho}{2}L^3 u^2 N_{prop} \tag{36}$$

4.1.7 Substitution terms

$$U_{cf}(x) = \sqrt{(v + xr)^2 + (w - xq)^2} \tag{37}$$

$$X_{prop} = C_{d0}(\eta|\eta| - 1); \quad \eta = 0.012\frac{n}{u}; \quad C_{d0} = 0.00385 \tag{38}$$

$$\epsilon(n) = -1 + \frac{\text{sign}(n)}{\text{sign}(u)} \cdot \frac{\sqrt{C_t + 1} - 1}{\sqrt{C_{t1} + 1} - 1} \tag{39}$$

$$C_t = 0.008\frac{L^2\eta|\eta|}{2.0}; \quad C_{t1} = 0.008\frac{L^2}{2.0} \tag{40}$$

4.2 Control
The *Aries*-precursor's low-level control encompasses three separate, distinctly designed decoupled control loops:

1. Forward speed control by the main propeller rate of revolution,
2. Heading control by the deflection of the stern rudder,
3. Combined control of the pitch and depth by the deflection of the stern elevator plates.

All of the controllers are sliding mode controllers, and the precise design procedure is presented in (Healey & Lienard, 1993). In the interest of brevity, final controller forms will be presented in the ensuing subsections.

4.2.1 Forward speed
The forward speed sliding mode controller is given in terms of a signed squared term for the propeller revolution signal, with parameters (α, β) dependent on the nominal operational parameters of the vehicle, and the coefficients presented in table 1:

$$n(t)|n(t)| = (\alpha\beta)^{-1}\left[\alpha u(t)|u(t)| + \dot{u}_c(t) - \eta_u\tanh\frac{\tilde{u}(t)}{\phi_u}\right] \tag{41}$$

$$\alpha = \frac{\rho L^2 C_d}{2m + \rho L^3 X_{\dot{u}}}; \quad C_d = 0.0034$$

$$\beta = \frac{n_0}{u_0}; \quad n_0 = 52.359\,\frac{\text{rad}}{\text{s}}; \quad u_0 = 1.832\,\frac{\text{m}}{\text{s}} \tag{42}$$

$W = 53.4$ kN	$B = 53.4$ kN	$L = 5.3$ m	$I_x = 13587$ Nms^2
$I_{xy} = -13.58$ Nms^2	$I_{yz} = -13.58$ Nms^2	$I_{xz} = -13.58$ Nms^2	$I_y = 13587$ Nms^2
$I_x = 2038$ Nms^2	$x_G = 0.0$ m	$x_B = 0.0$ m	$y_G = 0.0$ m
$y_B = 0.0$ m	$z_G = 0.061$ m	$z_B = 0.0$ m	$g = 9.81$m/s^2
$\rho = 1000.0$ kg/m^2	$m = 5454.54$ kg		
$X_{pp} = 7.0 \cdot 10^{-3}$	$X_{qq} = -1.5 \cdot 10^{-2}$	$X_{rr} = 4.0 \cdot 10^{-3}$	$X_{pr} = 7.5 \cdot 10^{-4}$
$X_{\dot{u}} = -7.6 \cdot 10^{-3}$	$X_{wq} = -2.0 \cdot 10^{-1}$	$X_{vp} = -3.0 \cdot 10^{-3}$	$X_{vr} = 2.0 \cdot 10^{-2}$
$X_{q\delta_s} = 2.5 \cdot 10^{-2}$	$X_{q\delta_b/2} = -1.3 \cdot 10^{-3}$	$X_{r\delta_r} = -1.0 \cdot 10^{-3}$	$X_{vv} = 5.3 \cdot 10^{-2}$
$X_{ww} = 1.7 \cdot 10^{-1}$	$X_{v\delta_r} = 1.7 \cdot 10^{-3}$	$X_{w\delta_s} = 4.6 \cdot 10^{-2}$	$X_{w\delta_b/2} = 0.5 \cdot 10^{-2}$
$X_{\delta_s\delta_s} = -1.0 \cdot 10^{-2}$	$X_{\delta_b\delta_b/2} = -4.0 \cdot 10^{-3}$	$X_{\delta_r\delta_r} = -1.0 \cdot 10^{-2}$	$X_{q\delta_{sn}} = 2.0 \cdot 10^{-3}$
$X_{w\delta_{sn}} = 3.5 \cdot 10^{-3}$	$X_{\delta_s\delta_{sn}} = -1.6 \cdot 10^{-3}$		
$Y_{\dot{p}} = 1.2 \cdot 10^{-4}$	$Y_{\dot{r}} = 1.2 \cdot 10^{-3}$	$Y_{pq} = 4.0 \cdot 10^{-3}$	$Y_{qr} = -6.5 \cdot 10^{-3}$
$Y_{\dot{v}} = -5.5 \cdot 10^{-2}$	$Y_p = 3.0 \cdot 10^{-3}$	$Y_r = 3.0 \cdot 10^{-2}$	$Y_{vq} = 2.4 \cdot 10^{-2}$
$Y_{wp} = 2.3 \cdot 10^{-1}$	$Y_{wr} = -1.9 \cdot 10^{-2}$	$Y_v = -1.0 \cdot 10^{-1}$	$Y_{vw} = 6.8 \cdot 10^{-2}$
$Y_{\delta_r} = 2.7 \cdot 10^{-2}$			
$Z_{\dot{q}} = -6.8 \cdot 10^{-3}$	$Z_{pp} = 1.3 \cdot 10^{-4}$	$Z_{pr} = 6.7 \cdot 10^{-3}$	$Z_{rr} = -7.4 \cdot 10^{-3}$
$Z_{\dot{w}} = -2.4 \cdot 10^{-1}$	$Z_q = -1.4 \cdot 10^{-1}$	$Z_{vp} = -4.8 \cdot 10^{-2}$	$Z_{vr} = 4.5 \cdot 10^{-2}$
$Z_w = -3.0 \cdot 10^{-1}$	$Z_{vv} = -6.8 \cdot 10^{-2}$	$Z_{\delta_s} = -7.3 \cdot 10^{-2}$	$Z_{\delta_b/2} = -1.3 \cdot 10^{-2}$
$Z_{qn} = -2.9 \cdot 10^{-3}$	$Z_{wn} = -5.1 \cdot 10^{-3}$	$Z_{\delta_{sn}} = -1.0 \cdot 10^{-2}$	
$K_{\dot{p}} = -1.0 \cdot 10^{-3}$	$K_{\dot{r}} = -3.4 \cdot 10^{-5}$	$K_{pq} = -6.9 \cdot 10^{-5}$	$K_{qr} = 1.7 \cdot 10^{-2}$
$K_{\dot{v}} = 1.3 \cdot 10^{-4}$	$K_p = -1.1 \cdot 10^{-2}$	$K_r = -8.4 \cdot 10^{-4}$	$K_{vq} = -5.1 \cdot 10^{-3}$
$K_{wp} = -1.3 \cdot 10^{-4}$	$K_{wr} = 1.4 \cdot 10^{-2}$	$K_v = 3.1 \cdot 10^{-3}$	$K_{vw} = -1.9 \cdot 10^{-1}$
$K_{\delta_b/2} = 0.0$	$K_{pn} = -5.7 \cdot 10^{-4}$	$K_{prop} = 0.0$	
$M_{\dot{q}} = -1.7 \cdot 10^{-2}$	$M_{pp} = 5.3 \cdot 10^{-5}$	$M_{pr} = 5.0 \cdot 10^{-3}$	$M_{rr} = 2.9 \cdot 10^{-3}$
$M_{\dot{w}} = -6.8 \cdot 10^{-2}$	$M_{uq} = -6.8 \cdot 10^{-2}$	$M_{vp} = 1.2 \cdot 10^{-3}$	$M_{vr} = 1.7 \cdot 10^{-2}$
$M_{uw} = 1.0 \cdot 10^{-1}$	$M_{vv} = -2.6 \cdot 10^{-2}$	$M_{\delta_s} = -4.1 \cdot 10^{-2}$	$M_{\delta_b/2} = 3.5 \cdot 10^{-3}$
$M_{qn} = -1.6 \cdot 10^{-3}$	$M_{wn} = -2.9 \cdot 10^{-3}$	$M_{\delta_{sn}} = -5.2 \cdot 10^{-3}$	
$N_{\dot{p}} = -3.4 \cdot 10^{-5}$	$N_{\dot{r}} = -3.4 \cdot 10^{-3}$	$N_{pq} = -2.1 \cdot 10^{-2}$	$N_{qr} = 2.7 \cdot 10^{-3}$
$N_{\dot{v}} = 1.2 \cdot 10^{-3}$	$N_p = -8.4 \cdot 10^{-4}$	$N_r = -1.6 \cdot 10^{-2}$	$N_{vq} = -1.0 \cdot 10^{-2}$
$N_{wp} = -1.7 \cdot 10^{-2}$	$N_{wr} = 7.4 \cdot 10^{-3}$	$N_v = -7.4 \cdot 10^{-3}$	$N_{vw} = -2.7 \cdot 10^{-2}$
$N_{\delta_r} = -1.3 \cdot 10^{-2}$	$N_{prop} = 0.0$		

Table 1. Parameters of the Model Dynamics

It is apparent from the above that the propeller rate of revolution command comprises a term that accelerates the vehicle in the desired measure ($\dot{u}_c(t)$), overcomes the linear drag ($u(t)|u(t)|$), and attenuates the perturbations due to disturbances and process noise ($\tilde{\sigma}_u(t)$).

4.2.2 Heading
The sliding surface for the subset of states governing the vehicle's heading is given below, in (43). The resulting sliding mode controller is contained in (44).

$$\sigma_r = -0.074\tilde{v}(t) + 0.816\tilde{r}(t) + 0.573\tilde{\varphi}(t) \tag{43}$$

$$\delta_r = 0.033v(t) + 0.1112r(t) + 2.58 \tanh \frac{0.074\tilde{v}(r) + 0.816\tilde{r}(t) + 0.573\tilde{\varphi}(t)}{0.1}$$

It should be noted that $\tilde{v}(r)$ seems to imply the possibility of defining some $v_c(t)$ for the vehicle to track. This is impractical. The *Aries*-precursor's thrust allocation and kinematics, nonholonomic in sway, would lead to severe degradation of this sliding mode controller's performance in its main objective – tracking the heading. Lienard (1990) provides a further detailed discussion of this and similar sliding mode controllers.

4.2.3 Pitch and depth

The main objective of the third of the three controllers onboard the *Aries*-precursor HIL simulator, that for the combination of pitch and depth, is to control *depth*. For a vehicle with the holonomic constraints and kinematics of the model used here, this is only possible by using the stern elevators δ_s to pitch the vehicle down and dive. Accordingly, the sliding surface is designed in (44), and the controller in (45).

$$\sigma_z(t) = \tilde{q}(t) + 0.520\tilde{\vartheta}(t) - 0.011\tilde{z}(t) \tag{44}$$

$$\delta_s(t) = -5.143q(t) + 1.070\vartheta(t) + 4\tanh\frac{\sigma_z(t)}{0.4}$$

$$= -5.143q(t) + 1.070\vartheta(t)$$

$$+4\tanh\frac{\tilde{q}(t) + 0.520\tilde{\vartheta}(t) - 0.011\tilde{z}(t)}{0.4} \tag{45}$$

5. Obstacle classification, state estimation and conditioning the control signals

In this section, the issues of obstacle classification will be addressed, giving the expressions for (d_i, n_i) of every type of obstacle considered, which are functions prerequisite to obtaining P_i-s through composition with one of (6, 8, 11). Also, full-state estimation of the AUV (modeled after the NPS *Aries*-precursor vehicle described in the preceding section), $\hat{x} = [\hat{u}\ \hat{v}\ \hat{w}\ \hat{p}\ \hat{q}\ \hat{r}\ \hat{x}\ \hat{y}\ \hat{z}\ \hat{\varphi}\ \hat{\vartheta}\ \hat{\psi}]^T$ will be explored. Realistic plant and measurement noise $(\tilde{n}, \tilde{y})$, which can be expected when transposing this control system from HILS to a real application will be discussed and a scheme for the generation of non-stationary stochastic noise given. Finally, the section will address a scheme for conditioning / clamping the low-level control signals to values and dynamic ranges realizable by the AUV with the *Aries* body-plan. The conditioning adjusts the values in the low-level command vector $c = [a_c\ u_c\ r_c\ \psi_c]^T$ to prevent unfeasible commands which can cause saturation in the actuators and temporary break-down of feedback.

5.1 Obstacle classification

The problem of classification in a 2D waterspace represented by $\mathbb{R}^2$ is a well studied topic. We have adopted an approach based on modeling real-world features after a sparse set of geometrical primitives – circles, rectangles and ellipses.

In the ensuing expressions, $\{x^{int}\}$ will be used for the closed, connected set comprising the interior of the obstacle being described. $\mathcal{T}_i$ shall be a homogeneous, isomorphic coordinate transform from the global reference coordinate system to the coordinate system attached to the obstacle, affixed to the *centroid* of the respective obstacle with a possible rotation by some ψ_i if applicable.

5.1.1 Circles

Circles are the simplest convex obstacles to formulate mathematically. The distance and normal vector to a circle defined by $(x_i \in \mathbb{R}^2, r_i \in \mathbb{R}^+)$, its center and radius respectively, are given below:

$$d_i : \quad \mathbb{R}^2 \setminus \left\{ x^{int} : \left\| x^{int} - x_i \right\| < r_i \right\} \to \mathbb{R}^+$$

$$d_i(x) = \| x - x_i \| - r_i \in \mathbb{R}^+ \tag{46}$$

$$n_i : \quad \mathbb{R}^2 \setminus \left\{ x^{int} : \left\| x^{int} - x_i \right\| < r_i \right\} \to SO^2$$

$$n_i(x) = \frac{x - x_i}{\| x - x_i \|} \tag{47}$$

Robust and fast techniques of classifying 2D point-clouds as circular features are very well understood both in theory and control engineering practice. It is easy to find solid algorithms applicable to hard-real time implementation. Good coverage of the theoretic and practical aspects of the classification problem, solved by making use of the circular Hough transform is given in (Haule & Malowany, 1989; Illingworth & Kittler, 1987; Maitre, 1986; Rizon et al., 2007).

5.1.2 Rectangles

The functions for the distance and normal vector $(d_i(x), n_i(x))$, of a point with respect to a rectangle in Euclidean 2-space defined by $(x_i \in \mathbb{R}^2, a_i, b_i \in \mathbb{R}^+, \psi_i \in [-\pi, \pi))$, the center of the rectangle, the half-length and half-breadth and the angle of rotation of the rectangle's long side w.r.t. the global coordinate system, respectively, are given below:

$$d_i : \quad \mathbb{R}^2 \setminus \left\{ x^{int} : \left\| \begin{bmatrix} a_i & 0 \\ 0 & b_i \end{bmatrix}^{-1} \cdot T_i(x^{int}) \right\|_{\infty} < 1 \right\} \to \mathbb{R}^+$$

$$d_i(x) = \begin{cases} |\hat{\imath} \cdot T_i(x)| < a_i : & |\hat{\jmath} \cdot T_i(x)| - b_i \\ |\hat{\jmath} \cdot T_i(x)| < b_i : & |\hat{\imath} \cdot T_i(x)| - a_i \\ \text{otherwise} : & \left\| |T_i(x)| - \begin{bmatrix} \frac{a_i}{2} & \frac{b_i}{2} \end{bmatrix}^{\mathrm{T}} \right\| \end{cases} \tag{48}$$

$$n_i : \quad \mathbb{R}^2 \setminus \left\{ x^{int} : \left\| \begin{bmatrix} a_i & 0 \\ 0 & b_i \end{bmatrix}^{-1} \cdot T_i(x^{int}) \right\|_{\infty} < 1 \right\} \to SO^2$$

$$n_i(x) = \begin{cases} |\hat{\imath} \cdot T_i(x)| < \frac{a_i}{2} : & T_i^{-1} \left\{ \mathrm{sign} \left[\hat{\jmath} \cdot T_i(x) \right] \hat{\jmath} \right\} \\ |\hat{\jmath} \cdot T_i(x)| < \frac{b_i}{2} : & T_i^{-1} \left\{ \mathrm{sign} \left[\hat{\imath} \cdot T_i(x) \right] \hat{\imath} \right\} \\ \text{otherwise} : & T_i^{-1} \left\{ T_i(x) - \left[a_i \, \mathrm{sign} \left[\hat{\imath} \cdot T_i(x) \right] \; b_i \, \mathrm{sign} \left[\hat{\jmath} \cdot T_i(x) \right] \right]^{\mathrm{T}} \right\} \end{cases} \tag{49}$$

There is a large amount of published work dedicated to the extraction of the features of rectangles from sensed 2D point-clouds. Most of these rely on Hough space techniques (Hough & Powell, 1960) and (Duda & Hart, 1972) to extract the features of distinct lines in an image and determine whether intersections of detected lines are present in the image (He & Li, 2008; Jung & Schramm, 2004; Nguyen et al., 2009).

5.1.3 Ellipses

The method of solving for a distance of a point to an ellipse involves finding the roots of the quartic (57). Therefore, it is challenging to find *explicit analytical solutions*, although some options include Ferrari's method (Stewart, 2003) or algebraic geometry (Faucette, 1996). A computer-based control system can, however, employ a good, numerically stable algorithm to obtain a precise enough solution. The rudimentary part of analytic geometry that formulates the quartic to be solved is given below in (50 - 57).

The equation of the ellipse with the center in the origin and axes aligned with the axes of the coordinate system is:

$$\left(\frac{x}{a}\right)^2 + \left(\frac{y}{b}\right)^2 = 1 \tag{50}$$

The locus of its solutions is the ellipse, $\{x_e = \begin{bmatrix} x_e & y_e \end{bmatrix}^T\}$. The analysis proceeds by considering those $x = \begin{bmatrix} x & y \end{bmatrix}^T \in \mathbb{R}^2$ for which $x - x_e$ is normal to the ellipse. The equation of such a normal is:

$$x_n(\tau) = k\tau + x_e \tag{51}$$

Where $\tau \in \mathbb{R}$ is an independent parameter, the degree of freedom along the line and k is the direction vector of the line, given below:

$$k = \nabla \left\{ \left(\frac{x_e}{a}\right)^2 + \left(\frac{y_e}{b}\right)^2 - 1 \right\} = \begin{bmatrix} \frac{x_e}{a^2} & \frac{y_e}{b^2} \end{bmatrix}^T \tag{52}$$

It follows that if $\tau = t = \arg x$, i.e. $x_n(t) \overset{id}{=} x$. Then, the following manipulation can be made:

$$\begin{bmatrix} x - x_e & y - y_e \end{bmatrix}^T = \begin{bmatrix} \frac{tx_e}{a^2} & \frac{ty_e}{b^2} \end{bmatrix}^T \tag{53}$$

$$\begin{bmatrix} x_e & y_e \end{bmatrix}^T = \begin{bmatrix} \frac{a^2 x}{t+a^2} & \frac{b^2 y}{t+b^2} \end{bmatrix}^T \tag{54}$$

Substituting the right-hand side of (54) into (50), the quartic discussed is obtained as:

$$\left(\frac{ax}{t+a^2}\right)^2 + \left(\frac{by}{t+b^2}\right)^2 = 1 \tag{55}$$

$$(t+b^2)^2 a^2 x^2 + (t+a^2)^2 b^2 y^2 = (t+a^2)^2 (t+b^2)^2 \tag{56}$$

$$(t+a^2)^2 (t+b^2)^2 - (t+b^2)^2 a^2 x^2 - (t+a^2)^2 b^2 y^2 = 0 \tag{57}$$

The greatest root of (57), $\bar{t}$, allows for the calculation of $(d_i(x), n_i(x))$ in (51, 54), as given below:

$$d_i : \quad \mathbb{R}^2 \setminus \left\{ x^{int} : x^T T_i \left\{ \begin{bmatrix} a & 0 \\ b & 0 \end{bmatrix} T_i^{-1}(x) \right\} < 1 \right\} \rightarrow \mathbb{R}^+$$

$$d_i(x) = \|x - x_e\| = \|k\bar{t}\| = \bar{t} \left\| \begin{bmatrix} \frac{a^2 x_e}{\bar{t}+a^2} & \frac{b^2 y_e}{\bar{t}+b^2} \end{bmatrix}^T \right\|$$

$$= \bar{t} \sqrt{\frac{(\bar{t}+b^2)^2 a^4 x_e^2 + (\bar{t}+a^2)^2 b^4 y_e^2}{(\bar{t}+a^2)^2 (\bar{t}+b^2)^2}} \tag{58}$$

$$= \bar{t} \sqrt{\frac{(\bar{t}+b^2)^2 a^4 [\hat{\imath} T_i(x)]^2 + (\bar{t}+a^2)^2 b^4 [\hat{\jmath} T_i(x))]^2}{(\bar{t}+a^2)^2 (\bar{t}+b^2)^2}} \tag{59}$$

$$n_i: \quad \mathbb{R}^2 \setminus \left\{ x^{int} : x^T \mathcal{T}_i \left\{ \begin{bmatrix} a & 0 \\ b & 0 \end{bmatrix} \mathcal{T}_i^{-1}(x) \right\} < 1 \right\} \rightarrow \mathrm{SO}^2$$

$$\mathcal{T}_i n_i^{(i)}(x) = \frac{k}{\|k\|} = \frac{\left[\dfrac{x_e}{\bar{t} + a^2} \quad \dfrac{y_e}{\bar{t} + b^2} \right]^T}{\left\| \left[\dfrac{x_e}{\bar{t} + a^2} \quad \dfrac{y_e}{\bar{t} + b^2} \right]^T \right\|} \tag{60}$$

$$\mathcal{T}_i n_i(x) = \frac{\left[\dfrac{\hat{\imath}\mathcal{T}_i(x)}{\bar{t} + a^2} \quad \dfrac{\hat{\jmath}\mathcal{T}_i(x)}{\bar{t} + b^2} \right]^T}{\left\| \left[\dfrac{\hat{\imath}\mathcal{T}_i(x)}{\bar{t} + a^2} \quad \dfrac{\hat{\jmath}\mathcal{T}_i(x)}{\bar{t} + b^2} \right]^T \right\|} \tag{61}$$

$$n_i(x) = \mathcal{T}_i^{-1} \left\{ \frac{\left[\dfrac{\hat{\imath}\mathcal{T}_i(x)}{\bar{t} + a^2} \quad \dfrac{\hat{\jmath}\mathcal{T}_i(x)}{\bar{t} + b^2} \right]^T}{\left\| \left[\dfrac{\hat{\imath}\mathcal{T}_i(x)}{\bar{t} + a^2} \quad \dfrac{\hat{\jmath}\mathcal{T}_i(x)}{\bar{t} + b^2} \right]^T \right\|} \right\} \tag{62}$$

With the advent of cheap solid-state perception sensors in service robotics and aerial photography in the last decade, publication on fast and robust ellipse-fitting of 2D point clouds has intensified – (Ahn et al., 1999; Jiang et al., 2007; Pilu et al., 1996).

5.2 State estimation

The full state $\mathbf{x} = \begin{bmatrix} \mathbf{v}^T & \boldsymbol{\omega}^T & x^T & \boldsymbol{\Theta}^T \end{bmatrix}^T = \begin{bmatrix} u\,v\,w | p\,q\,r | x\,y\,z | \varphi\,\vartheta\,\psi \end{bmatrix}^T$ of the AUV will be estimated using the Scaled Unscented Transform Sigma-Point Kalman Filter (SP-UKF) introduced by van der Merwe (2004).

The Extended Kalman Filter formulations that feature more prominently in marine control engineering state-of-the-art are capable of estimating the states of nonlinear model dynamics by taking into account only first-order statistics of the states (with possible addition of plant / process noise. EKFs use Jacobians of the nonlinear operator(s) evaluated at the current state estimate. The Unscented Kalman Filters (UKF), in contrast, use the original non-linear model dynamics to propagate *samples* – called *sigma-points*, which are characteristic of the *current estimate of the statistical distribution of the states*, influenced by process and measurement noise. The Kalman gain is evaluated based on the covariance of state hypotheses thus propagated vs. the covariance of the samples characteristic of the current estimate of the statistical distribution of *the measurements*. The Kalman gain will award a higher gain to those measurements for which a significant correlation is discovered between the state and measurement hypotheses and for which the covariance of the measurement hypotheses themselves is relatively small. The algorithm is listed in table 2.

5.3 Measurement and process noise

The AUV is assumed to carry a 4-beam DVL[4] which it can use to record the true 3D speed-over-ground measurement $\mathbf{v} = \begin{bmatrix} u\,v\,w \end{bmatrix}^T$. Furthermore, the AUV carries a 3-axial rate gyro package capable of measuring the body-fixed angular velocities $\boldsymbol{\omega} = \begin{bmatrix} p\,q\,r \end{bmatrix}^T$. A

---

[4] Doppler Velocity Logger.

Table 2. The Scaled Unscented Transform Sigma-Point Kalman Filter Algorithm

0. Parameterization

0.1. Set α, the scaling parameter for the Scaled Unscented Transform

0.2. Set β, the parameter of accentuation of the central estimate

0.3. Set κ, the scaling parameter of the set of sigma-point drawn from the underlying distribution

0.4. $L = 12$, the number of states

0.5. $\lambda = \alpha^2 (L + \kappa) - L$

0.6. $\mathbf{w}_c = \left\{ w_c^{(0)} \ w_c^{(1)} \ \cdots \ w_c^{(2L)} \right\}$,

$w_c^{(0)} = \frac{\lambda}{L+\lambda} + (1 - \alpha^2 + \beta)$,

$w_c^{(1..2L)} = w_m^{(1..2L)} = \frac{1}{2(L+\lambda)}$

0.7. $\mathbf{w}_m = \left\{ w_m^{(0)} \ w_m^{(1)} \ \cdots \ w_m^{(2L)} \right\}$,

$w_m^{(0)} = \frac{\lambda}{L+\lambda}$

1. Initialization

1.1. Set $\hat{x}(0|1) = \bar{x}$
- the initial a priori estimate

1.2. Set $\mathbf{P_x}(k) = \mathrm{E}\left[(\mathbf{x} - \bar{x})(\mathbf{x} - \bar{x})^\mathsf{T} \right]$
- the initial covariance matrix of the estimates

1.3. Set $\mathbf{R}_f$ - the process noise covariance

1.4. Set $\mathbf{R_n}$ - the measurement noise covariance

2. Iteration for $k = 1 \ldots \infty$

2.1. Sigma-points and hypotheses of the states

2.1.1. $\boldsymbol{\mathcal{X}}^-(k) = \{\mathbf{x}^-\} =$
$\left\{ \mathbf{x}(k|k-1) \ \mathbf{x}(k|k-1) + \gamma\sqrt{\mathbf{P_x}(k)} \ \mathbf{x}(k|k-1) - \gamma\sqrt{\mathbf{P_x}(k)} \right\}$

2.2. Time-update

2.2.1. $\boldsymbol{\mathcal{X}}^{-*}(k) = \{\mathbf{x}^{-*}\} = \{\mathbf{F}(\boldsymbol{\mathcal{X}}^-(k))\}$

2.2.2. $\hat{x}^-(k|k-1) = \sum_{i=0}^{2L} w_m^{(i)} \mathbf{x}^{-*(i)}$

2.2.3. $\mathbf{P_x^-} = \sum_i^{2L} w_c^{(i)} \left(\boldsymbol{\mathcal{X}}^{-*(i)} - \hat{x}^-(k|k-1) \right) \left(\boldsymbol{\mathcal{X}}^{-*(i)} - \hat{x}^-(k|k-1) \right)^\mathsf{T}$

2.2.4. Re-draw the hypotheses taking into account the process noise covariance
$X(k|k-1) = \{ \hat{x}^- \ \hat{x}^- + \gamma\sqrt{\mathbf{R_v}} \ \hat{x}^- - \gamma\sqrt{\mathbf{R_v}} \}$

2.2.5. $Y(k|k-1) = \mathbf{H}(X)$

2.2.6. $\hat{y}^- = \sum_{i=0}^{2L} w_m^{(i)} Y^{(i)}$

2.3. Measurement update

2.3.1. $\mathbf{P_y} = \sum_{i=0}^{2L} w_c^{(i)} \left(Y^{(i)} - \hat{y}^- \right) \left(Y^{(i)} - \hat{y}^- \right)^\mathsf{T}$

2.3.2. $\mathbf{P_{xy}} = \sum_{i=0}^{2L} w_c^{(i)} \left(X^{(i)} - \hat{x}^- \right) \left(Y^{(i)} - \hat{y}^- \right)^\mathsf{T}$

2.3.3. $\mathbf{K}(k) = \mathbf{P_{xy}} \mathbf{P_y}^{-1}$

2.3.4. $\hat{x}(k|k) = \hat{x}^-(k|k-1) + \mathbf{K}(k)\left(\mathbf{y} - \hat{y}^- \right)$

2.3.5. $\mathbf{P_x}(k) = \mathbf{P_x^-} - \mathbf{K}(k)\mathbf{P_y}\mathbf{K}(k)^\mathsf{T}$

low-grade commercial USBL[5] system is assumed to provide estimates of $[x \; y]^{\mathrm{T}}$. A fusion of the USBL estimate and the pressure gauge prior to the SP-UKF entry point is assumed to provide a relatively good-quality depth reading of z. A 3-axial middle-market strap-down AHRS[6] is assumed to provide the Tait-Bryan angle readings, $\Theta = [\varphi \; \vartheta \; \psi]^{\mathrm{T}}$.

In the proposed HILS framework, measurement noises should mimic the actual experience during AUV fieldwork operations. Therefore, a noise generator which can produce non-stationary, varying noises is required. It is intended that these noises include errors in the sensor readings whose sources cannot be simply identified by recourse to first-order statistics, and which therefore cannot be easily calibrated (de-biased) for. Additionally, we wish to be able to generate *sporadic irrecoverable faults* i.e. events during which a sensor reading cannot be relied on in any meaningful way.

For this reason, we propose the use of a bank of *Gaussian Markov models* – GMMs, for generating the additive measurement noise. Markov models are stochastic state-machines whose state-switching is governed by random number generators. GMMs ultimately output a normally distributed random number with the statistics dependent on the current state. Means and standard deviations (μ_i, σ_i) of each state i are designed into the GMM. In this chapter, a bank of $12n_{AUV}$ Gaussian Markov models, one for each state of each of the n_{AUV} AUVs is used. All of the GMM states contain separate *univariate rate-limited white noise generators* parameterized by (μ_i, σ_i, n_i), where n_i is the rate limit of the additive measurement noise in the i-th channel.

Relying on the MATLAB normally distributed random number generator invoked by the `randn` command, each state generates a number according to:

$$\tilde{y}_i^-(k) = \mu_i + \sigma_i \cdot \mathtt{randn} \tag{63}$$

$$\tilde{y}_i(k) = \mathrm{sign}(\tilde{y}^-(k) - \tilde{y}(k-1)) \cdot \min\left[\left|\tilde{y}^-(k) - \tilde{y}(k-1)\right|, \frac{n_i}{T}\right] \tag{64}$$

Where T is the sampling time.

To optimize between a realistic nature of the measurement noises and HILS complexity, each of the Markov models in the employed bank contains 6 states, $\{nominal, +reliable, -reliable + unreliable, -unreliable, fault\}$. The 6-state Gaussian Markov models are initialized by a 6-tuple $\mathcal{M} = ((\mu_1, \sigma_1, n_1), \cdots, (\mu_6, \sigma_6, n_6))$ and a 6×6 *transition matrix* $\mathbf{T} = [t_{ij}]$ with t_{ij} being a priori probabilities of switching from state i to state j. The actual parameters used in the HILS simulation are presented in (65 – 74). Before adding them to idealized state measurements, the noise channels are mixed together (correlated) as $\mathbf{y} \leftarrow \mathbf{S}_y\mathbf{y}$ using the matrix $\mathbf{S}_y$ in (75), to mimic the physics of the relevant sensors' interdependence of measurements. Notice that $\mathbf{S}_y$ has a pronounced block-diagonal structure, indicative of the fact that the mentioned sensors (DVL, USBL, AHRS, gyro-compass and rate gyros) output several state measurements each. The correlation between the states measured by a single instrument is more pronounced than the one between measurements of mutually dislocated sensors operating along different physical principles.

$$\mathcal{M}_{\mathbf{y_v}} = \mathcal{M}_{\mathbf{y_u}} = \mathcal{M}_{\mathbf{\bar{y}_v}} = \mathcal{M}_{\mathbf{y_w}} = \begin{cases} \text{State } nominal: & (\mu_1, \sigma_1, n_1) = (0, 0.06, 0.03815) \\ \text{State } \pm reliable: & (\mu_1, \sigma_1, n_1) = (\pm 0.09, 0.11, 0.05) \\ \text{State } \pm unreliable: & (\mu_1, \sigma_1, n_1) = (\pm 0.2981, 0.24, 0.09) \\ \text{State } faulty: & (\mu_1, \sigma_1, n_1) = (\mathtt{NaN}, \mathtt{NaN}, \mathtt{NaN}) \end{cases} \tag{65}$$

---

[5] Ultra-short baseline hydroacoustic localization.
[6] Attitude and heading reference system.

$$\mathbf{T_{yv}} = \mathbf{T_{y_u}} = \mathbf{T_{y_v}} = \mathbf{T_{y_w}} = \begin{bmatrix} 0.7542 & 0.1000 & 0.1000 & 0.0208 & 0.0208 & 0.0042 \\ 0.4739 & 0.3365 & 0.1043 & 0.0379 & 0.0284 & 0.0190 \\ 0.4739 & 0.1043 & 0.3365 & 0.0284 & 0.0379 & 0.0190 \\ 0.3825 & 0.1749 & 0.1749 & 0.0984 & 0.0984 & 0.0710 \\ 0.3825 & 0.1749 & 0.1749 & 0.0984 & 0.0984 & 0.0710 \\ 0.0270 & 0.1622 & 0.1622 & 0.2703 & 0.2703 & 0.1081 \end{bmatrix} \tag{66}$$

$$\mathcal{M}_{y\omega} = \mathcal{M}_{y_p} = \mathcal{M}_{y_q} = \mathcal{M}_{y_r} = \begin{cases} \text{State } nominal: & (\mu_1, \sigma_1, n_1) = \left(0, \frac{\pi}{85}, \frac{\pi}{227.608}\right) \\ \text{State } \pm reliable: & (\mu_1, \sigma_1, n_1) = \left(\pm\frac{\pi}{72}, \frac{\pi}{60}, \frac{\pi}{144.201}\right) \\ \text{State } \pm unreliable: & (\mu_1, \sigma_1, n_1) = \left(\pm\frac{\pi}{21.5}, \frac{\pi}{18.8}, \frac{\pi}{64.454}\right) \\ \text{State } faulty: & (\mu_1, \sigma_1, n_1) = (\texttt{NaN, NaN, NaN}) \end{cases} \tag{67}$$

$$\mathbf{T_{y\omega}} = \mathbf{T_{y_p}} = \mathbf{T_{y_q}} = \mathbf{T_{y_r}} = \begin{bmatrix} 0.5928 & 0.1596 & 0.1596 & 0.0423 & 0.0423 & 0.0033 \\ 0.4978 & 0.2489 & 0.1511 & 0.0356 & 0.0533 & 0.0133 \\ 0.4978 & 0.1511 & 0.2489 & 0.0533 & 0.0356 & 0.0133 \\ 0.5234 & 0.1963 & 0.0561 & 0.0935 & 0.0748 & 0.0561 \\ 0.5234 & 0.0561 & 0.1963 & 0.0748 & 0.0935 & 0.0561 \\ 0.0588 & 0.2941 & 0.2941 & 0.1176 & 0.1176 & 0.1176 \end{bmatrix} \tag{68}$$

$$\mathcal{M}_{yxy} = \mathcal{M}_{y_x} = \mathcal{M}_{y_y} = \begin{cases} \text{State } nominal: & (\mu_1, \sigma_1, n_1) = (0, 1.0, 0.012) \\ \text{State } \pm reliable: & (\mu_1, \sigma_1, n_1) = (\pm1.3, 1.5, 0.06) \\ \text{State } \pm unreliable: & (\mu_1, \sigma_1, n_1) = (\pm3.85, 4.0, 1.28) \\ \text{State } faulty: & (\mu_1, \sigma_1, n_1) = (\texttt{NaN, NaN, NaN}) \end{cases} \tag{69}$$

$$\mathbf{T_{yxy}} = \mathbf{T_{y_x}} = \mathbf{T_{y_y}} = \begin{bmatrix} 0.4809 & 0.1967 & 0.1967 & 0.0601 & 0.0601 & 0.0055 \\ 0.4160 & 0.3200 & 0.1440 & 0.0800 & 0.0320 & 0.0080 \\ 0.4160 & 0.1440 & 0.3200 & 0.0320 & 0.0800 & 0.0080 \\ 0.3689 & 0.2136 & 0.1359 & 0.1942 & 0.0777 & 0.0097 \\ 0.3689 & 0.1359 & 0.2136 & 0.0777 & 0.1942 & 0.0097 \\ 0.0102 & 0.2959 & 0.2959 & 0.1837 & 0.1837 & 0.0306 \end{bmatrix} \tag{70}$$

$$\mathcal{M}_{yz} = \begin{cases} \text{State } nominal: & (\mu_1, \sigma_1, n_1) = (0, 0.08, 0.012) \\ \text{State } \pm reliable: & (\mu_1, \sigma_1, n_1) = (\pm0.11, 0.1208, 0.06) \\ \text{State } \pm unreliable: & (\mu_1, \sigma_1, n_1) = (\pm0.55, 0.71, 1.28) \\ \text{State } faulty: & (\mu_1, \sigma_1, n_1) = (\texttt{NaN, NaN, NaN}) \end{cases} \tag{71}$$

$$\mathbf{T_{yz}} = \begin{bmatrix} 0.5198 & 0.1762 & 0.1762 & 0.0617 & 0.0617 & 0.0044 \\ 0.4020 & 0.4020 & 0.1106 & 0.0503 & 0.0302 & 0.0050 \\ 0.4020 & 0.1106 & 0.4020 & 0.0302 & 0.0503 & 0.0050 \\ 0.3704 & 0.2667 & 0.1481 & 0.1481 & 0.0593 & 0.0074 \\ 0.3704 & 0.1481 & 0.2667 & 0.0593 & 0.1481 & 0.0074 \\ 0.0667 & 0.2000 & 0.2000 & 0.2000 & 0.2000 & 0.1333 \end{bmatrix} \tag{72}$$

$$\mathcal{M}_{y\Theta} = \mathcal{M}_{y_\varphi} = \mathcal{M}_{y_\theta} = \mathcal{M}_{y_\psi} = \begin{cases} \text{State } nominal: & (\mu_1, \sigma_1, n_1) = \left(0, \frac{\pi}{220}, \frac{\pi}{98.05}\right) \\ \text{State } \pm reliable: & (\mu_1, \sigma_1, n_1) = \left(\pm\frac{\pi}{192}, \frac{\pi}{176}, \frac{\pi}{42.60}\right) \\ \text{State } \pm unreliable: & (\mu_1, \sigma_1, n_1) = \left(\pm\frac{\pi}{60}, \frac{\pi}{42}, \frac{\pi}{10}\right) \\ \text{State } faulty: & (\mu_1, \sigma_1, n_1) = (\texttt{NaN, NaN, NaN}) \end{cases} \tag{73}$$

$$\mathbf{T_{y\Theta}} = \mathbf{T_{y_\varphi}} = \mathbf{T_{y_\theta}} = \mathbf{T_{y_\psi}} = \begin{bmatrix} 0.4686 & 0.2301 & 0.2301 & 0.0335 & 0.0335 & 0.0042 \\ 0.4014 & 0.2721 & 0.1769 & 0.1020 & 0.0340 & 0.0136 \\ 0.4014 & 0.1769 & 0.2721 & 0.0340 & 0.1020 & 0.0136 \\ 0.3982 & 0.1403 & 0.0995 & 0.1719 & 0.0995 & 0.0905 \\ 0.3982 & 0.0995 & 0.1403 & 0.0995 & 0.1719 & 0.0905 \\ 0.0526 & 0.1579 & 0.1579 & 0.2105 & 0.2105 & 0.2105 \end{bmatrix} \tag{74}$$

$$S_y = \begin{bmatrix} 1 & 1.0\cdot10^{-4} & 1.0\cdot10^{-4} & 3.0\cdot10^{-5} & 3.0\cdot10^{-5} & 3.0\cdot10^{-5} & 0 & 0 & 0 & 5.0\cdot10^{-7} & 5.0\cdot10^{-7} & 5.0\cdot10^{-7} \\ 1.0\cdot10^{-4} & 1 & 1.0\cdot10^{-4} & 3.0\cdot10^{-5} & 3.0\cdot10^{-5} & 3.0\cdot10^{-5} & 0 & 0 & 0 & 5.0\cdot10^{-7} & 5.0\cdot10^{-7} & 5.0\cdot10^{-7} \\ 1.0\cdot10^{-4} & 1.0\cdot10^{-4} & 1 & 3.0\cdot10^{-5} & 3.0\cdot10^{-5} & 3.0\cdot10^{-5} & 0 & 0 & 0 & 5.0\cdot10^{-7} & 5.0\cdot10^{-7} & 5.0\cdot10^{-7} \\ 3.0\cdot10^{-5} & 3.0\cdot10^{-5} & 3.0\cdot10^{-5} & 1 & 6.0\cdot10^{-4} & 6.0\cdot10^{-4} & 0 & 0 & 0 & 0 & 0 & 0 \\ 3.0\cdot10^{-5} & 3.0\cdot10^{-5} & 3.0\cdot10^{-5} & 6.0\cdot10^{-4} & 1 & 6.0\cdot10^{-4} & 0 & 0 & 0 & 0 & 0 & 0 \\ 3.0\cdot10^{-5} & 3.0\cdot10^{-5} & 3.0\cdot10^{-5} & 6.0\cdot10^{-4} & 6.0\cdot10^{-4} & 1 & 0 & 0 & 0 & 0 & 0 & 0 \\ 0 & 0 & 0 & 0 & 0 & 0 & 1 & 3.0\cdot10^{-4} & 4.0\cdot10^{-6} & 2.0\cdot10^{-5} & 2.0\cdot10^{-5} & 2.0\cdot10^{-5} \\ 0 & 0 & 0 & 0 & 0 & 0 & 3.0\cdot10^{-4} & 1 & 4.0\cdot10^{-6} & 2.0\cdot10^{-5} & 2.0\cdot10^{-5} & 2.0\cdot10^{-5} \\ 0 & 0 & 0 & 0 & 0 & 0 & 4.0\cdot10^{-6} & 2.0\cdot10^{-5} & 1 & 0 & 3.0\cdot10^{-5} & 3.0\cdot10^{-5} \\ 5.0\cdot10^{-7} & 5.0\cdot10^{-7} & 5.0\cdot10^{-7} & 0 & 0 & 0 & 2.0\cdot10^{-5} & 4.0\cdot10^{-6} & 1 & 0 & 0 & 0 \\ 5.0\cdot10^{-7} & 5.0\cdot10^{-7} & 5.0\cdot10^{-7} & 0 & 0 & 0 & 2.0\cdot10^{-5} & 2.0\cdot10^{-5} & 0 & 3.0\cdot10^{-5} & 1 & 3.0\cdot10^{-5} \\ 5.0\cdot10^{-7} & 5.0\cdot10^{-7} & 5.0\cdot10^{-7} & 0 & 0 & 0 & 2.0\cdot10^{-5} & 2.0\cdot10^{-5} & 0 & 3.0\cdot10^{-5} & 3.0\cdot10^{-5} & 1 \end{bmatrix} \tag{75}$$

5.3.1 Sensor fault simulation

The last mentioned state, *"fault"* doesn't generate random additive measurement noises, but outputs NaN[7] values, which are ignored by the SP-UKF. On detecting the NaN value in a measurement channel, the SP-UKF sets the corresponding, i-th row of the column vector of measurements y_i to the value of $\hat{y}_i$. This results in the corresponding rows and columns of the P_y covariance matrix being zero, an update to the corresponding row of the estimate column vector $\hat{x}(k|k)$ being omitted, and the corresponding rows and columns of $P_x(k)$ growing rather than falling. The latter signifies a decrease in the trustworthiness of the estimate, and *compromises the stability of the SP-UKF* should a string of faulty readings continue overlong.

5.3.2 Outlier rejection

In order for the SP-UKF to remain stable and deliver trustworthy state estimates to the feedback of the relevant controllers, outlier measurements are rejected. Rejection of an outlier is dealt with identically to a faulty measurement, the appropriate row of the measurement column vector y being over-written with NaN as if there were a sensor fault occurring in the outlier measurement channel.

Outlier rows of y are considered those for which *any* values, inspected column-wise, fulfill:

$$R_y(k) = y(k)^T \cdot y(k) = \left[r_y^{(i,j)} \right] \tag{76}$$

$$i_{out} = \left\{ \arg_i \left(\exists j, \, r_y^{(i,j)} > 16 \cdot P_x(k) \right) \right\} \tag{77}$$

$$y\left[\{i_{out}\}\right] \overset{\text{redef}}{=} \text{NaN} \tag{78}$$

Where $P_x(k)$ is the covariance matrix of the estimate of the full-state vector rendered by the SP-UKF.

5.3.3 Process noise

The additive process noise is assumed to be multivariate rate-limited white noise without bias. The used covariance matrix is given in (79) and the rate limits in a vector in (80).

$$R_v^{(true)} = \begin{bmatrix} 0.2 & 0.01 & 0.01 & 0 & 0.005 & 0.005 \\ 0.01 & 0.1 & 0.0275 & 0.01 & 0.003 & 0.035 \\ 0.01 & 0.0275 & 0.1 & 0.01 & 0.003 & 0 \\ 0 & 0.01 & 0.01 & 0.0011 & 0.0001 & 0.00015 \\ 0.005 & 0.003 & 0.03 & 0.0001 & 0.002 & 0 \\ 0.005 & 0.035 & 0 & 0.00015 & 0 & 0.0022 \end{bmatrix} \tag{79}$$

$$n_v^{(true)} = \begin{bmatrix} 0.2 & 0.2 & 0.2 & \frac{\pi}{36} & \frac{\pi}{36} & \frac{\pi}{36} \end{bmatrix}^T \tag{80}$$

---

[7] Not a Number.

5.4 Control signals conditioning

The solution of $f_i(k) \leftarrow \mathbf{E}(\hat{x}_{AUV}(k))$, obtained by evaluating (2), or more precisely (29), at the estimate $\hat{x}_{AUV}(k)$ rendered by the SP-UKF, will be used for the formation of commands accepted by the forward speed and heading controllers in (41, 44), $\mathbf{c} = [\dot{u}_c \, u_c \, \dot{\psi}_c \, \psi_c]$.

The task of the low-level control system of the AUV is to try to recreate a motion that would result from applying $f(k)$ to an unconstrained point unit mass, i.e. a holonomic 2D double integrator model, up to the thrust allocation and kinematic and dynamic constraints of the actual vehicle. In the following equations, the sampling with time T of the integration over the $\mathbb{R}^2$ space is assumed in the form of *the Euler backwards formula*. So naively:

$$u(k) = \sqrt{u(k-1)^2 + T^2 f^2 + 2Tu f_\|} \tag{81}$$

$$\dot{u}(k) = \frac{1}{T}(u(k) - u(k-1))$$

$$= \frac{1}{T}\sqrt{u(k-1)^2 + T^2 f^2 + 2Tf_\| u(k-1)} - \frac{u(k-1)}{T} \tag{82}$$

$$\dot{\psi}(k) = \frac{1}{T} \arctan\left(\frac{Tf_\perp}{Tf_\| + u(k-1)}\right) \tag{83}$$

$$\ddot{\psi}(k) = \frac{1}{T}\left[\frac{1}{T}\arctan\left(\frac{Tf_\perp}{Tf_\| + u(k-1)}\right) - \dot{\psi}(k-1)\right] \tag{84}$$

Where:
- $f \overset{\text{id}}{=} \|f\|$ is the norm of the total controlling force, admitting decomposition into $[f_\| \, f_\perp]^T$, the components parallel and perpendicular to the direction of heading of the AUV (given by $\psi(k-1)$, notwithstanding possible *sideslip* resulting from $x_2 = v_2 = v \neq 0$).

At this point, it is assumed that an AUV has a specified *performance envelope* of $(\bar{u}, \bar{\dot{u}}, \bar{\dot{\psi}}, \bar{\ddot{\psi}})$. With these as given independent parameters, the manipulation of (81 - 84) results in the constraints that dictate the admissible ranges to which f needs to be clamped to avoid forcing the low-level controllers of the AUV beyond their normal operating range.

5.4.1 The constraint inequalities

The locus of solutions of (81) for f in easier to visualize in an AUV-fixed coordinate system (one with the origin in x_i, with the x-axis aligned with $u(k-1)\angle\psi(k-1)$). In it, the *admissible solution locus* is a disc offset along the x-axis by $u(k-1)/T$, given by the implicit expression:

$$\left\| \left[f_\| + \frac{u(k-1)}{T} \, f_\perp \right]^T \right\| \leq \frac{\bar{u}}{T} \tag{85}$$

The locus of solutions of (82) is, similarly to the preceding case, an offset *annulus* concentric to the disc described by (85):

$$\frac{u(k-1)}{T} - \bar{\dot{u}} \leq \left\| \left[f_\| + \frac{u(k-1)}{T} \, f_\perp \right]^T \right\| \leq \frac{u(k-1)}{T} + \bar{\dot{u}} \tag{86}$$

The locus of solutions of (83) is an *angular sector of an infinite disk* (a 1-cone) concentric to the previous two loci, expressed in terms of:

$$\left| \arctan \left(\frac{a_\perp}{a_\| + \frac{u(k-1)}{T}} \right) \right| \leq T\overline{\dot{\psi}} \tag{87}$$

The locus of solutions of (84) is likewise an *angular sector of an infinite disk* (a 1-cone) concentric to all the other loci, which satisfies the inequality:

$$T\dot{\psi}(k-1) - \frac{T^2}{2}\overline{\dot{\psi}} \leq \arctan \left(\frac{f_\perp}{f_\| + \frac{u(k-1)}{T}} \right)$$
$$\leq T\dot{\psi}(k-1) + \frac{T^2}{2}\overline{\dot{\psi}} \tag{88}$$

A solution for $f(k)$ is legal if it meets *all* of the criteria stated in (85–88), i.e. if it belongs to a subset of $\mathbb{R}^2$ shaped as an annular sector.

5.4.2 Clamping the total controlling force
If (85 – 88) are not fulfilled, a non-linear procedure for the clamping f to an admissible range is employed. Consequently, the low-level controllers' operating point(s) remain within a quasi-linear vicinity of the sliding surfaces. The procedure is pseudo-coded in Algorithm 3. After clamping by the presented algorithm, (81 – 84) are used to form the commands $\mathbf{c} = \begin{bmatrix} a_c & u_c & r_c & \psi_c \end{bmatrix}^T$ for the low-level controllers 41, 44).

6. Simulation results

Combining the presented virtual potentials framework and the HILS presented in the previous chapters, a full simulation is presented for a group of 4 simulated *Aries*-precursor AUVs cruising in formation.

6.1 Simulation 1
The first simulation presents a cruise in formation down an unobstructed channel in between two obstacles towards the way-point. Figure 10 presents the actual paths traveled by the AUVs. Figure 11 presents the speeds of all four vehicles. The initial dips in the path occur due to the non-holonomic nature of the vehicles' kinematics, due to which they cannot initialize the manoeuvres from zero starting speed that would preserve the initial formation perfectly and still commence navigation to the way-point. This is especially exacerbated by the fact that at near-zero speeds, the control surfaces (δ_r, δ_s) are terribly ineffective. The final dips in the area of the paths around the way-point occur after the AUVs have parked in the stable formation configuration. Near the waypoint and at low speeds, the drift in the state estimates is accentuated by a lack of passive stability provided by AUVs' streamlining at higher speeds. This, in hand with non-holonomic kinematics, causes the vehicles to momentarily break formation. It is only after accumulating enough speed that vehicles can turn within a small enough radius to re-establish the formation. Dips in the path correspond to the dips in the speed graphs for the vehicles, as their commanded speed shoots up again in order to re-establish the formation.

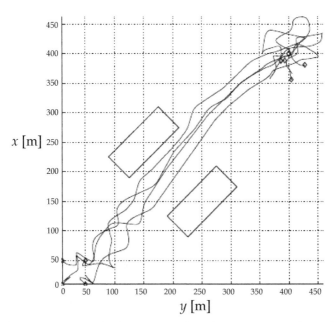

Fig. 10. Paths of the 4 HILS models of AUVs based on the precursor to the NPS *Aries* vehicle cruising in an uncluttered environment.

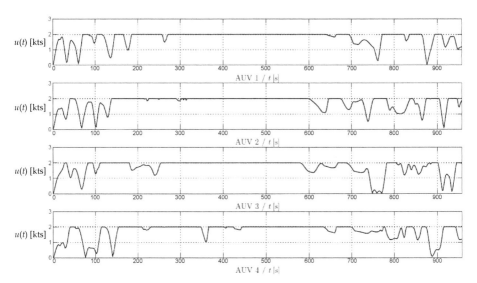

Fig. 11. Speeds of the 4 HILS models of AUVs based on the precursor to the NPS *Aries* vehicle cruising in an uncluttered environment.

if not (85) or (86)
  if $f_\parallel > 0$
   — *this indicates that the required* dominant *behaviour produced*
    by the virtual potentials framework is to accelerate the AUV further in
    the forward direction; Trying to follow through on this behaviour, having in mind
    the operational wisdom concerning real-world missions and experiences,
    it is much safer to give priority to the component that manoeuvres the AUV to the side,
    in order for the AUV to be able to circumnavigate the obstacles it might be heading towards.

$$f_\perp = \text{sign}(f_\perp) \min\{|f_\perp|,\ \bar{u}/T,\ u(k-1)/T + \bar{u}\}$$
$$f_\parallel = \sqrt{\min\{\bar{u}/T,\ u(k-1)/T + \bar{u}\}^2 - f_\perp^2} - u(k-1)/T$$

    — *therefore, require the* $f_\perp$ *component to be as large as admissible,*
     and adjust the $f_\parallel$ *accordingly*

  else
   — *otherwise the required* dominant *behaviour produced*
    by the virtual potentials framework is to decelerate or break the AUV;
    Trying to follow through on this behaviour, it is much safer
    to give priority to the component that decreases the forward speed of
    the AUV, in order for the AUV to avoid colliding with possible obstacles
    it is heading towards.

$$f_\parallel = \max\{f_\parallel,\ -u(k-1)/T,\ -\bar{u}\}$$
$$f_\perp = \text{sign}(f_\perp) \min\{|f_\perp|,\ \sqrt{(\bar{u}/T)^2 - (f_\parallel + u(k-1)/T)^2}\}$$

    — *therefore, require that* $f_\parallel$ *component to be as large as*
     admissible, and adjust the $f_\perp$ *accordingly*
  end if
end if

$\psi_{crit} \overset{\text{def}}{=}$ minimum right-hand side of (87) and (88)

if either of the left-hand sides of (87) or (88) $> \psi_{crit}$
  — *if, after preceding adjustments,* $\angle (f - f \cdot \mathbf{u}(k) / \|\mathbf{u}(k)\|)$ *is inadmissible*
  if $f_\parallel > 0$

$$f_\parallel = f_\parallel \cot \psi_{crit} + f_\perp \cos^2 \psi_{crit} - (u(k-1)/T) \sin^2 \psi_{crit}$$

    — *again, pay attention to the intention of the manoeuvre, and if the dominant behavior is*
     acceleration, decrease the accelerating component of f, $f_\parallel$, *further*

  end if
  $f_\perp = \text{sign}(f_\perp) \cdot (f_\parallel + u(k-1)/T) \tan \psi_{crit}$
  — *clamp the* $f_\perp$ *component to an admissible value*
end if

Table 3. Controlling Force Clamping / Saturation

6.2 Simulation 2

The second simulation presents a cruise in formation down a heavily cluttered corridor defined by two larger rectangular obstacles. Figure 12 presents the actual paths traveled. Figure 13 presents the speeds of the four vehicles.

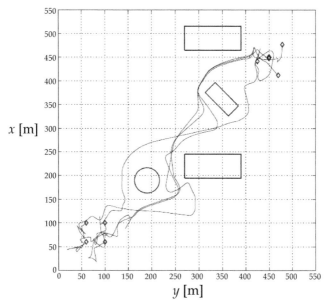

Fig. 12. Paths of the 4 HILS models of AUVs based on the precursor to the NPS *Aries* vehicle cruising in a cluttered environment.

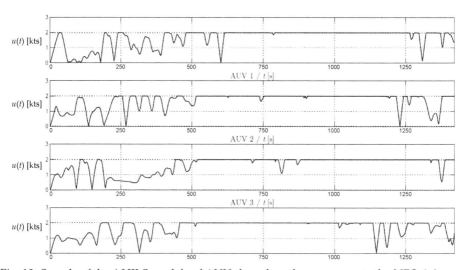

Fig. 13. Speeds of the 4 HILS models of AUVs based on the precursor to the NPS *Aries* vehicle cruising in a cluttered environment.

6.2.1 Cruise phase 1

A cruise is started in a formation. However, very soon the formation encounters the first obstacle. As the leading formation-members are momentarily slowed down before they circumnavigate to either side of the obstacle, the trailing members "pile up" in front of this artificial potential barrier (especially the AUV closest to the origin). This is evident in the dips and temporary confusion before the first, circular obstacle. However, the operational safety approach that is implicitly encapsulated by the *cross-layer design* is preserved. The vehicles break formation, so that one of the vehicle circumnavigates the first obstacle on the left, and the others on the right. Since this produces a significantly different trajectory from the rest of the group, vehicle 1 isn't able to rejoin the formation until much later.

6.2.2 Cruise phase 2

The other 3 vehicles (2, 3 and 4), before being able to restore a formation encounter the first of the two large rectangular obstacles. Note that vehicles 2 and 3 remain in the leader-follower arrangement as evidenced by their closely matching trajectories in this phase of the cruise. The "outrigger" vehicle 4, trying to keep in formation with 2 and 3, encounters the large rectangular obstacle at a bearing much closer to head-on. Therefore, it executes a significant course change manoeuvre, during which it cannot satisfactorily compromise between safe avoidance of the obstacle and staying in formation with 2 and 3. As vehicles 2 and 3 navigate in formation through the strait in between the circular and the first large rectangular obstacle, the leader vehicle 3 starts to manoeuvre to starboard towards the way-point. This manoeuvre causes the formation cell vertex trailing behind vehicle 3 that represents the dominant navigation goal for vehicle 2 to start accumulating speed in excess of what 2 is able to match. This is due to the fact that as 3 swings to starboard, the formation cell vertex "sweeps" through water with a velocity that consists of the sum of linear velocity of vehicle 3 and the tangential velocity contributed by the "arm" of the formation cell f. Therefore, the formation is temporarily completely broken.

6.2.3 Cruise phase 3

However, the breaking of formation between 2 and 3 occurs at such a time that 1 catches up with 2 before 2 gets much farther afield, presenting its trailing cell vertex as a local navigation goal to 2. That is why 2 exhibits a hard break to starboard, trying to form itself up as a follower of vehicle 1. However, just as 2 is completing its formation, vehicle 1, manoeuvres around the final obstacle – the small diagonally presented rectangle. As the trailing cell vertex of 1 is, from 2's viewpoint, shadowed by the obstacle's repulsion, it reorients towards what until then is a secondary navigation goal in its vicinity – the cell vertex of the "latecomer" of Phase 2, vehicle 4. This reorientation is what contributes to 2's "decision" to circumnavigate the diagonal rectangle to starboard, rather than to port, as would be optimal if no formation influences were present. Phase 3 finishes as vehicle 2 is trying to pursue vehicle 4, and vehicle 4 corners the diagonal rectangle, getting away from vehicle 2.

6.2.4 Cruise phase 4

Phase 4 is entered into without formations. This phase is characterized by converging on the way-point, which all the vehicles reach independently, followed by re-establishing the formation. However, an ideal formation is impossible due to operational safety, as no vehicle is "willing" to approach the second large rectangle (towards the top of the figure). This is exacerbated by reduced manoeuvring capabilities, as all vehicles reduce speed in the vicinity of the way-point.

7. Conclusion

The chapter has presented a virtual potentials-based decentralized formation guidance framework that operates in 2D. The framework guarantees the stability of trajectories, convergence to the way-point which is the global navigation goal, and avoidance of salient, hazardous obstacles. Additionally, the framework offers a *cross-layer* approach to subsuming two competing behaviours that AUVs in a formation guidance framework need to combine – a priority of formation maintenance, opposed by operational safety in avoiding obstacles while cruising amidst clutter.

Additionally to the theoretical contribution, a well-rounded functional hardware-in-the-loop system (HILS) for realistic simulative analysis was presented. Multiple layers of realistic dynamic behaviour are featured in the system:

1. A full-state coupled model dynamics of a seaworthy, long-autonomy AUV model based on rigid-body physics and hydrodynamics of viscous fluids like water,

2. An unbiased rate-limited white noise model of the process noise,

3. A non-stationary generator of measurement noise based on Gaussian Markov models with an explicitly included fault-mode,

4. An outlier-elimination scheme based on the evaluation of the state estimate covariance returned by the employed estimator,

5. A Scaled Unscented Transform Sigma-Point Kalman Filter (SP-UKF) that can work either in the filtering mode, or a combination of filtering and pure-prediction mode when faulty measurements are present, utilizing a full-state non-linear coupled AUV model dynamics,

6. A command signal adaptation mechanism that accents operational safety concerns by prioritizing turning manoeuvres while accelerating, and "pure" braking / shedding forward speed when decelerating.

7.1 Further work

Several distinct areas of research, based on the developed HILS framework, remain to ascertain the quality of the presented virtual potential-based decentralized cooperative framework. These are necessary in order to clear the framework for application in costly and logistically demanding operations in the real Ocean environment.

1. Realistically model the representation of knowledge of the other AUVs aboard each AUV locally.

   This can be approached on several fronts:

   (a) Exploring the realistic statistics of the sensing process when applied to sensing other AUVs as opposed to salient obstacles in the waterspace. Exploring and modeling the beam-forming issues arising with mechanically scanning sonars vs. more complex and costlier multi-beam imaging sonars,

   (b) Exploring the increases in complexity (and computer resource management), numerical robustness and stability issues of AUV-local estimation of other AUVs in the formation,

(c) Dealing with the issues of the instability of the "foreign" AUVs' state estimates covariance matrix by one of three ways: *(i)* using synchronous, pre-scheduled hydroacoustic communication. Communication would entail improved estimates coming from on-board the AUVs, where the estimates are corrected by collocated measurement; *(ii)* exploring an on-demand handshake-based communication scheme. Handshaking would be initiated by an AUV polling a team-member for a correction to the local estimate featuring unacceptably large covariance; *(iii)* exploring a predictive communication scheme where the AUVs themselves determine to broadcast their measurements without being polled. This last option needs to involve each AUV continually predicting how well other AUVs are keeping track of its own state estimates.

2. Explore the applicability of the framework to non-conservative, energetic manoeuvring in 3D, i.e. use the same framework to generate commands for the depth / pitch low-level controllers. Explore the behaviour of 3D-formations based on the *honeycombs* (3D tesselations) of the vector space of reals.

8. References

Abkowitz, M. (1969). *Stability and Motion Control of Ocean Vehicles*, MIT Press Cambridge MA USA.

Ahn, S., Rauh, W. & Recknagel, M. (1999). Ellipse fitting and parameter assessment of circular object targets for robot vision, *Intelligent Robots and Systems 1999. IROS '99 Proceedings. 1999 IEEE/RSJ International Conference on*, Vol. 1, pp. 525 –530.

Allen, B., Stokey, R., Austin, T., Forrester, N., Goldborough, R., Purcell, M. & von Alt, C. (1997). REMUS A Small Low Cost AUV system description field trials and performance results, *OCEANS 1997 Conference Record (IEEE)*, Vol. 2, pp. 994–100.

Allen, T., Buss, A. & Sanchez, S. (2004). Assessing obstacle location accuracy in the REMUS unmanned underwater vehicle, *Proceedings - Winter Simulation Conference*, Vol. 1, pp. 940–948.

An, P., Healey, A., Park, J. & Smith, S. (1997). Asynchronous data fusion for auv navigation via heuristic fuzzy filtering techniques, *OCEANS '97. MTS/IEEE Conference Proceedings*, Vol. 1, pp. 397 –402 vol.1.

Barisic, M., Vukic, Z. & Miskovic, N. (2007a). Kinematic simulative analysis of virtual potential field method for AUV trajectory planning, *in* K. Valavanis & Z. Kovacic (eds), *15th Mediterranean Conference on Control and Automation, 2007. Proceedings of the*, p. on CD.

Barisic, M., Vukic, Z. & Miskovic, N. (2007b). A kinematic virtual potentials trajectory planner for AUV-s, *in* M. Devy (ed.), *6th IFAC Symposium on Intelligent Autonomous Vehicles, 2007. Proceedings of the*, p. on CD.

Bildberg, R. (2009). Editor's foreword, *in* B. R. (ed.), *Proceedings of the 16th International Symposium on Unmanned Untethered Submersibles Technology*, Autonomous Undersea Systems Institute. on CD.

Boncal, R. (1987). *A Study of Model Based Maneuvering Controls for Autonomous Underwater Vehicles*, Master's thesis, Naval Postgraduate School Monterey CA USA.

Carder, K., Costello, D., Warrior, H., Langebrake, L., Hou, W., Patten, J. & Kaltenbacher, E. (2001). Ocean-science mission needs: Real-time AUV data for command control and model inputs, *IEEE Journal of Oceanic Engineering* 26(4): 742–751.

Clegg, D. & Peterson, M. (2003). User operational evaluation system of unmanned underwater vehicles for very shallow water mine countermeasures, *OCEANS 2003 Conference Record (IEEE)*, Vol. 3, pp. 1417–1423.

Curtin, T., Bellingham, J., Catipovic, J. & Webb, D. (1993). Autonomous oceanographic sampling networks, *Oceanography* 6(3): 86–94.

Duda, R. O. & Hart, P. E. (1972). Use of the Hough transformation to detect lines and curves in pictures, *Communications of the ACM* 15(1): 11–15.

Eisman, D. (2003). Navy ships seize boats carrying mines in Iraqi port, *The Virginian-pilot* .

Farrell, J., S., P. & Li, W. (2005). Chemical plume tracing via an autonomous underwater vehicle, *IEEE Journal of Oceanic Engineering* 30(2): 428–442.

Faucette, W. M. (1996). A geometric interpretation of the solution of the general quartic polynomial, *The American Mathematical Monthly* 103(1): 51–57.

Gertler, M. & Hagen, G. (1967). Standard Equations of Motion for Submarine Simulation, *Technical report*, David W. Taylor Naval Ship Research and Development Center Bethesda MD USA.

Haule, D. & Malowany, A. (1989). Object recognition using fast adaptive Hough transform, *Communications Computers and Signal Processing 1989. Conference Proceedings IEEE Pacific Rim Conference on*, pp. 91 –94.

He, Y. & Li, Z. (2008). An effective approach for multi-rectangle detection, *Young Computer Scientists 2008. ICYCS. The 9th International Conference for*, pp. 862 –867.

Healey, A. & Lienard, D. (1993). Multivariable sliding mode control for autonomous diving and steering of unmanned underwater vehicles, *Oceanic Engineering IEEE Journal of* 18(3): 327 –339.

Hough, P. & Powell, B. (1960). A method for faster analysis of bubble chamber photographs, Vol. 18, Italian Physical Society, pp. 1184 –1191.

Illingworth, J. & Kittler, J. (1987). The adaptive Hough transform, *Pattern Analysis and Machine Intelligence IEEE Transactions on* PAMI-9(5): 690 –698.

Jiang, X., Huang, X., Jie, M. & Yin, H. (2007). Rock detection based on 2D maximum entropy thresholding segmentation and ellipse fitting, *Robotics and Bioimetics 2007. ROBIO 2007. IEEE International Conference on*, pp. 1143 –1147.

Jung, C. & Schramm, R. (2004). Rectangle detection based on a windowed Hough transform, *Computer Graphics and Image Processing 2004. Proceedings, 17th Brazilian Symposium on*, pp. 113 – 120.

Lienard, D. (1990). *Sliding Mode Control for Multivariable AUV Autopilots*, Master's thesis, Naval Postgraduate School Monterey CA USA.

Maitre, H. (1986). Contribution to the prediction of performances of the Hough transform, *Pattern Analysis and Machine Intelligence IEEE Transactions on* PAMI-8(5): 669 –674.

Marco, D. & Healey, A. (2000). Current developments in underwater vehicle control and navigation: the NPS Aries AUV, *OCEANS Conference record (IEEE)*, Vol. 2, pp. 1011–1016.

Marco, D. & Healey, A. (2001). Command control and navigation experimental results with the NPS Aries AUV, *IEEE Journal of Oceanic Engineering* 26(4): 466–476.

Nguyen, T., Pham, X. & Jeon, J. (2009). Rectangular object tracking based on standard Hough transform, *Robotics and Biomimetics 2008. ROBIO IEEE International Conference on*, pp. 2098 –2103.

Pang, S. (2006). Development of a guidance system for AUV chemical plume tracing, *OCEANS 2006 Conference Record (IEEE)*, pp. 1–6.

Pang, S., Arrieta, R., Farell, J. & Li, W. (2003). AUV reactive planning: Deepest Point, *OCEANS 2009 Conference Record (IEEE)*, Vol. 4, pp. 2222–2226.

Papoulias, F., Cristi, R., Marco, D. & Healey, A. (1989). Modeling, sliding mode control design, and visual simulation of auv dive plane dynamic response, *Unmanned Untethered Submersible Technology, 1989. Proceedings of the 6th International Symposium on*, pp. 536 –547.

Pilu, M., Fitzgibbon, A. & Fisher, R. (1996). Ellipse-specific direct least-square fitting, *Image Processing 1996. Proceedings. International Conference on*, Vol. 3, pp. 599 –602.

Plueddemann, A., Packard, G., Lord, J. & Whelan, S. (2008). Observing Arctic Coastal Hydrography Using the REMUS AUV, *Autonomous Underwater Vehicles 2008. AUV 2008. IEEE/OES*, pp. 1–4.

Rizon, M., Yazid, H. & Saad, P. (2007). A comparison of circular object detection using Hough transform and chord intersection, *Geometric Modeling and Imaging 2007. GMAI '07*, pp. 115 –120.

Stewart, I. (2003). *Galois Theory*, Chapman & Hall / CRC Press, 6000 Broken Sound Pkwy. Suite 300 Boca Raton FL 33487 USA.

Tuohy, S. T. (1994). A simulation model for AUV navigation, *Proceedings of the 1994 Symposium on Autonomous Underwater Vehicle Technology*, pp. 470–478.

US Navy (2004). The US Navy Unmanned Underwater Vehicles Master Plan, `http://www.navy.mil/navydata/technology/uuvmp.pdf`.

van der Merwe, R. (2004). *Sigma-Point Kalman Filters for Probabilistic Inference in Dynamic State-Space Models*, PhD thesis, OGI School of Science & Engineering, Oregon Health & Science University, Portland, OR, USA.

6

Real-Time Optimal Guidance and Obstacle Avoidance for UMVs

Oleg A. Yakimenko and Sean P. Kragelund
Naval Postgraduate School Monterey, CA
USA

1. Introduction

The single most important near-term technical challenge of developing an autonomous capability for unmanned vehicles is to assess and respond appropriately to near-field objects in the path of travel. For unmanned aerial vehicles (UAVs), that near field may extend to several nautical miles in all directions, whereas for unmanned ground and maritime vehicles, the near field may only encompass a few dozen yards directly ahead of the vehicle. Nevertheless, when developing obstacle avoidance (OA) manoeuvres it is often necessary to implement a degree of deliberative planning beyond simply altering the vehicle's trajectory in a reactive fashion. For unmanned maritime vehicles (UMVs) the ability to generate near-optimal OA trajectories in real time is especially important when conducting sidescan sonar surveys in cluttered environments (e.g., a kelp forest or coral reef), operations in restricted waterways (e.g., rivers or harbours), or performing feature-based, terrain-relative navigation, to name a few. For example, a primary objective of sidescan sonar surveys is 100% area coverage while avoiding damage to the survey vehicle. Ideally, a real-time trajectory generator should minimize deviations from the pre-planned survey geometry yet also allow the vehicle to retarget areas missed due to previous OA manoeuvres. Similarly, for operations in restricted waterways, effective OA trajectories should incorporate all known information about the environment including terrain, bathymetry, water currents, etc.

In the general case, this OA capability should be incorporated into an onboard planner or trajectory generator computing optimal (or near-optimal) feasible trajectories faster than in real time. For unmanned undersea vehicles (UUVs) the planner should be capable of generating full, three-dimensional (3D) trajectories, however some applications may require limiting the planner's output to two-dimensions (2D) for vertical-plane or horizontal-plane operating modes. For unmanned surface vehicles (USVs) the latter case is the only mode of operations.

Consider a typical hardware setup consisting of a UUV augmented with an autopilot (Fig.1). The autopilot not only stabilizes the overall system, but also enables vehicle control at a higher hierarchical level than simply changing a throttle setting $\delta_T(t)$, or deflecting stern plane $\delta_s(t)$ or rudder $\delta_r(t)$ angles.

In Fig.1, $\mathbf{x}^{WP}$, $\mathbf{y}^{WP}$, $\mathbf{z}^{WP}$ are the vectors defining x, y, and z coordinates of some points in the local tangent (North-East-Down (NED)) plane for waypoint navigation. Alternatively a typical autopilot may also accept some reference flight-path angle $\gamma(t)$ (or altitude/depth) command and heading $\Psi(t)$ (or yaw angle $\psi(t)$), respectively. The motion sensors, accelerometers, and rate gyros measure the components of inertial acceleration, $\ddot{x}_I(t)$, $\ddot{y}_I(t)$ and $\ddot{z}_I(t)$, and angular velocity – roll rate $p(t)$, pitch rate $q(t)$, and yaw rate $r(t)$.

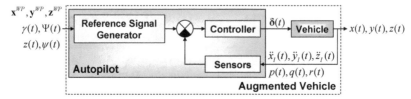

Fig. 1. A. UUV augmented with an autopilot

A trajectory generator would consider an augmented UUV as a new plant (Fig.2) and provide this plant with the necessary inputs based on the mission objectives (final destination, time of arrival, measure of performance, etc.). Moreover, the reference signals, $\gamma(t)$ and $\Psi(t)$, are to be computed dynamically (once every few seconds) to account for disturbances (currents, etc.) and newly detected obstacles.

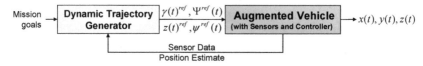

Fig. 2. Providing an augmented UUV with a reference trajectory

Ideally, the trajectory generator software should also produce the control inputs $\delta^{ref}(t)$ corresponding to the feasible reference trajectory (Fig.3) (Basset et al., 2008). This enhanced setup assures that the inner-loop controller deals only with small errors. (Of course this setup is only viable if the autopilot accepts these direct actuator inputs.)

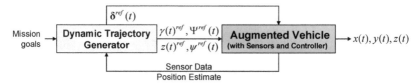

Fig. 3. Providing an augmented UUV with the reference trajectory and reference controls

The goal of this chapter is to present the dynamic trajectory generator developed at the Naval Postgraduate School (NPS) for the UMVs of the Center for Autonomous Vehicle Research (CAVR) and show how the OA framework is built upon it. Specifically, Section 2 formulates a general feasible trajectory generation problem, followed by Section 3, which introduces the general ideas behind the proposed framework for solving this problem that utilizes the inverse dynamics in the virtual domain (IDVD) method. Section 4 considers simplifications that follow from reducing the general spatial problem to two planar subcases. Section 5 describes the REMUS UUV and SeaFox USV and their forward looking sonar (FLS) systems employed for OA research at NPS. Section 6 addresses path-following considerations and practical implementation details for tracking nonlinear trajectories with conventional vehicle autopilots. Section 7 presents results from computer simulations and field experiments for several different scenarios which benefit from faster-than-real-time computation of near-optimal trajectories.

2. Problem formulation

Let us consider the most general case and formulate an optimization problem for computing collision-free trajectories in 3D (it can always be reduced to a 2D problem by eliminating two states). We will be searching within a set of admissible trajectories described by the state vector

$$\mathbf{z}(t) = \left[x(t), y(t), z(t), u(t), v(t), w(t)\right]^T \in S, \ S = \left\{\mathbf{z}(t) \in Z^6 \subset E^6\right\}, \ t \in \left[t_0, t_f\right] \tag{1}$$

where the components of the velocity vector – surge u, sway v, and heave w, defined in the body frame $\{b\}$ – are added to the UUV NED coordinates x, y and z ($z = 0$ at the surface and increases in magnitude with depth). While many UUVs are typically programmed to operate at a constant altitude above the ocean floor, it is still preferable to generate vertical trajectories in the NED local tangent plane because the water surface is a more reliable absolute reference datum than a possibly uneven sea floor. In general, however, it is a trivial matter to convert the resulting depth trajectory $z(t)$ to an altitude trajectory $h(t)$ for vehicles equipped with both altitude and depth sensors. Section 6.2 describes such practical considerations in detail.

The admissible trajectories should satisfy the set of ordinary differential equations describing the UUV kinematics

$$\begin{bmatrix} \dot{x}(t) \\ \dot{y}(t) \\ \dot{z}(t) \end{bmatrix} = {}_b^u R \begin{bmatrix} u(t) \\ v(t) \\ w(t) \end{bmatrix} \tag{2}$$

In (2) ${}_b^u R$ is the rotation matrix from the body frame $\{b\}$ to the NED frame $\{u\}$, defined using two Euler angles, pitch $\theta(t)$ and yaw $\psi(t)$, and neglecting a roll angle as

$$ {}_b^u R(t) = \begin{bmatrix} \cos\psi(t)\cos\theta(t) & -\sin\psi(t) & \cos\psi(t)\sin\theta(t) \\ \sin\psi(t)\cos\theta(t) & \cos\psi(t) & \sin\psi(t)\sin\theta(t) \\ -\sin\theta(t) & 0 & \cos\theta(t) \end{bmatrix} \tag{3}$$

Although we are not going to exploit it in this study, admissible trajectories should also obey UUV dynamic equations describing translational and rotational motion. This means that the following linearized system holds for the vector $\varsigma(t)$, which includes speed components u, v, w (being a part of our state vector $\mathbf{z}(t)$) and angular rates p, q, r:

$$\dot{\varsigma}(t) = \mathbf{A}\varsigma(t) + \mathbf{B}\delta(t) \tag{4}$$

Here $\mathbf{A}$ and $\mathbf{B}$ are the state and control matrices and $\delta = [\delta_T, \delta_s, \delta_r]^T$ is the control vector (Healey, 2004).

Next, the admissible trajectories (1) should satisfy the initial and terminal conditions

$$\mathbf{z}(t_0) = \mathbf{z}_0, \ \mathbf{z}(t_f) = \mathbf{z}_f \tag{5}$$

Finally, certain constraints should be obeyed by the state variables, controls and their derivatives. For example, in the case of a UUV these can include obvious constraints on the UUV depth:

$$z_{min} \le z(t) \le z_{max} \tag{6}$$

where $z_{max}(x, y)$ describes a programmed operational depth limit. For vehicles programmed to operate at some nominal altitude above the sea floor, the $z_{max}(x, y)$ constraint can be converted into a minimum altitude $h_{min}(x, y)$ constraint as described in Section 6.2.

A 3D OA requirement can be formulated as

$$[x(t); y(t); z(t)] \cap \Re = 0 \tag{7}$$

where $\Re$ is the set of all known obstacle locations. The constraints are usually imposed not only on the controls themselves $|\delta| \le \delta_{max}$ but on their time derivatives as well $|\dot{\delta}| \le \dot{\delta}_{max}$ to account for actuator dynamics. Knowing the system's dynamics (4) (or simply complying with the autopilot specifications), these latter constraints can be elevated to the level of the reference signals, for instance

$$|\theta(t)| \le \theta_{max} \ \text{and} \ |\dot{\psi}(t)| \le \dot{\psi}_{max} \tag{8}$$

The objective is to find the best trajectory and corresponding control inputs that minimize some performance index J. Typical performance index specifications include: i) minimizing time of the manoeuvre $t_f - t_0$, ii) minimizing the distance travelled to avoid the obstacle(s), and iii) minimizing control effort or energy consumption. In addition, the performance index may include some "exotic" constraints dictated by a sensor payload. For example, a UUV may require vehicle trajectories which point a fixed FLS at specified terrain features or minimize vehicle pitch motion in order to maintain level, horizontal flight along a survey track line for accurate synthetic aperture sonar imagery (Horner et al., 2009).

Before we proceed with the development of the control algorithm, it should be noted that quite often the UUV surge velocity is assumed to be constant, $u(t) \equiv U_0$, to provide enough control authority in two other channels. This uniquely defines a throttle setting $\delta_T(t)$, and leaves only two control inputs, $\delta_s(t)$ and $\delta_r(t)$, for altering the vehicle's trajectory. It also allows us to consider matrices $\mathbf{A}$ and $\mathbf{B}$ in (4) to be constant (time- and states-independent). If this assumption is not required, inverting kinematic and dynamic equations will differ slightly from the examples presented in the next section. However, the general ideas of the proposed approach remain unchanged.

3. Real-time near-optimal guidance

For the dynamic trajectory generator shown in Figs. 2 and 3, it is advocated to use the direct-method-based IDVD (Yakimenko, 2000). The primary rationale is that this approach features several important properties required for real-time implementations: i) the boundary conditions including high-order derivatives are satisfied *a priori*; ii) the resulting control commands are smooth and physically realizable, iii) the method is very robust and is not sensitive to small variations in input parameters, iv) any compound performance index can be used during optimization. Moreover, this specific method uses only a few variable parameters, thus ensuring that the iterative process during optimization converges very fast compared to other direct methods. The IDVD-based trajectory generator consists of several blocks. The goal of the first block, to be discussed next, is to produce a candidate trajectory, which satisfies the boundary conditions.

3.1 Generating a candidate trajectory

Again, consider the most general case of a UUV operating in 3D (as opposed to a USV). Suppose that each coordinate x_i, $i = 1,2,3$ of the candidate UUV trajectory is represented as a polynomial of degree M of some abstract argument τ, the virtual arc

$$x_i(\tau) = \sum_{k=0}^{M} a_{ik} \tau^k, \tag{9}$$

(for simplicity of notation we assume $x_1(\tau) \equiv x(\tau)$, $x_2(\tau) \equiv y(\tau)$ and $x_3(\tau) \equiv z(\tau)$). In general, analytic expressions for the trajectory coordinates can be constructed from any combination of basis functions to produce a rich variety of candidate trajectories. For example, a combination of monomials and trigonometric functions was utilized in (Yakimenko & Slegers, 2009).

As discussed in (Yakimenko, 2000; Horner & Yakimenko, 2007) the degree M is determined by the number of boundary conditions that must be satisfied. Specifically, it should be greater or equal to the number of preset boundary conditions but one. In general the desired trajectory includes constraints on the initial and final position, velocity and acceleration: x_{i0}, x_{if}, x'_{i0}, x'_{if}, x''_{i0}, x''_{if}. In this case the minimal order of polynomials (9) is 5, because all coefficients in (9) will be uniquely defined by these boundary conditions, leaving the "length" of the virtual arc τ_f as the only varied parameter. For more flexibility in the candidate trajectory, additional varied parameters can be obtained by increasing the order of the polynomials (9). For instance, using seventh-order polynomials will introduce two more varied parameters for each coordinate expression. Rather than varying two coefficients in these extended polynomials directly, we will vary the initial and final jerk, x'''_{i0} and x'''_{if}, respectively. In this case, coefficients a_{ik} in (9) can be determined by solving the obvious system of linear algebraic equations equating polynomials (9) to x_{i0}, x_{if}, x'_{i0}, x'_{if}, x''_{i0}, x''_{if}, x'''_{i0} and x'''_{if} at two endpoints ($\tau = 0$ and $\tau = \tau_f$) (Yakimenko, 2000, 2008).

By construction, the boundary conditions (5) will be satisfied unconditionally for any value of the final arc τ_f. However, varying τ_f will alter the shape of the candidate trajectory. Figure 4 demonstrates a simple example whereby a UUV operating 2m above the sea floor at 1.5m/s must perform a pop-up manoeuvre to avoid some obstacle. Even with a single varied parameter, changing the value of τ_f allows the UUV to avoid obstacles of different heights. Similar trajectories could be produced solely in the horizontal plane or in all three dimensions. It should be pointed out that even at this stage infeasible candidate trajectories will be ruled out. (In Fig.4 the trajectory requiring the UUV to jump out of the water is infeasible because it violates the constraint (6).)

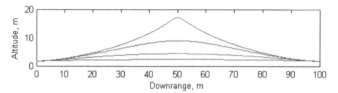

Fig. 4. Varying the candidate trajectory while changing the value of τ_f

With six free parameters, which in our case are components of the initial and final jerk (x_{i0}''' and x_{if}''', $i = 1,2,3$) the trajectory generator can change the overall shape of the trajectory even further. To this end, Fig.5 illustrates candidate trajectories for a UUV avoiding a 10m obstacle located between its initial and final points. These trajectories were generated by varying just two components of the jerk, x_{30}''' and x_{3f}''', and minimizing τ_f. This additional flexibility can produce trajectories which satisfy operational constraints (6), as well as OA constraints (7).

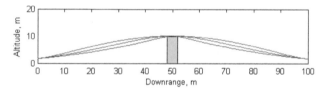

Fig. 5. Candidate trajectories obtained by varying the terminal jerks

The selection of a specific trajectory will be based upon whether the trajectory is feasible (satisfies constraints (8)) and if so, whether it assures the minimum value of the performance index calculated using the values of the vehicle states (and controls) along that trajectory. As an example, Fig.6 presents collision-free solutions for two different locations of a 10m-tall obstacle when five varied parameters, x_{10}''', x_{1f}''', x_{30}''', x_{3f}''' and τ_f, are optimized to assure feasible minimum-path-length trajectories

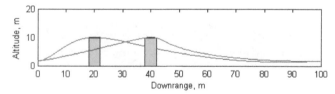

Fig. 6. Examples of minimum-path-length trajectories

Now, let us address the reason for choosing some abstract parameter τ as an argument for the reference functions (9) rather than time or path length, which are commonly used. Assume for a moment that $\tau \equiv t$. In this case, once we determine the trajectory we unambiguously define a speed profile along this trajectory as well, since

$$V(t) = \sqrt{u(t)^2 + v(t)^2 + w(t)^2} = \sqrt{\dot{x}(t)^2 + \dot{y}(t)^2 + \dot{z}(t)^2} \tag{10}$$

Obviously, we cannot allow this to happen because we want to vary the speed profile independently. With the abstract argument τ this becomes possible via introduction of a speed factor λ such that

$$\lambda(\tau) = \frac{d\tau}{dt}. \tag{11}$$

Now instead of (10) we have

$$V(\tau) = \lambda(\tau)\sqrt{x'(\tau)^2 + y'(\tau)^2 + z'(\tau)^2} \tag{12}$$

and by varying $\lambda(\tau)$ we can achieve any desired speed profile.

The capability to satisfy higher-order derivatives at the trajectory endpoints, specifically at the initial point, allows continuous regeneration of the trajectory to accommodate sudden changes like newly discovered obstacles. As an example, Fig.7 demonstrates a scenario whereby a UUV executing an OA manoeuvre discovers a second obstacle and must generate a new trajectory beginning with the current vehicle states and control values (up to the second-order derivatives of the states). The suggested approach enables this type of continuous trajectory generation and ensures smooth, non-shock transitions.

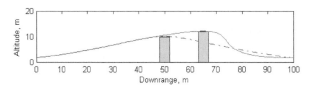

Fig. 7. Example of dynamic trajectory reconfiguration

3.2 Inverse dynamics

The second key block inside the dynamic trajectory generator in Figs. 2 and 3 accepts the candidate trajectory as an input and computes the components of the state vector and control signals required to follow it. In this way we can ensure that each candidate trajectory does not violate any constraints (including those of (8)).

First, using the following relation for any parameter ζ,

$$\dot{\zeta}(\tau) = \frac{d\zeta}{d\tau}\frac{d\tau}{dt} = \zeta'(\tau)\lambda(\tau) \tag{13}$$

we convert kinematic equations (2) into the τ domain

$$\lambda(\tau)\begin{bmatrix} x'(\tau) \\ y'(\tau) \\ z'(\tau) \end{bmatrix} = {}_b^u R \begin{bmatrix} U_0 \\ v(\tau) \\ w(\tau) \end{bmatrix} \tag{14}$$

Next, we assume the pitch angle to be small enough to let $\sin\theta(t) \approx 0$ and $\cos\theta(t) \approx 1$, so that the rotation matrix (3) becomes

$$
{}_b^u R(\tau) = \begin{bmatrix} \cos\psi(\tau) & -\sin\psi(\tau) & 0 \\ \sin\psi(\tau) & \cos\psi(\tau) & 0 \\ 0 & 0 & 1 \end{bmatrix} \tag{15}
$$

While this step is not required, it simplifies the expressions in the following development. Inverting (14) via the rotation matrix (15) yields

$$
\begin{bmatrix} U_0 \\ v \\ w \end{bmatrix} = \lambda \begin{bmatrix} \cos\psi & \sin\psi & 0 \\ -\sin\psi & \cos\psi & 0 \\ 0 & 0 & 1 \end{bmatrix} \begin{bmatrix} x' \\ y' \\ z' \end{bmatrix} \tag{16}
$$

Hereafter each variable's explicit dependence on τ will be omitted from the notation.
Now the three equations of system (16) must be resolved with respect to three unknown parameters, v, w and ψ. While the last one readily yields

$$
w = \lambda z' \tag{17}
$$

the first two require more rigorous analysis.
Consider Fig.8. Geometrically, a scalar product of two vectors on the right-hand-side of the first equation in (16) represents the length of the longest side of the shaded rectangle. Similarly, the second equation expresses the length of the shortest side of this rectangle. From here it follows that the square of the length of the diagonal vector can be expressed in two ways: $v^2\lambda^{-2} + U_0^2\lambda^{-2} = x'^2 + y'^2$. This yields

$$
v = \sqrt{\lambda^2(x'^2 + y'^2) - U_0^2} \tag{18}
$$

From the same figure it follows that

$$
\psi = \Psi - \tan^{-1}\frac{v\lambda^{-1}}{U_0\lambda^{-1}} = \Psi - \tan^{-1}\frac{v}{U_0}, \quad \Psi = \tan^{-1}\frac{y'}{x'} \tag{19}
$$

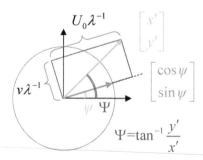

Fig. 8. Kinematics of horizontal plane parameters

Now, using these inverted kinematic equations, we can check whether each candidate trajectory obeys the constraints imposed on it (constraints (8)).

3.3 Discretization

We proceed with computing the remaining states along the reference trajectory over a fixed set of N points (for instance, $N=100$) spaced evenly along the virtual arc $[0;\tau_f]$ with the interval

$$\Delta\tau = \tau_f(N-1)^{-1} \tag{20}$$

so that

$$\tau_j = \tau_{j-1} + \Delta\tau, \quad j = 2,...,N, \; (\tau_1 = 0) \tag{21}$$

In order to determine coefficients for polynomials (9) we will have to guess on the values of the varied parameters τ_f, x_{i0}''', x_{i0}''', x_{if}''', and x_{if}'''. These guesses will be used along with the known or desired boundary conditions x_{i0}, x_{i0}', x_{i0}'', x_{if}, x_{if}', and x_{if}''. The boundary conditions on coordinates x_{i0} and x_{if} come directly from (5). According to (14), the given boundary conditions on surge, sway, and heave velocities define the first-order time derivatives of the coordinates as

$$\begin{bmatrix} x_{0;f}' \\ y_{0;f}' \\ z_{0;f}' \end{bmatrix} = \lambda_{0;f}^{-1} \, {}_b^u R_{0;f} \begin{bmatrix} U_0 \\ v_{0;f} \\ w_{0;f} \end{bmatrix} \tag{22}$$

They also define the initial and final pitch and yaw angles used to compute ${}_b^u R_{0;f}$ in (22) as

$$\theta_{0;f} = \gamma_0 + \tan^{-1}\frac{-w_{0;f}}{\sqrt{U_0^2 + v_{0;f}^2}} \quad \text{and} \quad \psi_{0;f} = \Psi_0 - \tan^{-1}\frac{v_{0;f}}{U_0} \tag{23}$$

In equation (22) we may use any value for the initial and final speed factor λ, for example, $\lambda_{0;f} = 1$. This value simply scales the virtual domain; the higher the values for λ, the larger the values for τ_f. This follows directly from equations (11) and (12) that $\lambda_{0;f}\tau_f^{-1} = U_0 s_f^{-1}$, where s_f is the physical path length.

Finally, initial values for the second-order derivatives are provided by the UUV motion sensors (see Figs. 1-3) (after conversion to the τ domain), while final values for the second-order derivatives are usually set to zero for a smooth arrival at the final point. Having an analytical representation of the candidate trajectory (9) defines the values of x_{ij}, and x_{ij}', $i = 1,2,3$, $j = 1,...,N$.

Now, for each node $j = 1,...,N$ we compute

$$\lambda_j = \Delta\tau\Delta t_{j-1}^{-1} \tag{24}$$

where

$$\Delta t_{j-1} = \frac{\sqrt{(x_j - x_{j-1})^2 + (y_j - y_{j-1})^2 + (z_j - z_{j-1})^2}}{\sqrt{U_0^2 + v_{j-1}^2 + w_{j-1}^2}} \tag{25}$$

and then use (17)-(19) to compute w, v, ψ and Ψ at each timestamp. The vertical plane parameters, flight path angle γ and pitch angle θ, can be computed using the following relations:

$$\gamma_j = \tan^{-1} \frac{-z'_j}{\sqrt{x'^2_j + y'^2_j}} \; , \quad \theta_j \approx \gamma_j + \tan^{-1} \frac{-w_j}{\sqrt{U^2_0 + v^2_j}} \tag{26}$$

In order to check the yaw rate constraints (8) we must first numerically differentiate the expression for Ψ in (19).

3.4 Optimization
When all parameters (states and controls) are computed in each of N points, we can compute the performance index J and the penalty function. For example, we can combine constraints (6) and (8) into the joint penalty

$$\Delta = \left[k^{z_{\min}}, k^{z_{\max}}, k^{\theta}, k^{\psi} \right] \begin{bmatrix} \min_j(0; z_j - z_{\min})^2 \\ \max_j(0; z_j - z_{\max})^2 \\ \max_j(0; |\theta_j| - \theta_{\max})^2 \\ \max_j(0; |\dot{\psi}_j| - \dot{\psi}_{\max})^2 \end{bmatrix} \tag{27}$$

with $k^{z_{\min}}$, $k^{z_{\max}}$, k^{θ} and $k^{\dot{\psi}}$ being scaling (weighting) coefficients. Now the problem can be solved using numerical methods such as the built-in *fmincon* function in the Mathworks' MATLAB development environment. Alternatively, by combining the performance index J with the joint penalty Δ we may exploit MATLAB's non-gradient *fminsearch* function. For real-time applications, however, the authors prefer to use a more robust optimization routine based on the gradient-free Hooke-Jeeves pattern search algorithm (Yakimenko, 2011).

4. Planar cases

This section presents two simplified cases for a vehicle manoeuvring exclusively in either the horizontal or vertical plane.

4.1 Horizontal plane guidance
For the case of a UUV manoeuvring in the horizontal plane or a USV, the computational procedure is simplified. The trajectory is represented by only two reference polynomials for coordinates x_1 and x_2. Hence, we end up having only five varied parameters, which are: τ_f, x'''_{10}, x'''_{20}, x'''_{1f} and x'''_{2f}. The remaining kinematic formulas are identical to those presented above with $z \equiv 0$, $z' \equiv 0$ and $\gamma \equiv 0$. Figure 9 shows an example of a planar scenario in which a USV has to compute a new trajectory twice. First, after detecting an obstacle blocking its original path, a new trajectory is generated to steer right and pass safely in front of the object (dotted line). Second, while executing the first avoidance manoeuvre the USV detects that the object has moved south into its path. It therefore

generates a new trajectory to steer left and pass safely behind the object's stern. The complete trajectory is shown as a solid line.

4.2 Vertical plane guidance

For the case of a UUV manoeuvring in the vertical plane, the 3D algorithm can be reduced to the 2D case in a manner similar to the horizontal case. Specifically, using five varied parameters, τ_f, x_{10}''', x_{30}''', x_{1f}''' and x_{3f}''', we can develop reference trajectories for x_1 and x_3, and then use the same general equations developed in Section 3, assuming $y \equiv 0$, $y' \equiv 0$, and $\Psi \equiv 0$.

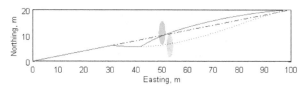

Fig. 9. Moving obstacle avoidance in a horizontal plane

Alternatively, we can use a single reference polynomial to approximate just x_3 and then integrate the third equation of (4) to get the heave velocity w. That allows computation of the time period Δt_{j-1} using

$$\Delta t_{j-1} = (z_j - z_{j-1})w_{j-1}^{-1} \tag{28}$$

instead of (25).

Another way of dealing with vertical plane manoeuvres is to invert the dynamic equations (4) (Horner & Yakimenko, 2007). After developing the reference functions for two coordinates, x_1 and x_3, the stern plane δ_s control input is computed subject to five variable parameters: τ_f, x_{10}''', x_{30}''', x_{1f}''', and x_{3f}'''.

In this case, the corresponding time period Δt_{j-1} is computed similarly to (28):

$$\Delta t_{j-1} = t_j - t_{j-1} = \frac{z_j - z_{j-1}}{w_{j-1}\cos\theta_{j-1} + u_0\sin\theta_{j-1}} \approx \frac{z_j - z_{j-1}}{w_{j-1}} \tag{29}$$

and the heave velocity is calculated using the third equation of system (4) as

$$\begin{aligned} x_{j-1}' &= \lambda_{j-1}^{-1}(w_{j-1}\sin\theta_{j-1} + U_0\cos\theta_{j-1}) & x_j &= x_{j-1} + \Delta\tau x_{j-1}' \\ w_{j-1}' &= \lambda_{j-1}^{-1}\left(A_{33}w_{j-1} + A_{35}q_{j-1} + B_{32}\delta_{s;j-1}\right) & w_j &= w_{j-1} + \Delta\tau w_{j-1}' \end{aligned} \tag{30}$$

The next step involves computing the pitch angle, pitch rate and pitch acceleration as

$$\theta_j = \cos^{-1}\left(\lambda_j \frac{u_0 x_j' + w_j z_j'}{w_j^2 + U_0^2}\right), \quad q_j = \dot{\theta}_j \approx \frac{\theta_j - \theta_{j-1}}{\Delta t_{j-1}}, \quad \dot{q}_j = \ddot{\theta}_j \approx \frac{q_j - q_{j-1}}{\Delta t_{j-1}} \tag{31}$$

Finally, we can compute the dive plane deflection required to follow the trajectory using the 5th equation of system (4) as

$$\delta_{s;j} = (\dot{q}_j - A_{53}w_j - A_{55}q_j)B_{52}^{-1} \tag{32}$$

In this case the last two terms in the joint penalty Δ, similar to that of (27) but developed for the new controls, enforce $|\delta_s| \le \delta_{s\max}$ and $|\dot{\delta}_s| \le \dot{\delta}_{s\max}$.

5. Test vehicles and sensing architecture

The preceding trajectory generation framework has been implemented on several UMVs. Before presenting simulated and experimental results with specific vehicle platforms at sea, we first introduce two such vehicles in use at CAVR - the REMUS UUV and SeaFox USV. Both vehicles utilize FLS to detect and localize obstacles in their environment and employ the suggested direct method to generate real-time OA trajectories. This section provides system-level descriptions of both platforms including their sensors, and proposes a way of building the OA framework on top of the trajectory generation framework, i.e. enhancing the architecture of Figs. 2 and 3 even further.

5.1 REMUS UUV and SeaFox USV

Remote Environmental Monitoring UnitS (REMUS) are UUVs developed by Woods Hole Oceanographic Institute and sold commercially by Hydroid, LLC (Hydroid, 2011). The NPS CAVR owns and operates two REMUS 100 vehicles in support of various navy-sponsored research programs. The REMUS 100 is a modular, 0.2m diameter UUV designed for operations in coastal environments up to 100m deep. Typical configurations measure less than 1.6m in length and weigh less than 45kg, allowing the entire system to be easily transported worldwide and deployed by a two-man team (Fig.10a). Designed primarily for hydrographic surveys, REMUS comes equipped with sidescan sonar and sensors for collecting oceanographic data such as conductivity, temperature, depth or optical backscatter. The REMUS 100 system navigates using a pair of external transponders for long baseline acoustic localization or ultra-short baseline terminal homing, as well as an Acoustic Doppler Current Profiler/Doppler Velocity Log (ADCP/DVL). The ADCP/DVL measures vehicle altitude, ground- or water-relative vehicle velocity, and current velocity profiles in body-fixed coordinates.

a)   b)

Fig. 10. NPS REMUS 100 UUV (a) and FLS arrays (b)

To support ongoing CAVR research into sonar-based OA, terrain-relative navigation, and multi-vehicle operations in cluttered environments, each NPS REMUS vehicle has been modified to incorporate a FLS, multi-beam bathymetric sonar, acoustic communications modem, navigation-grade inertial measurement system, and fore/aft horizontal/vertical cross-body thrusters for hovering or precise manoeuvring. (Figure 10b provides a close up of the NPS REMUS FLS arrays with nose cap removed.) To maximize the REMUS system's utility as a research platform, Hydroid developed the RECON communications interface so that sensor and computer payloads can interact with the REMUS autopilot. Using this interface, NPS payloads receive vehicle sensor data and generate autopilot commands based on NPS sonar processing, trajectory generation, and path-following algorithms.

The SeaFox USV was designed and manufactured by Northwind Marine (Seattle, WA) as a remote-controlled platform for intelligence, surveillance, reconnaissance, anti-terrorist force protection, and maritime interdiction operations (Northwind Marine, 2011). SeaFox is a 4.88m long, aluminium, rigid-hulled inflatable boat with a 1.75m beam; 0.25m draft; fold-down communications mast; and fully-enclosed electronics and engine compartments. SeaFox's water jet propulsion system is powered by a JP5-fueled, 185-HP V-6 Mercury Racing engine, and can deliver a top speed of 74km/h. Standard sensing systems include three daylight and three low light navigation cameras for remote operation, as well as twin daylight and infrared gyro-stabilized camera turrets for video surveillance. All video is accessible over a wireless network via two onboard video servers.

The NPS SeaFox was modified to enable fully-autonomous operations by integrating a payload computer with the primary autopilot (Fig.11). Meanwhile, the original remote control link was retained to provide an emergency stop function. NPS algorithms running on the payload computer generate rudder and throttle commands that are sent directly to the SeaFox autopilot. Recent navigational upgrades include a satellite compass that uses differential Global Positioning System (GPS) navigation service for accurate heading information, a tactical-grade inertial measurement unit for precise attitude estimation, and an optional ADCP/DVL for water velocity measurements. To support ongoing CAVR research into autonomous riverine navigation, the NPS SeaFox was further upgraded to deploy a retractable, pole-mounted FLS system for underwater obstacle detection and avoidance (Gadre et al., 2009). Figure 12 shows the SeaFox USV operating on a river with its sonar system deployed below the waterline.

5.2 Sonar system

The NPS REMUS and SeaFox vehicles rely on FLS to detect and localize obstacles in their environment. Both platforms utilize commercial blazed array sonar systems manufactured by BlueView Technologies (BlueView Technologies, 2011). These sonar systems comprise one or more pairs of arrays grouped into sonar "heads." Each sonar head generates a 2D cross-sectional image of the water column in polar coordinates, typically plotted as the image plane's field of view angle vs. range. Due to the sonar arrays' beam width, the resulting FLS imagery has a 12-degree out-of-plane ambiguity. The REMUS FLS system consists of two fixed sonar heads, which provide a 90-degree horizontal field of view (FOV) and a 45-degree vertical FOV. Similarly, the SeaFox FLS system is comprised of twin sonar heads mounted on port and starboard pan/tilt actuators, providing each side with a 45-degree FOV image at an adjustable mounting orientation that can be swept through the water column for increased sensor coverage.

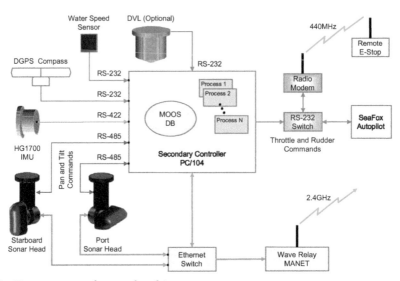

Fig. 11. SeaFox sensors and control architecture

Fig. 12. SeaFox USV navigating on the Sacramento River near Rio Vista, CA

5.3 Obstacle avoidance framework

The proposed OA framework built into the architecture of Figs. 2 and 3 is shown in Fig.13. It consists of an environmental map, a planning module, a localization module, sensors and actuators (Horner & Yakimenko, 2007). The environmental map can include *a priori* knowledge, such as the positions of charted underwater obstacles, but also incorporate unexpected threats discovered by sonar. The positions of all obstacles are eventually resolved in the vehicle-centred coordinate frame with the help of the localization module. The planning module is responsible for generating collision-free trajectories the vehicle should follow. This reference trajectory, possibly with reference controls, is then used to excite actuators.

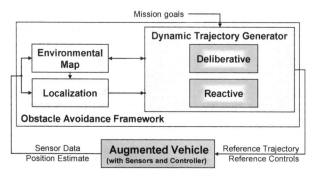

Fig. 13. Components of the NPS OA framework

The proposed OA framework supports both deliberative and reactive obstacle avoidance behaviours. Deliberative OA involves the ability to generate and follow a trajectory that avoids all known obstacles between an arbitrary start location and some desired goal location, whereas reactive OA involves the ability to avoid any previously unknown obstacles detected while following this trajectory. Since the sonar system continuously resamples the environment, this reactive behaviour can be achieved by a deliberative planner as long as i) it executes fast enough to incorporate all new obstacle information from the sonar, and ii) it generates feasible trajectories which begin with the vehicle's current state vector. Specifically, since the REMUS and SeaFox FLS have limited range and limited fields of view in both image planes, new trajectories must be generated continuously (e.g. on some fixed time interval or upon detection of a new obstacle) during execution of the current manoeuvre to ensure reactive avoidance of new obstacles.

As an example of deliberative OA, assume a REMUS vehicle is mapping a minefield with sidescan sonar prior to a mine clearance operation. For this mission, the goal locations are provided by the sequence of waypoints making up a typical lawn-mowing survey pattern. If an obstacle is detected along a specified track line, the preferred OA manoeuvre for this mission would be one that also minimizes the cumulative deviation from this track line, since we desire 100% sensor coverage of the survey area. Hence, deliberative OA implies the optimization of some performance index. Likewise, while digital nautical charts or previous vehicle surveys can be used to identify some obstacles *a priori*, this data is usually incomplete or outdated. Vehicles should be capable of storing in memory the locations of any uncharted obstacles discovered during their mission so that subsequent trajectories can avoid them — even when they are no longer in the sonar's current field of view. Deliberative OA, therefore, also entails the creation and maintenance of obstacle maps.

5.4 Obstacle detection and mapping

Detecting obstacles from sonar imagery is challenging because several factors affect the intensity of sonar reflections off objects in the water column. These factors include the size, material, and geometry of an object relative to the sonar head; interference from other acoustic sensors; and the composition of the acoustic background (e.g. bottom type, amount of sediment, etc.) to name a few (Masek, 2008). Once an obstacle has been detected, other image processing algorithms must measure its size and compute its location within the navigational reference frame. While localizing obstacles via the range and bearing data embedded in the sonar imagery is straightforward, computing their true size is very difficult. First, for the REMUS FLS, an obstacle's height and width can be measured directly by both sonar heads only when it is located within a narrow 12-degree by 12-degree "window" directly ahead of the vehicle. Due to this narrow beam width, most obstacles are not imaged by both the horizontal and vertical sonar at the same time. Moreover, FLS images do not contain information in the region behind an obstacle's ensonified leading edge; this portion of the image is occluded. Therefore, the true horizontal and vertical extent of each obstacle must be deduced from multiple views of the same object. For a vehicle with a fixed sensor like the REMUS, this may be accomplished by deliberately inducing vehicle motion to vary the sonar angle (Furukawa, 2006) or by generating trajectories that will image the object from a different location at a later time. For these scenarios, it is desirable to balance OA behaviours with exploration behaviours in order to maximize sensor coverage and generate more complete obstacle maps. In this way, the proposed trajectory generation framework can be adapted to produce exploratory trajectories which more accurately measure the size and extent of detected obstacles (Horner et al., 2009). Nevertheless, due to the uncertainty in sonar images arising from environmental factors, sensor geometry, or obstacle occlusion, it is prudent to make conservative assumptions about an obstacle's boundaries until other information becomes available.

For the remainder of this section, we highlight different representations for incorporating obstacle size, location, and uncertainty into an obstacle map for efficient collision detection during the trajectory optimization phase. These representations can be tailored to the working environment. For operations in a kelp forest, for example, kelp stalks often appear as point-like features in horizontal-plane sonar imagery (Fig.14) but seldom appear in vertical-plane images. By making the reasonable assumption (for this environment) that these obstacles extend vertically from the sea floor to the surface, it may be simpler to perform horizontal-plane OA through this type of obstacle field. Nevertheless, when building an obstacle map comprised primarily of point features, mapping algorithms must account for the uncertainty inherent in sonar imagery. One simple but effective technique adds spherical (3D) or circular (2D) uncertainty bounds to each point feature stored in the obstacle map. Candidate OA trajectories which penetrate these boundaries violate constraint (7). Under this construct, collision detection calculations are reduced to a simple test to determine whether line segments in a discretized trajectory intersect with the uncertainty circle (2D) or sphere (3D) for each obstacle in the map. In general, when checking for line segment intersections with a circle or sphere there are five different test cases to consider (Bourke, 1992). Our application, however, requires only two computationally efficient tests to determine: i) which line segment along a discretized trajectory contains the closest point of approach (CPA) to an obstacle, and ii) whether this CPA is located inside the obstacle's uncertainty bound.

Most objects appear in sonar imagery not as point features, but as complex shapes. Unlike point features, it is difficult and computationally expensive to determine exhaustively whether or not a candidate vehicle trajectory will collide with these shapes. Instead, we can bound an arbitrary shape with a minimal area rectangle (or box, in 3D) aligned with the shape's principle axes (Fig.15). This type of object, called an oriented bounding box, is widely used in collision-detection algorithms for video games. One technique, based on the Separating Axis Theorem from complex geometry, results in an extremely fast test for line segment intersections with an oriented bounding box (Kreuzer, 2006). With slight modification, this test can also be used to detect when a trajectory passes directly above a bounding box. In our application, we use the OpenCV computer vision library to generate a bounding box around each object detected in the horizontal image plane. For each box, we then compute its centre point, length extent, and angle relative to the vehicle's navigation frame. Due to occlusion, the width extent produced from this rectangle does not accurately convey the true size of the obstacle, so we assume a constant value for this parameter. To create a 3D (actually 2.5D) bounding box around the object, we compute its vertical extents from vertical sonar imagery. At this time, the assumption is that obstacles extend from the ocean floor to its measured height above bottom, but this method can be generalized to obstacles suspended in the water column or extending from the surface to a measured depth (i.e. ships in a harbour).

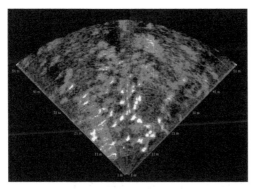

Fig. 14. Horizontal FLS image of a kelp forest

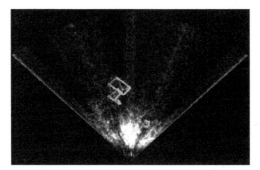

Fig. 15. Example of the bounding boxes used in conservative collision detection calculations

While oriented bounding boxes work well for mapping discrete obstacles in open-water environments, they require an additional image processing step and are not easily adapted to operations in restricted waterways. For these environments, a probabilistic occupancy grid is preferable for robustly mapping large continuous obstacles (e.g. harbour breakwaters) or natural terrain (e.g., a river's banks). Occupancy grids divide the environment into a grid of cells and assign each cell a probability of being occupied by an obstacle. Given a probabilistic sensor model, Bayes' Theorem is used to compute the probability that a given cell is occupied, based upon current sensor data. By extension, an estimate for the occupancy state of each cell can be continually updated using an iterative technique that incorporates all previous measurements (Elfes, 1989). Figure 16a shows an occupancy grid map of a river generated by the SeaFox FLS system. In this image, each pixel corresponds to a 1-metre square grid cell whose colour represents the cell's probability of being occupied (red) or empty (green). For comparison, the inset portion of the occupancy grid map has been overlaid with an obstacle map of oriented bounding boxes in Fig.16b. Clearly, using discrete bounding boxes to represent a long, continuous shoreline quickly becomes intractable as more and more boxes are required. The occupancy grid framework is a much more efficient obstacle map representation for wide area operations in restricted waterways.

a)   b)

Fig. 16. Occupancy grid for a river as generated by the SeaFox FLS system

NPS has developed probabilistic sonar models for the BlueView FLS and has successfully combined separate 2D occupancy grids in order to reconstruct the 3D geometry of an obstacle imaged by the REMUS UUV's horizontal and vertical sonar arrays (Horner et al., 2009). Using this occupancy grid framework, each candidate trajectory's risk of obstacle collision is computed using the occupancy probabilities (a direct lookup operation) of the grid cells it traverses. Trajectory optimization for OA entails minimizing the cumulative risk of collision along the entire trajectory.

6. Path following

While the REMUS UUV and SeaFox USV are both commercial vehicles with proprietary autopilots, both provide communications interfaces that allow experimental sensor and computer payloads to override the primary autopilot via high-level commands. The REMUS RECON interface, for example, closely resembles the augmented autopilot depicted in Figs. 2 and 3 (although direct actuator inputs are only available for propeller and cross-body thrusters settings). For full overriding control, a payload module must periodically send valid commands containing all of the following: i) desired depth or altitude, ii) desired vehicle or propeller speed, and iii) desired heading, turn rate, or waypoint location. The developed trajectory generator (described in Section 3) outputs reference trajectories as parameterized expressions for each coordinate in a spatial curve plus a speed factor to use while traversing that curve. Using these expressions as reference trajectories, the 3D path following controller developed earlier (Kaminer et al., 2007) can compute turn rate and pitch rate commands required to drive a vehicle onto (and along) the desired trajectory. The

RECON interface, however, does not accept pitch rate commands (for vehicle safety reasons). Therefore, in order to use the aforementioned path following controller to track 3D trajectories with the REMUS UUV, controller outputs must be partitioned into horizontal (turn rate) and vertical (depth or altitude) commands as described in the following section (obviously, the SeaFox USV only uses the turn rate commands).

6.1 Horizontal plane
Consider the 2D problem geometry depicted in Fig.17, which defines an inertial $\{I\}$ frame, Serret-Frenet $\{F\}$ error frame and body-fixed reference frame $\{b\}$. The kinematic model of the vehicle (2)-(3) reduces to

$$\begin{bmatrix} \dot{x}(t) \\ \dot{y}(t) \end{bmatrix} = \begin{bmatrix} U_0 \cos\psi(t) \\ U_0 \sin\psi(t) \end{bmatrix} \tag{33}$$

with dynamics described by

$$\dot{\psi} = r \tag{34}$$

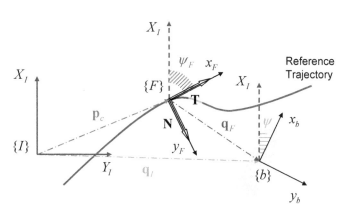

Fig. 17. Horizontal path-following kinematics

By construction, the local trajectory planner produces an analytic expression for each component of the spatial trajectory as a function of virtual arc length, $\mathbf{p}_c(\tau)$. We can also compute analytic expressions for $\mathbf{p}'_c(\tau)$ and $\mathbf{p}''_c(\tau)$, the first and second derivatives, respectively, of the spatial trajectory. Using the relationships in Fig.17, the errors can be expressed in the Serret-Frenet frame $\{F\}$ as

$$\mathbf{q}_F = \begin{bmatrix} x_F \\ y_F \end{bmatrix} = {}^F_I R (\mathbf{q}_I - \mathbf{p}_c) \tag{35}$$

where ${}^F_I R = [\mathbf{T}, \mathbf{N}, \mathbf{B}]^T$ is a rotation matrix constructed from the tangent, normal, and binormal vectors of the Serret-Frenet error frame $\{F\}$. The tangent vector is computed from the expression for the trajectory's first derivative as:

$$\mathbf{T} = \frac{\mathbf{p}'_c(\tau)}{\left\| \mathbf{p}'_c(\tau) \right\|} \tag{36}$$

For the 2D problem, the normal vector components can be computed directly from the tangent vector components: $N_x = -T_y$ and $N_y = T_x$. Additionally, the signed curvature of the trajectory can be computed using the expressions for the trajectory's first and second derivatives as:

$$\kappa(\tau) = \frac{p''_{cy}(\tau)p'_{cx}(\tau) - p''_{cx}(\tau)p'_{cy}(\tau)}{\left\| \mathbf{p}'_c(\tau) \right\|^3} \tag{37}$$

Taking the time derivative of $\mathbf{q}_F$, we obtain the following state space representation for the error kinematics (i.e. the position and heading of the vehicle's body-fixed frame $\{b\}$ relative to the Serret-Frenet frame $\{F\}$, which follows the desired trajectory):

$$\begin{aligned}
\dot{x}_F &= -\dot{l}(1 - \kappa y_F) + U_0 \cos\psi_e \\
\dot{y}_F &= -\dot{l}(\kappa x_F) + U_0 \sin\psi_e \\
\dot{\psi}_e &= \dot{\psi} - \dot{\psi}_F = u_\psi - \dot{\psi}_F = u_\psi - \kappa \dot{l}
\end{aligned} \tag{38}$$

where l is the path length of the desired spatial curve and $\dot{l}$ describes the speed at which a virtual target travels along this curve.

The goal is to drive the vehicle's position error ($\mathbf{q}_F$) and heading error (ψ_e) to zero. This will drive the vehicle to the commanded trajectory location ($\mathbf{p}_c$) and align its velocity vector with the trajectory's tangent vector ($\mathbf{T}$). The control signal u_ψ must now be chosen to asymptotically drive the vehicle position and velocity vectors onto the commanded trajectory. We choose the candidate Lyapunov function

$$V = \frac{1}{2}\left(x_F^2 + y_F^2 + (\psi_e - \delta_\psi)^2\right) \tag{39}$$

where δ_ψ is a shaping function that controls the manner in which the vehicle approaches the path

$$\delta_\psi = \sin^{-1}\left(\frac{-y_F}{|y_F| + d}\right) \tag{40}$$

with $d > 0$ an arbitrary constant.
Using some algebra, we choose the following control laws to ensure that $\dot{V} < 0$:

$$\begin{aligned}
\dot{l} &= K_1 x_F + U_0 \cos\psi_e \\
u_\psi &= \kappa \dot{l} + \dot{\delta}_\psi + K_2(\psi_e - \delta_\psi) - \frac{\sin\psi_e - \sin\delta_\psi}{\psi_e - \delta_\psi} U_0 y_F
\end{aligned} \tag{41}$$

In these expressions K_1, K_2, and d can be used as gains to tune the closed-loop performance of the path following controller.

6.2 Vertical plane

Now consider the REMUS UUV manoeuvring in the vertical plane using altitude commands. For survey operations, REMUS is typically programmed to follow a lawnmower pattern at a constant altitude above the sea floor determined by the desired sidescan sonar range. Since the ADCP/DVL sensor continuously measures vehicle altitude above the bottom, this operating mode ensures safe operation over undulating bottoms with slopes of up to 45 degrees (Healey, 2004). Obstacle avoidance manoeuvres are required to safely negotiate steeper slopes, step-like terrain features (e.g. sand bars or coral heads), or objects proud of the ocean floor. As described earlier, since the REMUS FLS is mounted in a fixed orientation, it may detect new obstacles while executing a manoeuvre to avoid the current obstacle threat. Periodic or detection-based replanning can handle these situations. This scenario was illustrated conceptually in Fig.7.

When negotiating a step-up ridge or sand bar whose extent is not known due to sonar occlusion at the time of detection, it may not be desirable to follow the planned vertical trajectory to its completion. Between planning iterations, a simple yet safer approach is to revert back to constant altitude control once the vehicle reaches a position directly above the detected object boundaries. This condition can be checked using a 2.5D version of the 3D bounding box intersection test described above. Figure 18 illustrates a simulation whereby the REMUS FLS detects the leading edge of a ridge at maximum sonar range. Image processing algorithms compute range to the object (80m) as well as its width (W, into the page) and its height above the seafloor (5.5m) but cannot determine the length of the ridge since it is occluded by its own leading edge. Therefore, the obstacle detection algorithm generates a 3D bounding box measuring W m wide x 1.0m long (assumed) x 5.5m high.

While the IDVD-method planner generates a vertical trajectory in NED coordinates, in shallow water it is safer to operate the vehicle in an altitude control mode. Therefore, the vertical coordinate trajectory is converted from a depth plan into an altitude plan by assuming constant water depth over the planning horizon and using the relationships

$$D = h_{nom} + z(0) \qquad h_{plan}(\tau) = D - z_{plan}(\tau)$$
$$\Delta h_{plan}(\tau) = h_{plan}(\tau) - h_{nom} \qquad h_{cmd}(\tau) = h_{nom} + \Delta h_{plan}(\tau) \tag{42}$$

where D is the water depth computed at planner initialization time, h_{nom} is the nominal altitude set point, and $h_{cmd}(\tau)$ is the altitude command sent to the autopilot via the RECON interface. The resulting altitude plan $\Delta h_{plan}(\tau)$ is shown in Fig.18 as a deviation from the nominal mission segment altitude. As seen, sending this altitude command directly to the vehicle autopilot would cause an undesirable jump in the altitude profile once the ADCP/DVL sensor measures the vehicle's true altitude above the ridge. Instead, switching the altitude command to the nominal mission segment altitude once the vehicle reaches the ridge will produce the desired altitude profile). Note that even though Fig.18 depicts a sudden drop in altitude when the vehicle passes the ridge while commanding the nominal mission segment altitude, in practice the vehicle dynamics will ensure that the UUV executes a smooth transition back to its nominal survey altitude.

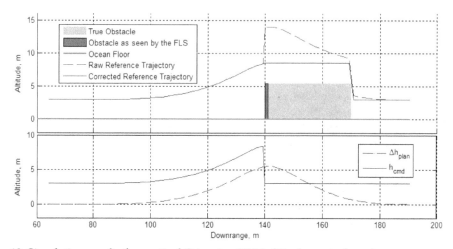

Fig. 18. Simulation results for vertical OA using UUV altitude control mode

7. Computer simulations and sea trials

The proposed path-planning method can be tailored to a specific vehicle or operational domain by modifying the performance index J to incorporate vehicle or actuator dynamics (feasibility constraints) and mission objectives such as OA or underwater rendezvous. This section presents simulated and in-water experimental results for four different applications which use the proposed trajectory optimization framework for UMV guidance: i) underwater docking of a UUV with a mobile underwater recovery system (MURS); ii) optimal exploitation of a terrain-relative feature map to improve UUV self-localization accuracy; iii) 2D or 3D OA in cluttered environments; and iv) specific USV implementations for sonar-based OA in riverine operations.

7.1 Underwater recovery

The goal of underwater recovery (Yakimenko et al., 2008) is to be able to compute a rendezvous trajectory from any point on the UUV holding pattern to any point on the MURS holding pattern as shown in Fig.19 (note hereinafter that depth values are shown as negative numbers).

While the stochastic simulation shown in Fig.19 employs circular race tracks, in practice the MURS would establish a race track that allowed it to travel back and forth along two long track legs (see Fig.20). These legs are needed to allow sufficient time to contact the UUV (which is assumed to be in its holding pattern somewhere within communication range) and allow it to transit from its holding pattern to the rendezvous point. The proposed sequence of events is to have the MURS (at position 1 in Fig.20) signal the UUV (at position 2) and command it to proceed to a rendezvous point by a certain time. The UUV computes the trajectory required to comply with the command. If the commanded rendezvous is feasible, the UUV sends an acknowledgement message. Otherwise (i.e. the request violated some constraint) the UUV sends a denial message (stage A in Fig.20) and requests that the MURS command a different rendezvous point or time. The final point of the trajectory is located in the approximate location of the MURS docking station at a given time. Knowing the geometry of the MURS allows the planner to construct a "keep out" zone corresponding to the MURS propeller and aft control surfaces. The UUV rendezvous trajectory must avoid this area. Once the rendezvous plan has been agreed upon and acknowledged, both the UUV and the MURS proceed to position 3 for rendezvous (stage B). Finally, at position 4 the recovery operation (stage C) is completed.

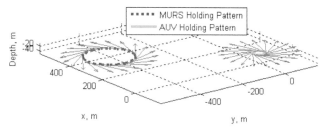

Fig. 19. Manifold of initial and final conditions

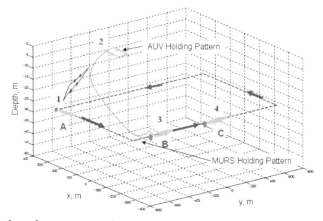

Fig. 20. Proposed rendezvous scenario

The simulated rendezvous scenario assumes three stages: communication (A), execution (B), and recovery (C), respectively. From the trajectory generation standpoint we are primarily concerned with optimizing the path that would bring the UUV from its current position (point 2) to a certain rendezvous state (point 3) in the preset time T_r proposed by the MURS, while obeying all possible real-life constraints and avoiding the MURS keep out zone.

Figures 21 and 22 present a computer simulation in which a MURS is moving due east at $1m/s$ (1.94kn) with the docking station at a depth of 15m. A UUV is located 800 meters away. The MURS wishes to conduct a rendezvous operation T_r minutes later and sends the corresponding information to the UUV. This information includes the proposed final position x_f, y_f, z_f rendezvous course, speed, and time. Figure 21 shows several generated trajectories, which meet the desired objectives for this scenario and also avoid an obstacle located along the desired path to MURS. These trajectories differ by the arrival time T_r.

During handshaking communications with the MURS, the UUV determines whether the suggested T_r is feasible. Of the four trajectories shown, the trajectory generated for

$T_r = 450s$ happens to be infeasible (the constraints on controls are violated). The solution of the minimum-time problem for this scenario yielded 488 seconds as the soonest possible rendezvous time.

The other three trajectories shown in Fig.21 are feasible. That means that the boundary conditions are met (by construction) and all constraints including OA are satisfied (via optimization). As an example, Fig.22 shows the time histories for the yaw rate $\dot{\psi}_c$ and flight path angle γ_c vehicle control parameters as well as the UUV's speed as it followed the trajectory for $T_r = 600s$.

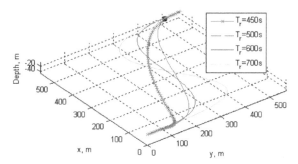

Fig. 21. Examples of rendezvous trajectories

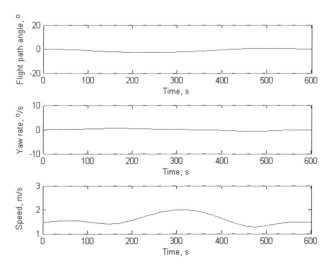

Fig. 22. Constrained vehicle parameters for $T_r = 600s$

Stochastic simulations of the manifolds shown in Fig.21 illustrate that a successful rendezvous can take place in all cases as long as T_r is greater than a certain value. Furthermore, they show that minimization of the performance index using the IDVD method ensures that a smooth, realizable trajectory is calculated in just a few seconds, regardless of the initial guess. Converting code to an executable file in lieu of using an interpretative programming language reduces execution time down to a fraction of a second.

7.2 Feature-based navigation

In the last decade, several different UUVs have been developed to perform a variety of underwater missions. Survey-class vehicles carry highly accurate navigational and sonar payloads for mapping the ocean floor, but these payloads make such vehicles very expensive. Vehicles which lack these payloads can perform many useful missions at a fraction of the cost, but their performance will degrade over time from inaccurate self-localization unless external navigation aids are available. Therefore, it is interesting to consider collaborative operations via a team of vehicles for maximum utility at reasonable cost. The NPS CAVR has been investigating one such concept of operations called feature-based navigation. This technique allows vehicles equipped only with a GPS receiver and low cost imaging sonar to exploit an accurate sonar map generated by a survey vehicle. This map is comprised of terrain or bottom object features that have utility as future navigational references. This sonar map is downloaded to the low-cost follow-on vehicles before launch. Starting from an initial GPS position fix obtained at the surface, these vehicles then navigate underwater by correlating current sonar imagery with the sonar features from the survey vehicle's map. The localization accuracy of vehicles performing feature-based navigation can be improved by maximizing the number of times navigational references are detected with the imaging sonar. The following simulation demonstrates how the IDVD trajectory generation framework can be tailored to this application. By incorporating a simple geometric model of an FLS having a range of 60m, 30-degree horizontal FOV and operating at a nominal ping rate of 1Hz, a new performance index was designed to favour candidate trajectories, which point the sonar toward navigational references in the *a priori* feature map. For this example, we sought trajectories that could obtain at least three sonar images of each feature in the map. Figure 23 shows results of a computer simulation in which the number of times each target was imaged by the sonar has been annotated. The resulting trajectory is feasible (i.e. satisfies turn rate constraints) and yields three or more sonar images of all but two targets.

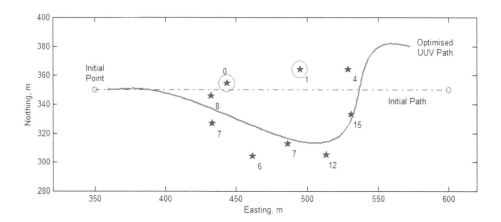

Fig. 23. Simulation results for a feature-based navigation application

7.3 Obstacle avoidance in cluttered environments

Another application which benefits from the aforementioned trajectory generation algorithm is real-time OA in a highly cluttered environment. Figure 24 illustrates simulated trajectories for avoiding a field of point-like objects in the 2D horizontal plane (e.g. a kelp forest) and in all three dimensions (e.g. a mine field). In both simulations, the performance index was designed to minimize deviations from a predefined survey track line while avoiding all randomly generated obstacles via a CPA calculation. Terminal boundary conditions for the OA manoeuvre were chosen to ensure the UUV rejoined the desired track line before reaching the next waypoint (i.e. the manoeuvre terminated at a position 95% along the track segment). Initial boundary conditions were chosen to simulate a random obstacle detection which triggers an avoidance manoeuvre after the UUV has completed about 10% of the predefined track segment. For illustration purposes, Fig.24 includes several candidate trajectories evaluated during the optimization process although the algorithm ultimately converged to the trajectory depicted with a thicker (red) line (CPA distances to each obstacle appear as dashed lines).

Figure 25 shows the results from an initial sea trial of 3D OA that took place in Monterey Bay on 9 December 2008. This experiment tested periodic trajectory generation and replanning on the REMUS UUV using a simulated obstacle map comprised of oriented bounding boxes. As seen in Fig.25, initially the REMUS UUV follows a predefined track segment (dash-dotted line) at 4 meters altitude. At some point the vehicle's FLS simulator "detects" an obstacle (i.e. the current REMUS position and orientation place the virtual obstacle within the range and aperture limits of the FLS). This activates the OA mode, and the planner generates an initial trajectory (green) from the current vehicle position to the final waypoint. REMUS follows this trajectory until the next planning cycle (4 seconds later) when the vehicle generates a new trajectory and continues this path planning-path following cycle.

7.4 Obstacle avoidance in restricted waterways

The NPS CAVR in collaboration with Virginia Tech (VT) is developing technologies to enable safe, autonomous navigation by USVs operating in unknown riverine environments. This project involves both surface (laser) and subsurface (sonar) sensing for obstacle detection, localization, and mapping as well as global-scale (wide area) path planning, local-scale trajectory generation, and robust vehicle control. The developed approach includes a hybrid receding horizon control framework that integrates a globally optimal path planner with a local, near-optimal trajectory generator (Xu et al., 2009).

The VT global path planner uses a Fast Marching Method (Sethian, 1999) to compute the optimal path between a start location and a desired goal location based on all available map information. While resulting paths are globally optimal, they do not incorporate vehicle dynamics and thus cannot be followed accurately by the USV autopilot. Moreover, since level set calculations are computationally expensive, global plans are recomputed only when necessary and thus do not always incorporate recently detected obstacles. Therefore, a complimentary local path planner operating over a short time horizon is required to incorporate current sensor information and generate feasible OA trajectories. The IDVD-based trajectory generator described above is ideally suited for this purpose. VT has developed a set of matching conditions which guarantee the asymptotic stability of this

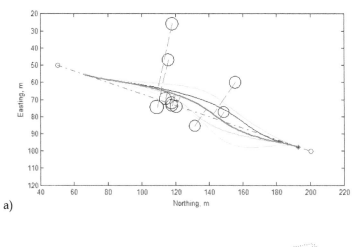

a)

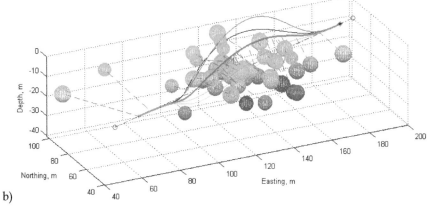

b)

Fig. 24. Simulated 2D (a) and 3D (b) near-optimal OA trajectories

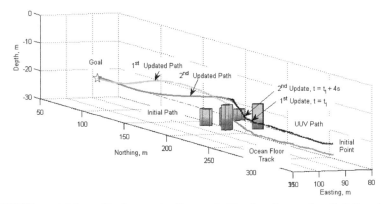

Fig. 25. REMUS sea trial results demonstrating periodic planning and path following

framework. When these matching conditions are satisfied, the sequence of local trajectories will converge to the global path's goal location. If the local trajectories no longer satisfy these conditions (usually because the global path is no longer compatible with recently detected obstacles), the global path is recomputed.

Simulation results demonstrate the need for local trajectories that incorporate vehicle dynamics and real-time sensor data (Fig.26). For this simulation, an initial level set map was computed using an occupancy grid created by masking land areas as occupied and water areas as unoccupied in an aerial image of the Sacramento River operating area. Performing gradient descent on the level set from the USV's initial position produces an optimal path shown in blue. To simulate local trajectory generation with a stale global plan, the initial level set map was not updated during the entire simulation. Meanwhile, to simulate access to real-time sensor data, the local planner was provided with a complete sonar map generated during a previous SeaFox survey of the area. In Fig.26, this sonar map has been overlaid on the *a priori*

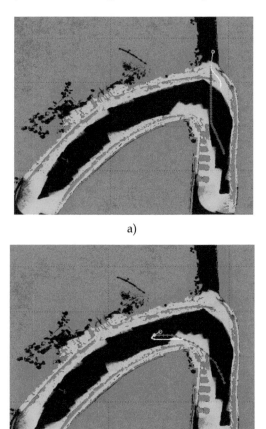

a)

b)

Fig. 26. Simulated local OA trajectories

map with red and green colour channels representing the probability that a cell is occupied or unoccupied, respectively. Black pixels represent cells with unknown status. A short green line segment depicts the USV's orientation when the local planner is invoked, and the resulting trajectory is shown in yellow. The first simulation (Fig.26a) shows a local trajectory which deviates from the stale global plan to avoid a sand bar detected with sonar. In the second simulation (Fig.26b) the USV is initially heading in a direction opposite from the global path, but the local planner generates a dynamically feasible trajectory to turn around and rejoin the global path later.

To track these local trajectories, the 2D controller described in Section 6.1 was implemented on the SeaFox USV by mapping the controller's turn rate commands into rudder commands understood by the SeaFox autopilot. After validating the turn rate controller design during sea trials on Monterey Bay, the direct method trajectory generator and closed-loop path following controller were tested on the Pearl River in Mississippi on 22 May 2010 (Fig.27). For this test, the local planner used a sonar map of the operating area to generate the trajectory (the cyan line) from an initial orientation (depicted by the yellow arrow) to a

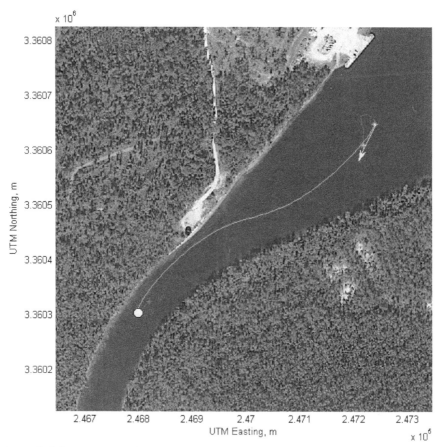

Fig. 27. Path-following controller test on the Pearl River

desired goal point (depicted by a circle). The SeaFox USV then followed it almost precisely (the magenta line). As seen from Fig.27 the trajectory generator was invoked at an arbitrary location while the USV was performing a clockwise turn. Since the USV was commanded to return to its start location upon completion of this manoeuvre, the magenta line includes a portion of this return trajectory as well (otherwise, the actual USV track would be nearly indistinguishable from the reference trajectory on this plot).

8. Conclusion

An onboard trajectory planner based on the Inverse Dynamics in the Virtual Domain direct method presented in this chapter is an effective means of augmenting an unmanned maritime vehicle's autopilot with smooth, feasible trajectories and corresponding controls. It also facilitates incorporation of sophisticated sensors such as forward-looking sonar for deliberative and reactive obstacle avoidance. This approach has been implemented on both unmanned undersea and surface vehicles and has demonstrated great potential. Beyond its ability to compute near-optimal collision-free trajectories much faster than in real time, the proposed approach supports the utilization of any practically-sound compound performance index. This makes the developed control architecture quite universal, yet simple to use in a variety of applied scenarios, as demonstrated in several simulations and preliminary sea trials. This chapter presented results from only a few preliminary sea trials. Future research will continue development of the suggested trajectory framework in support of other tactical scenarios.

9. Acknowledgements

The authors wish to gratefully acknowledge the support of Doug Horner, Co-Director of the CAVR and Principle Investigator for the REMUS UUV and SeaFox USV research programs at NPS. In addition, Sean Kragelund would like to thank his CAVR colleagues Tad Masek and Aurelio Monarrez. Mr. Masek's outstanding software development work to implement obstacle detection and mapping with forward looking sonar made possible the OA applications described herein. Likewise, the tireless efforts of Mr. Monarrez to continually upgrade, maintain, and operate CAVR vehicles in support of field experimentation have made a lasting contribution to this Center.

10. References

Basset, G., Xu, Y. & Yakimenko, O. (2010). Computing short-time aircraft maneuvers using direct methods," *Journal of Computer and Systems Sciences International*, 49(3), 145-176

BlueView Technologies, Inc. (2011). 2D Imaging sonar webpage. Available from: www.blueview.com/2d-Imaging-Sonar.html

Bourke, P. (1992). Intersection of a line and a sphere (or circle). Professional webpage. Available from: http://paulbourke.net/geometry/sphereline

Elfes, A. (1989). Using occupancy grids for mobile robot perception and navigation. *Computer*, 22(6), 46-57

Furukawa, T. (2006). *Reactive obstacle avoidance for the REMUS underwater autonomous vehicle using a forward looking sonar*. MS Thesis, NPS, Monterey, CA, USA

Gadre, A., Kragelund, S., Masek, T., Stilwell, D., Woolsey, C. & Horner, D. (2009). Subsurface and surface sensing for autonomous navigation in a riverine environment. In: *Proceedings of the Association of Unmanned Vehicle Systems International (AUVSI) Unmanned Systems North America convention, Washington, DC, USA*

Healey, A. J. (2004). Obstacle avoidance while bottom following for the REMUS autonomous underwater vehicle. In: *Proceedings of the IFAC conference, Lisbon, Portugal*

Horner, D. & Yakimenko, O. (2007). Recent developments for an obstacle avoidance system for a small AUV. In: *Proceedings of the IFAC conference on Control Applications in Marine Systems, Bol, Croatia*

Horner, D., McChesney, N., Kragelund, S. & Masek, T. (2009). 3D reconstruction with an AUV-mounted forward-looking sonar. In: *Proceedings of the International symposium on Unmanned Untethered Submersible Technology (UUST09), Durham, NH, USA*

Hydroid, Inc. (2011). REMUS 100 webpage. Available from: www.hydroidinc.com/remus100.html

Kaminer, I., Yakimenko, O., Dobrokhodov, V., Pascoal, A., Hovakimyan, N., Cao, C., Young, A. & Patel, V. (2007). Coordinated path following for time-critical missions of multiple UAVs via L_1 adaptive output feedback controllers. In: *Proceedings of the AIAA Guidance, Navigation, and Control conference, Hilton Head, SC, USA*

Kreuzer, J. (2006). 3D programming – weekly: Bounding boxes. Collision detection tutorial webpage. Available from: www.3dkingdoms.com/weekly/weekly.php?a=21

Masek, T. (2008). *Acoustic image mModels for navigation with forward-looking sonars*. MS Thesis, NPS, Monterey, CA, USA

Northwind Marine. (2011). SeaFox webPage. Available from: www.northwindmarine.com/military-boats

Sethian, J. (1999). Fast marching method. *SIAM Review*, 41(2), 199-235

Xu, B., Kurdila, A. J. & Stilwell, D. J. (2009). A hybrid receding horizon control method for path planning uncertain environments. In: *Proceedings of the IEEE/RSJ International conference on Intelligent Robots and Systems, St. Louis, MO, USA*

Yakimenko, O. & Slegers, N. (2009). Optimal control for terminal guidance of autonomous parafoils. In: *Proceedings of the 20th AIAA Aerodynamic Decelerator Systems Technology conference, Seattle, WA, USA*

Yakimenko, O. (2000). Direct method for rapid prototyping of near optimal aircraft trajectories. *Journal of Guidance, Control, and Dynamics*, 23(5), 865-875

Yakimenko, O. (2011). *Engineering computations and modeling in MATLAB/Simulink*. AIAA Education Series, ISBN 978-1-60086-781-1, Arlington, VA, USA

Yakimenko, O. A. (2008). Real-time computation of spatial and flat obstacle avoidance trajectories for AUVs. In: *Proceedings of the 2nd IFAC workshop on Navigation, Guidance and Control of Underwater Vehicles (NGCUV'08), Killaloe, Ireland*

Yakimenko, O.A., Horner, D.P. & Pratt, D.G. (2008). AUV rendezvous trajectories generation for underwater recovery, In: *Proceedings of the 16th Mediterranean conference on Control and Automation, Corse, France*

7

Fully Coupled 6 Degree-of-Freedom Control of an Over-Actuated Autonomous Underwater Vehicle

Matthew Kokegei, Fangpo He and Karl Sammut
Flinders University
Australia

1. Introduction

Unmanned underwater vehicles (UUVs) are increasingly being used by civilian and defence operators for ever more complex and dangerous missions. This is due to the underlying characteristics of safety and cost effectiveness when compared to manned vehicles. UUVs require no human operator be subject to the conditions and dangers inherent in the underwater environment that the vehicle is exposed to, and therefore the risk to human life is greatly minimised or even removed. Cost effectiveness, in both time and financial respects, comes from a much smaller vehicle not containing the various subsystems required to sustain life whilst underwater, as well as smaller, less powerful actuators not placing the same levels of stress and strain on the vehicles as compared to a manned vehicle. This leads to a much smaller team required to undertake the regular maintenance needed to keep a vehicle operational. Taking these two main factors into account, the progression from manned vehicles to unmanned vehicles is a logical step within the oceanographic industry.

Within the broad class of UUVs are the remotely operated vehicles (ROVs) and the autonomous underwater vehicles (AUVs). Both of these types of vehicles have been successfully used in industry, and their fundamental differences determine which type of vehicle is suited to a particular mission. The key difference between the two is that an ROV requires a tether of some description back to a base station, whereas an AUV does not. This tether connects the ROV to a human operator who can observe the current state of the vehicle and therefore provide the control for the vehicle while it executes its mission. This tether, depending on the configuration of the vehicle, can also provide the electricity to power the vehicles actuators, sensors and various internal electronic systems.

AUVs have an advantage over ROVs of not requiring this tether, which leads to two main benefits. Firstly, an AUV requires little or no human interaction while the vehicle is executing its mission. The vehicle is pre-programmed with the desired mission objectives and, upon launch, attempts to complete these objectives without intervention from personnel located at the base station. This minimises the effect of human error while the vehicle is operational. The second benefit is the increased manoeuvrability that is possible without a cable continuously attached to the AUV. This tether has the potential to become caught on underwater structures, which could limit the possible working environments of an ROV, as well as cause drag on the motion of the vehicle, thus affecting its manoeuvring performance. The possible range of the vehicle from the base station is also restricted, depending on the length of this tether. These two main benefits of AUVs over ROVs lead, in principle, to autonomous vehicles being selected for survey tasks in complex, dynamic and dangerous underwater environments, and therefore AUVs are the subject of this chapter. Furthermore, combining the desired performance characteristics of AUVs, with the aforementioned complex operation environment, leads to the conclusion that controllers implemented within AUVs must be precise and accurate, as well as robust to disturbances and uncertainties. Hence, the focus of this chapter will primarily be on the precise and robust control of AUVs.

Within the autonomy architecture of AUVs are three main systems. These are:- the *guidance system*, which is responsible for generating the trajectory for the vehicle to follow; the *navigation system*, which produces an estimate of the current state of the vehicle; and the *control system*, which calculates and applies the appropriate forces to manoeuvre the vehicle (Fossen, 2002). This chapter will focus on the control system and its two principal subsystems, namely the *control law* and the *control allocation*.

The chapter will be divided into three main parts with the first part focusing on the design and analysis of the control law, the second looking at control allocation, and the third providing an example of how these two systems combine to form the overall control system. Within the control law design and analysis section, the requirements of how the various systems within an AUV interact will be considered, paying particular attention to how these systems relate to the control system. An overview of the underwater environment will be given, which depicts the complexity of the possible disturbances acting on a vehicle. This will be followed by an analysis of the equations of motion, namely the kinematic and kinetic equations of motion that determine how a rigid body moves through a fluid. A summary of the relevant frames of reference used within the setting of underwater vehicles will be included. A review of the control laws that are typically used within the context of underwater vehicles will be conducted to conclude this section of the chapter.

The second section will look at the role of control allocation in distributing the desired control forces across a vehicle's actuators. An analysis of the principal types of actuators currently available to underwater vehicles will be conducted, outlining their useful properties, as well as their limitations. This section will conclude with an overview of various techniques for performing control allocation, with varying degrees of computational complexity.

The third and final section of this chapter will present an example of an overall control system for implementation within the architecture of AUVs. This example will demonstrate how the control law and control allocation subsystems interact to obtain the desired trajectory tracking performance while making use of the various actuators on the vehicle.

2. Control law design and analysis

Before delving into the laws governing how a particular control system produces a correcting signal, it is necessary to look at the requirements of the various systems within an AUV. This will provide an understanding of how the control system fits in with respect to the overall autonomy architecture of an AUV. The different types of disturbances must be acknowledged such that the effect of these disturbances is minimised, and the equations related to the dynamic motion of the vehicle must be analysed. Only after reviewing these factors can the control law be designed and analysed.

2.1 Requirements

As previously stated, the various components that make up the autonomy architecture of AUVs are the guidance system, navigation system, and control system. All three of these systems have their own individual tasks to complete, yet must also work cooperatively in order to reliably allow a vehicle to complete its objectives. Figure 1 shows a block diagram of how these various systems interact.

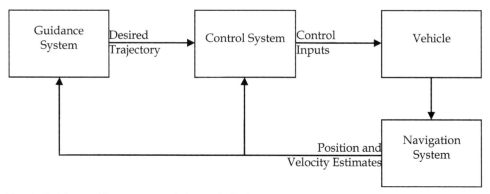

Fig. 1. Guidance, Navigation and Control Block Diagram.

2.1.1 Guidance

The guidance system is responsible for producing the desired trajectory for the vehicle to follow. This task is completed by taking the desired waypoints defined pre-mission and, with the possible inclusion of external environmental disturbances, generates a path for the vehicle to follow in order to reach each successive waypoint (Fossen, 1994, 2002). Information regarding the current condition of the vehicle, such as actuator configuration and possible failures, can also be utilised to provide a realistic trajectory for the vehicle to follow. This trajectory then forms the desired state of the vehicle, as it contains the desired position, orientation, velocity, and acceleration information.

2.1.2 Navigation

The navigation system addresses the task of determining the current state of the vehicle. For surface, land and airborne vehicles, global positioning system (GPS) is readily available and is often used to provide continuous accurate positioning information to the navigation system. However, due to the extremely limited propagation of these signals through water, GPS is largely unavailable for underwater vehicles. The task of the navigation system is then to compute a best estimate of the current state of the vehicle based on multiple measurements from other proprioceptive and exteroceptive sensors, and to use GPS only when it is available. This is completed by using some form of sensor fusion technique, such as Kalman filtering or Particle filtering (Lammas et al., 2010, 2008), to obtain a best estimate of the current operating condition, and allow for inclusion of a correction mechanism when GPS is available, such as when the vehicle is surfaced. Overall, the task of the navigation system is to provide a best estimate of the current state of the vehicle, regardless of what sensor information is available.

2.1.3 Control
The control system is responsible for providing the corrective signals to enable the vehicle to follow a desired path. This is achieved by receiving the desired state of the vehicle from the guidance system, and the current state of the vehicle from the navigation system. The control system then calculates and applies a correcting force, through use of the various actuators on the vehicle, to minimise the difference between desired and current states (Fossen, 1994, 2002). This allows the vehicle to track a desired trajectory even in the presence of unknown disturbances. Even though each of the aforementioned systems is responsible for their own task, they must also work collaboratively to fully achieve autonomy within an underwater vehicle setting.

2.2 Environment
Underwater environments can be extremely complex and highly dynamic, making the control of an AUV a highly challenging task. Such disturbances as currents and waves are ever present and must be acknowledged in order for an AUV to traverse such an environment.

2.2.1 Currents
Ocean currents, the large scale movement of water, are caused by many sources. One component of the current present in the upper layer of the ocean is due to atmospheric wind conditions at the sea surface. Differing water densities, caused by combining the effect of variation of salinity levels with the exchange of heat that occurs at the ocean surface, cause additional currents known as thermohaline currents, to exist within the ocean. Coriolis forces, forces due to the rotation of the Earth about its axis, also induce ocean currents, while gravitational forces due to other planetary objects, such as the moon and the sun, produce yet another effect on ocean currents (Fossen, 1994, 2002). Combining all of these sources of water current, with the unique geographic topography that are present within isolated coastal regions, leads to highly dynamic and complex currents existing within the world's oceans.

2.2.2 Wind generated waves
There are many factors that lead to the formation of wind generated waves in the ocean. Wind speed, area over which the wind is blowing, duration of wind influencing the ocean surface, and water depth, are just some of the elements that lead to the formation of waves. Due to the oscillatory motion of these waves on the surface, any vehicle on the surface will experience this same oscillatory disturbance. Moreover, an underwater vehicle will experience both translational forces and rotational moments while at or near the surface due to this wave motion.

2.3 Dynamics
All matter that exists in our universe must adhere to certain differential equations determining its motion. By analysing the physical properties of an AUV, a set of equations can be derived that determine the motion of this vehicle through a fluid, such as water. To assist in reducing the complexities of these equations, certain frames of reference are utilised depending on the properties that each frame of reference possesses. In order to make use of

these different reference frames for different purposes, the process of transforming information from one frame to another must be conducted.

2.3.1 Frames of reference
Within the context of control systems, the two main reference frames used are the n-frame and the b-frame. Both contain three translational components and three rotational components, yet the origin of each frame differs. This difference in origin can lead to useful properties which contain certain advantages when designing a control system.

2.3.1.1 N-Frame
The n-frame is a co-ordinate space usually defined as a plane oriented at a tangent to the Earth's surface. The most common of these frames for underwater vehicle control design is the North-East-Down (NED) frame. As its name suggests, the three axes of the translational components of this frame have the x-axis pointing towards true North, the y-axis pointing towards East, and the z-axis in the downward direction perpendicular to the Earth's surface. In general, waypoints are defined with reference to a fixed point on the earth, and therefore it is convenient to conduct guidance and navigation in this frame.

2.3.1.2 B-Frame
The b-frame, also known as the body frame, is a moving reference frame that has its origin fixed to the body of a vehicle. Due to various properties that exist at different points within the body of the vehicle, it is convenient to place the origin of this frame at one of these points to take advantage of, for example, body symmetries, centre of gravity, or centre of buoyancy. As a general rule, the x-axis of this frame points from aft to fore along the longitudinal axis of the body, the y-axis points from port to starboard, and the z-axis points from top to bottom. Due to the orientation of this frame, it is appropriate to express the velocities of the vehicle in this frame.

2.3.2 Kinematic equation
As mentioned in 2.3.1, both the NED and body frames have properties that are useful for underwater vehicle control design. Because both are used for different purposes, a means of converting information from one frame to the other is required. The kinematic equation, (1), achieves this task (Fossen, 1994, 2002).

$$\dot{\eta} = J(\eta)\upsilon \tag{1}$$

Here, the 6 degree-of-freedom (DoF) position and orientation vector in (1), decomposed in the NED frame, is denoted by (2).

$$\eta = \begin{bmatrix} p^n & \Theta \end{bmatrix}^T \tag{2}$$

Within (2), the three position components are given in (3),

$$p^n = \begin{bmatrix} x_n & y_n & z_n \end{bmatrix}^T \tag{3}$$

and the three orientation components, also known as Euler angles, are given in (4).

$$\Theta = \begin{bmatrix} \phi & \theta & \psi \end{bmatrix}^{T} \tag{4}$$

The 6 DoF translational and rotational velocity vector in (1), decomposed in the body frame, is denoted by (5).

$$v = \begin{bmatrix} v_o^b & \omega_{nb}^b \end{bmatrix}^{T} \tag{5}$$

Here, the three translational velocity components are given in (6),

$$v_o^b = \begin{bmatrix} u & v & w \end{bmatrix}^{T} \tag{6}$$

and the three rotational velocity components are given in (7).

$$\omega_{nb}^b = \begin{bmatrix} p & q & r \end{bmatrix}^{T} \tag{7}$$

In order to rotate from one frame to the other, a transformation matrix is used in (1). This transformation matrix is given in (8).

$$J(\eta) = \begin{bmatrix} R_b^n(\Theta) & 0_{3\times3} \\ 0_{3\times3} & T_\Theta(\Theta) \end{bmatrix} \tag{8}$$

Here, the transformation of the translational velocities from the body frame to the NED frame are achieved by rotating the translational velocities in the body frame, (6), using the Euler angles (4). Three principal rotation matrices are used in this operation, as shown in (9).

$$R_{x,\phi} = \begin{bmatrix} 1 & 0 & 0 \\ 0 & \cos\phi & -\sin\phi \\ 0 & \sin\phi & \cos\phi \end{bmatrix}, R_{y,\theta} = \begin{bmatrix} \cos\theta & 0 & \sin\theta \\ 0 & 1 & 0 \\ -\sin\theta & 0 & \cos\theta \end{bmatrix}, R_{z,\psi} = \begin{bmatrix} \cos\psi & -\sin\psi & 0 \\ \sin\psi & \cos\psi & 0 \\ 0 & 0 & 1 \end{bmatrix} \tag{9}$$

The order of rotation is not arbitrary, due to the compounding effect of the rotation order. Within guidance and control, it is common to use the zyx-convention where rotation is achieved using (10).

$$R_b^n(\Theta) = R_{z,\psi} R_{y,\theta} R_{x,\phi} \tag{10}$$

Overall, this yields the translational rotation matrix (11).

$$R_b^n(\Theta) = \begin{bmatrix} \cos\psi\cos\theta & -\sin\psi\cos\phi + \cos\psi\sin\theta\sin\phi & \sin\psi\sin\phi + \cos\psi\sin\theta\cos\phi \\ \sin\psi\cos\theta & \cos\psi\cos\phi + \sin\psi\sin\theta\sin\phi & -\cos\psi\sin\phi + \sin\psi\sin\theta\cos\phi \\ -\sin\theta & \cos\theta\sin\phi & \cos\theta\cos\phi \end{bmatrix} \tag{11}$$

The transformation of rotational velocities from the body frame to the NED frame is achieved by again applying the principal rotation matrices of (9). For ease of understanding, firstly consider the rotation from the NED frame to the body frame in which ψ is rotated by $R_{y,\theta}$, added to $\dot{\theta}$, and this sum then rotated by $R_{x,\phi}$ and finally added to $\dot{\phi}$. This process is given in (12).

$$\omega_{nb}^{b} = \begin{bmatrix} \dot{\phi} \\ 0 \\ 0 \end{bmatrix} + R_{x,\phi}^{T} \left(\begin{bmatrix} 0 \\ \dot{\theta} \\ 0 \end{bmatrix} + R_{y,\theta}^{T} \begin{bmatrix} 0 \\ 0 \\ \dot{\psi} \end{bmatrix} \right) \tag{12}$$

By expanding (12), the matrix for transforming the rotational velocities from the NED frame to the body frame is defined as (13),

$$T_{\Theta}^{-1}(\Theta) := \begin{bmatrix} 1 & 0 & -\sin\theta \\ 0 & \cos\phi & \cos\theta\sin\phi \\ 0 & -\sin\phi & \cos\theta\cos\phi \end{bmatrix} \tag{13}$$

and therefore the matrix for transforming the rotational velocities from the body frame to the NED frame is given in (14).

$$T_{\Theta}(\Theta) = \begin{bmatrix} 1 & \sin\phi\tan\theta & \cos\phi\tan\theta \\ 0 & \cos\phi & -\sin\phi \\ 0 & \sin\phi/\cos\theta & \cos\phi/\cos\theta \end{bmatrix} \tag{14}$$

Overall, (1) achieves rotation from the body frame to the NED frame, and by taking the inverse of (8), rotation from the NED frame to the body frame can be achieved, as shown in (15).

$$v = J^{-1}(\eta)\dot{\eta} \tag{15}$$

2.3.3 Kinetic equation
The 6 DoF nonlinear dynamic equations of motion of an underwater vehicle can be conveniently expressed as (16) (Fossen, 1994, 2002).

$$M\dot{v} + C(v)v + D(v)v + g(\eta) = \tau + \omega \tag{16}$$

Here, M denotes the 6×6 system inertia matrix containing both rigid body and added mass, as given by (17).

$$M = M_{RB} + M_{A} \tag{17}$$

Similar to (17), the 6×6 Coriolis and centripetal forces matrix, including added mass, is given by (18).

$$C(v) = C_{RB}(v) + C_{A}(v) \tag{18}$$

Linear and nonlinear hydrodynamic damping are contained within the 6×6 matrix $D(v)$, and given by (19).

$$D(v) = D + D_{n}(v) \tag{19}$$

Here, D contains the linear damping terms, and $D_{n}(v)$ contains the nonlinear damping terms.

The 6×1 vector of gravitational and buoyancy forces and moments are represented in (16) by $g(\eta)$, and determined using (20).

$$g(\eta) = \begin{bmatrix} (W-B)\sin\theta \\ -(W-B)\cos\theta\sin\phi \\ -(W-B)\cos\theta\cos\phi \\ -\left(y_gW - y_bB\right)\cos\theta\cos\phi + \left(z_gW - z_bB\right)\cos\theta\sin\phi \\ \left(z_gW - z_bB\right)\sin\theta + \left(x_gW - x_bB\right)\cos\theta\cos\phi \\ -\left(x_gW - x_bB\right)\cos\theta\sin\phi - \left(y_gW - y_bB\right)\sin\theta \end{bmatrix} \tag{20}$$

Here, W is the weight of the vehicle, determined using $W=mg$ where m is the dry mass of the vehicle and g is the acceleration due to gravity. B is the buoyancy of the vehicle which is due to how much fluid the vehicle displaces while underwater. This will be determined by the size and shape of the vehicle. Vectors determining the locations of the centre of gravity and the centre of buoyancy, relative to the origin of the body frame, are given by (21) and (22) respectively.

$$r_g^b = \begin{bmatrix} x_g & y_g & z_g \end{bmatrix}^T \tag{21}$$

$$r_b^b = \begin{bmatrix} x_b & y_b & z_b \end{bmatrix}^T \tag{22}$$

The 6×1 vector of control input forces is denoted by τ, and is given by (23).

$$\tau = \begin{bmatrix} X & Y & Z & K & M & N \end{bmatrix}^T \tag{23}$$

Here, the translational forces affecting surge, sway and heave are X, Y, and Z respectively, and the rotational moments affecting roll, pitch and yaw are K, M and N respectively.
The 6×1 vector of external disturbances is denoted by ω.
Overall, (16) provides a compact representation for the nonlinear dynamic equations of motion of an underwater vehicle, formulated in the body frame. By applying the rotations contained within (8), (16) can be formulated in the NED frame as given in (24).

$$M_\eta(\eta)\ddot{\eta} + C_\eta(v,\eta)\dot{\eta} + D_\eta(v,\eta)\dot{\eta} + g_\eta(\eta) = \tau_\eta(\eta) + \omega \tag{24}$$

Within (24), the equations in (25) contain the rotations of the various matrices from the body frame to the NED frame.

$$\begin{aligned} M_\eta(\eta) &= J^{-T}(\eta)MJ^{-1}(\eta) \\ C_\eta(v,\eta) &= J^{-T}(\eta)\left[C(v) - MJ^{-1}(\eta)\dot{J}(\eta)\right]J^{-1}(\eta) \\ D_\eta(v,\eta) &= J^{-T}(\eta)D(v)J^{-1}(\eta) \\ g_\eta(\eta) &= J^{-T}(\eta)g(\eta) \\ \tau_\eta(\eta) &= J^{-T}(\eta)\tau \end{aligned} \tag{25}$$

The presence of nonlinearities contained within $D_n(v)$, combined with the coupling effect of any non-zero off-diagonal elements within all matrices, can lead to a highly complex model containing a large number of coefficients.

2.4 Control laws

Various control strategies, and therefore control laws, have been implemented for AUV systems.

The benchmark for control systems would be the classical proportional-integral-derivative (PID) control that has been used successfully to control many different plants, including autonomous vehicles. PID schemes are, however, not very effective in handling the nonlinear AUV dynamics with uncertain models operating in unknown environments with strong wave and current disturbances. PID schemes are therefore only generally used for very simple AUVs working in environments without any external disturbances.

An alternative control scheme known as sliding mode control (SMC), which is a form of variable structure control, has proven far more effective and robust at handling nonlinear dynamics with modelling uncertainties and nonlinear disturbances. SMC is a nonlinear control strategy which uses a nonlinear switching term to obtain a fast transient response while still maintaining a good steady-state response. Consequently, SMC has been successfully applied by many researchers in the AUV community. One of the earliest applications of using SMC to control underwater vehicles was conducted by Yoerger and Slotine wherein the authors demonstrated through simulations studies on an ROV model, the SMC controller's robustness properties to parametric uncertainties (Yoerger & Slotine, 1985). A multivariable sliding mode controller based on state feedback with decoupled design for independently controlling velocity, steering and diving of an AUV is presented in Healey and Lienard (1993). The controller design was successfully implemented on NPS ARIES AUV as reported in Marco and Healey (2001).

2.4.1 PID control

The fundamentals of PID control is that an error signal is generated that relates the desired state of the plant to the actual state (26),

$$e(t) = x_d(t) - x(t) \tag{26}$$

Where $e(t)$ is the error signal, $x_d(t)$ is the desired state of the plant, and $x(t)$ is the current state of the plant, and this error signal is manipulated to introduce a corrective action, denoted $\tau(t)$, to the plant.

PID control is named due to the fact that the three elements that make up the corrective control signal are: *proportional* to the error signal by a factor of K_P, a scaled factor, K_I, of the *integral* of the error signal, and a scaled factor, K_D, of the *derivative* of the error signal, respectively (27).

$$\tau(t) = K_P e(t) + K_I \int_0^t e(\lambda) d\lambda + K_D \frac{d}{dt} e(t) \tag{27}$$

PID control is best suited to linear plants, yet has also been adopted for use on nonlinear plants even though it lacks the same level of performance that other control systems possess.

However, due to the wide use and acceptance of PID control for use in controlling a wide variety of both linear and nonlinear plants, it is very much employed as the "gold standard" that control systems are measured against. An example of a PID-based control strategy applied to underwater vehicles is given in Jalving (1994).

2.4.2 Sliding mode control

As mentioned previously, sliding mode control is a scheme that makes use of a discontinuous switching term to counteract the effect of dynamics that were not taken into account at the design phase of the controller.

To examine how to apply sliding mode control to AUVs, firstly (24) is compacted to the form of (28),

$$M_\eta(\eta)\ddot{\eta} + f(\dot{\eta},\eta,t) = \tau_\eta(\eta) \tag{28}$$

where $f(\dot{\eta},\eta,t)$ contains the nonlinear dynamics, including Coriolis and centripetal forces, linear and nonlinear damping forces, gravitational and buoyancy forces and moments, and external disturbances.

If a sliding surface is defined as (29),

$$s = \dot{\eta} + c\eta \tag{29}$$

where c is positive, it can be seen that by setting s to zero and solving for η results in η converging to zero according to (30)

$$\eta(t) = \eta(t_0)e^{ct_0}e^{-ct} \tag{30}$$

regardless of initial conditions. Therefore, the control problem simplifies to finding a control law such that (31) holds.

$$\lim_{t \to \infty} s(t) = 0 \tag{31}$$

This can be achieved by applying a control law in the form of (32),

$$\tau = -T(\dot{\eta},\eta)\text{sign}(s), T(\dot{\eta},\eta) > 0 \tag{32}$$

with $T(\dot{\eta},\eta)$ being sufficiently large. Thus, it can be seen that the application of (32) will result in η converging to zero.

If η is now replaced by the difference between the current and desired states of the vehicle, it can be observed that application of a control law of this form will now allow for a reference trajectory to be tracked.

Two such variants of SMC are the *uncoupled* SMC and the *coupled* SMC.

2.4.2.1 Uncoupled SMC

Within the kinetic equation of an AUV, (16), simplifications can be applied that will reduce the number of coefficients contained within the various matrices. These simplifications can be applied due to, for example, symmetries present in the body of the vehicle, placement of centres of gravity and buoyancy, and assumptions based on the level of effect a particular coefficient will have on the overall dynamics of the vehicle. Thus, the assumption of body symmetries allows reduced level of coupling between the various DoFs. An uncoupled SMC therefore assumes that no coupling exists between the various DoFs, and that simple manoeuvring is employed such that it does not excite these coupling dynamics (Fossen, 1994). The effect this has on (16) is to remove all off-diagonal elements within the various matrices which significantly simplifies the structure of the mathematical model of the vehicle (Fossen, 1994), and therefore makes implementation of a controller substantially easier.

2.4.2.2 Coupled SMC

Although the removal of the off-diagonal elements reduces the computational complexity of the uncoupled SMC, it also causes some limitation to the control performance of AUVs, particularly those operating in highly dynamic environments and required to execute complex manoeuvres. Taking these two factors into account, these off-diagonal coupling terms will have an influence on the overall dynamics of the vehicle, and therefore cannot be ignored at the design phase of the control law.

Coupled SMC is a new, novel control law that retains more of the coupling coefficients present in (16) compared to the uncoupled SMC (Kokegei et al., 2008, 2009). Furthermore, even though it is unconventional to design a controller in this way, the body frame is selected as the reference frame for this controller. This selection avoids the transformations employed in (24) and (25) used to rotate the vehicle model from the body frame to the NED frame although it does require that guidance and navigation data be transformed from the NED frame to the body frame. By defining the position and orientation error in the NED frame according to (33),

$$\tilde{\eta} = \hat{\eta} - \eta_d \tag{33}$$

where $\hat{\eta}$ represents an estimate of the current position and orientation provided by the navigation system, and η_d represents the desired position and orientation provided by the guidance system, a single rotation is required to transform this error from the NED frame to the body frame.

In general, desired and current velocity and acceleration data are already represented in the body frame, and as such, no further rotations are required here for the purposes of implementing a controller in the body frame.

By comparing the number of rotations required to transform the vehicle model into the NED frame, as seen in (25), for the uncoupled control scheme with the single rotation required by the coupled control scheme to transform the guidance and navigation data into the body frame, it can be seen that the latter has less rotations involved, and is therefore less computationally demanding.

3. Control allocation

The role of the control law is to generate a generalised force to apply to the vehicle such that a desired state is approached. This force, τ, for underwater vehicles consists of six components, one for each DoF, as seen in (23). The control allocation system is responsible for distributing this desired force amongst all available actuators onboard the vehicle such that this generalised 6 DoF force is realised. This means that the control allocation module must have apriori knowledge of the types, specifications, and locations, of all actuators on the vehicle.

3.1 Role

The role of the control allocation module is to generate the appropriate signals to the actuators in order for the generalised force from the control law to be applied to the vehicle. Since the vehicle under consideration is over-actuated, which means multiple actuators can apply forces to a particular DoF, the control allocation is responsible for utilising all available actuators in the most efficient way to apply the desired force to the vehicle. Power consumption is of particular importance for all autonomous vehicles, as it is a key factor in determining the total mission duration. The control allocation is therefore responsible for applying the desired forces to the vehicle, while minimising the power consumed.

3.2 Actuators

The force applied to a vehicle due to the various actuators of a vehicle can be formulated as (34),

$$\tau = TKu \tag{34}$$

where, for an AUV operating with 6 DoF with n actuators, T is the actuator configuration matrix of size $6 \times n$, K is the diagonal force coefficient matrix of size $n \times n$, and u is the control input of size $n \times 1$. Actuators are the physical components that apply the desired force to the vehicle, and the particular configuration of these actuators will determine the size and structure of T, K and u, with each column of T, denoted t_i, in conjunction with the corresponding element on the main diagonal of K, representing a different actuator.

A vast array of actuators are available to underwater vehicle designers, the more typical of which include propellers, control fins and tunnel thrusters, and each has their own properties that make them desirable for implementation within AUVs. For all the following actuator descriptions l_x defines the offset from the origin of the actuator along the x-axis, l_y defines the offset along the y-axis, and l_z defines the offset along the z-axis.

3.2.1 Propellers

Propellers are the most common actuators implemented to provide the main translational force that drives underwater vehicles. These are typically located at the stern of the vehicle and apply a force along the longitudinal axis of the vehicle. The structure of t_i for a propeller is given in (35).

$$t_i = \begin{bmatrix} 1 & 0 & 0 & 0 & l_z & -l_y \end{bmatrix}^T \tag{35}$$

As can be seen from (35), if the propeller is positioned such that there is no y-axis or z-axis offset, the force produced will be directed entirely along the x-axis of the vehicle, with no rotational moments produced.

3.2.2 Control surfaces

Control surfaces, or control fins, are actuators that utilise Newton's Third Law of motion to apply rotational moments to the vehicle. These surfaces apply a force to the water which causes a deflection in the water's motion. Hence, the water must also apply a force to the control surface. Due to this force being applied at a distance from the centre of gravity of the vehicle, a rotational moment is produced that acts on the vehicle. The typical configuration for control surfaces on an AUV is to have four independently controlled fins arranged in two pairs orientated horizontally and vertically at the stern of the vehicle. The structure of t_i for the horizontal fins is given in (36),

$$t_i = \begin{bmatrix} 0 & 0 & 1 & l_y & -l_x & 0 \end{bmatrix}^T \tag{36}$$

and for vertical fins given in (37).

$$t_i = \begin{bmatrix} 0 & 1 & 0 & -l_z & 0 & l_x \end{bmatrix}^T \tag{37}$$

These structures of t_i show that horizontal surfaces produce a heave force as well as roll and pitch moments, while vertical surfaces produce a sway force as well as roll and yaw moments. What must be considered here is that the force being produced by control surfaces relies on the vehicle moving relative to the water around it. If the vehicle is stationary compared to the surrounding water, control surfaces are ineffective. However, if the vehicle is moving relative to the surrounding water, these actuators are capable of applying forces and moments to the vehicle while consuming very little power.

3.2.3 Tunnel thrusters

The previously mentioned limitation of control surfaces can be overcome by the use of tunnel thrusters. These thrusters are usually implemented by being placed in tunnels transverse to the longitudinal axis of the vehicle. Similar to control surfaces, the typical arrangement is to position two horizontal tunnel thrusters equidistant fore and aft of the centre of gravity, and two vertical tunnel thrusters also equidistant fore and aft of the centre of gravity.

The structure of t_i for horizontal thrusters is given in (38),

$$t_i = \begin{bmatrix} 0 & 1 & 0 & -l_z & 0 & l_x \end{bmatrix}^T \tag{38}$$

while the structure for vertical thrusters is given in (39).

$$t_i = \begin{bmatrix} 0 & 0 & 1 & l_y & -l_x & 0 \end{bmatrix}^T \tag{39}$$

What can be observed here is that horizontal thrusters provide a sway force as well as a roll and yaw moment, while vertical thrusters provide a heave force as well as roll and pitch moment. In general, horizontal tunnel thrusters are located such that l_z is zero, and vertical thrusters are located such that l_y is zero. The result of this choice is that no roll moment is produced by these actuators.

The advantage of tunnel thrusters is that forces and moments can be produced even if the vehicle is stationary with respect to the surrounding water. This greatly increases the manoeuvrability of the vehicle, as control of the vehicle when moving at low speeds is possible. However, there are limitations associated with the use of these actuators. Firstly, these actuators consume more power when activated compared to control surfaces. This is due to force being produced by the thrusters only when the thruster itself is activated. In contrast, control surfaces consume power when the deflection angle is altered, but require very little power to hold the surface in place once the desired angle has been achieved.

Secondly, thruster efficiency is reduced when the vehicle is moving. Under certain conditions, an area of low pressure is produced at the exit of the tunnel, which has the effect of applying a force to the vehicle in the opposite direction to which the water jet from the tunnel thruster is attempting to provide. The result is less total force being applied to the vehicle, and therefore reduced performance when moving at non-zero forward speeds (Palmer et al., 2009).

3.3 Allocation methods

As previously mentioned when reviewing the various actuators for underwater vehicles, if the vehicle is stationary, control surfaces are ineffective while tunnel thrusters are very useful. Conversely, if the vehicle is moving, control surfaces are very efficient in providing force relative to power consumption compared to tunnel thrusters. The role of the control allocation is therefore to find a compromise between all actuators that both applies the desired generalised forces and moments to the vehicle while minimising power consumption. This balancing act allows the vehicle to maintain the manoeuvrability provided by all actuators, while at the same time allowing for as long a mission duration as possible.

3.3.1 Non-optimal scheme

One of the most straightforward methods for control allocation is application of the inverse of (34), i.e., (40) (Fossen, 2002).

$$u = (TK)^{-1} \tau \tag{40}$$

This method is very simple to implement as it consists of a single matrix multiplication. Therefore it is easy and efficient to implement within the computational processing constraints of an AUV. However, due to its simplicity, no attempt is made to minimise power consumption. If the vehicle contains both control surfaces and tunnel thrusters, both of these types of actuators will be utilised equally, even when the vehicle is moving at maximum velocity with respect to the surrounding fluid. However, since it is much more efficient to utilise control surfaces rather than tunnel thrusters while the vehicle is moving with respect to the water, a more intelligent approach is desired for implementing control allocation to minimise power consumption.

3.3.2 Quadratic programming

The limitations of the aforementioned non-optimal scheme can be overcome by formulating a quadratic programming optimisation problem to solve for actuator inputs (Fossen, 2002, Fossen et al., 2009). By introducing a weighting matrix into the problem statement, actuator usage can be biased towards utilising control surfaces over tunnel thrusters. Therefore, the generalised force desired from the control law can be realised by the actuators while minimising power consumption. There are however, limitations associated with this scheme. Firstly, although it is possible to calculate an explicit solution to this problem, in the event of actuator reconfiguration, such as an actuator failure, this explicit solution would need to be recalculated, which can be computationally intensive. Iterative approaches, such as sequential quadratic programming, can be implemented that allow for actuator failures, but this method has the potential to require several iterations of the programming problem be solved at each control sample interval. Again, this can be a computationally intensive task.

3.3.3 2-Stage scheme

A third scheme that is proposed here for the implementation of the control allocation is to break the control allocation problem into two smaller sub-problems, as seen in Figure 2, where the first sub-problem addresses control allocation to the main propeller and control surfaces, and the second addresses control allocation to the tunnel thrusters. Using this type of scheme, the control surfaces can firstly be used to their full extent in order to realise the generalised force as closely as possible. Only after full utilisation of the control surfaces occurs will the tunnel thrusters be introduced to provide forces and moments that the surfaces alone cannot produce. Using this methodology, the low power consumption control surfaces will be used as much as possible, while the higher power consumption tunnel thrusters will be called upon only when required to provide the extra manoeuvring capabilities that they possess. Furthermore, in a situation where accurate manoeuvring is not required, such as traversing from one waypoint to the next with no concern for what trajectory the vehicle follows, the second stage can be disabled such that no tunnel thruster is used, and control is performed entirely by the main propeller and control surfaces. However, if trajectory tracking is desired, the thruster allocation module can still be enabled. This will allow for the situation when the control surfaces provide inadequate force to the vehicle, and therefore the thrusters can assist in providing the extra force required to maintain the vehicle tracking the desired trajectory.

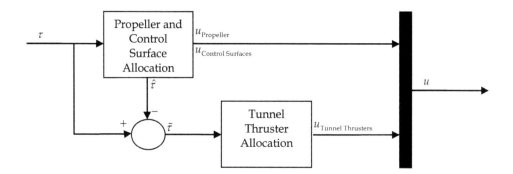

Fig. 2. 2-Stage Control Allocation Scheme Block Diagram.

Implementation of this scheme would look somewhat like a 2-stage non-optimal scheme, as seen in Figure 2. The first stage would require the matrix operation $(TK)^{-1}\tau$ for the main propeller and control surfaces in order to obtain as much force required from these actuators. An estimate of the force produced for this particular set of control values would then be calculated such that this force estimate can be subtracted from the total force required. Any residual force requirement would then become the input to the second stage of the control allocation, which would perform the matrix operation $(TK)^{-1}\tilde{\tau}$ for the tunnel thrusters in order for these actuators to provide any extra force that the control surfaces cannot deliver alone. Therefore, the computational requirement for this scheme is quite minimal compared to the quadratic programming scheme, yet still heavily biases the use of control surfaces over tunnel thrusters.

4. Overall control system

Overall, the control system of any autonomous underwater vehicle consists of a control law module and a control allocation module. The former is responsible for generating the generalised force in 6 DoF based on current and desired states, while the latter is responsible for distributing this generalised force amongst the actuators of the vehicle. Due to the modularity of such a control system, each of these subsystems can be designed and implemented independently, yet both must work cooperatively in order to accurately control the vehicle.

As discussed, the inputs to the control system are the desired state from the guidance system, and the current state from the navigation system. These are also the inputs to the control law. The outputs from the control system are the control signals that are sent to the individual actuators on the vehicle, which are the outputs generated by the control allocation. The intermediate signals connecting the control law to the control allocation are the 6 DoF generalised forces which are the outputs from the control law and therefore the inputs to the control allocation. Figure 3 shows the overall control system. As can be seen, the modularity of the system allows for control laws and control allocation schemes to be easily replaced, provided the inputs and outputs are of the same dimensions.

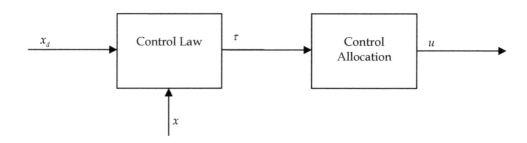

Fig. 3. Overall Control System Block Diagram.

For comparison purposes, both the uncoupled SMC law and the coupled SMC law are simulated here. Both simulations consist of identical control allocation modules that contain the previously described 2-stage scheme. The vehicle model used for these simulations is based on that which was proposed by Prestero (2001a, 2001b). This is a mathematical model for the REMUS (Remote Environmental Monitoring Unit) underwater vehicle developed by the Woods Hole Oceanographic Institute's Oceanographic Systems Laboratory. Alterations have been made to this model, based on current technology, which enable the vehicle to become fully actuated. These changes include the placement of four tunnel thrusters, two fore and two aft, such that sway, heave, pitch, and yaw motions are possible without any water flow over the control surfaces. Furthermore, the position of the vertical thrusters along the y-axis of the body, and the position of the horizontal thrusters along the z-axis of the body are set to 0m. Hence, there is no roll moment applied to the vehicle when these thrusters are activated. The thrusters introduced to the model are based on the 70mm IntegratedThruster™ produced by TSL Technology Ltd. This particular device can provide a maximum thrust of 42N, and due to its compact size, is well suited to this particular application. Also, the propulsion unit has been altered within the model. The simulation model used contains a propulsion unit based on the Tecnadyne Model 540 thruster. This device is able to provide approximately 93.2N of thrust and, with a propeller diameter of 15.2cm, is therefore well suited for this application.

The trajectory that the vehicle is asked to follow here consists of a series of unit step inputs applied to each DoF. The translational DoFs experience a step input of 1 metre whereas the rotational DoFs experience a step input of 1 radian, or approximately 60°. All inputs are applied for a period of 20 seconds, such that both transient and steady state behaviour can be observed. These unit step inputs excite the vehicle in all combinations of DoFs, from a single DoF, through to all DoFs being excited at once. This simulation assumes no water current, and therefore no water flow over the vehicle when it is stationary. Hence, roll cannot be compensated for when stationary and for this reason this simulation does not excite the roll component of the model. All other DoFs are excited, however.

Due to the way the aforementioned trajectory is supplied to the vehicle model, no guidance system is present. The step inputs will be applied at set times regardless of the state of the vehicle, and therefore observation of the control performance can be observed without influence from unnecessary systems. Therefore total execution time is constant for all simulations. Performance metrics used here in evaluating each control system is the accumulated absolute error between the desired translation/rotation and actual translation/rotation for each DoF. By looking at the following plots, observations can be made regarding such time-domain properties as rise time, settling time and percentage overshoot.

Figures 4-13 show the desired and actual trajectories for each individual DoF when this complex set of manoeuvres is applied to both the uncoupled SMC and the coupled SMC. Figures 4, 6, 8, 10 and 12 show the complete trajectory for each DoF, while Figures 5, 7, 9, 11 and 13 show these same trajectories with the focus being on the last 300 seconds of the mission. This latter section of the mission is when multiple DoFs, particularly the rotational DoFs, are excited simultaneously.

Figure 4 shows the surge motion of the vehicle for the two different control systems. As can be seen, both systems exhibit desired properties for the first 180 seconds. During this period, all manoeuvring is exciting only the translational DoFs which the main propeller and tunnel thrusters can handle independently. After 180 seconds, the other DoFs are also excited, and the effect of this combined motion produces significant overshoots, especially for the uncoupled system, within the surge motion of the vehicle. This is more easily seen in Figure 5 where the larger overshoots can be seen for the uncoupled system compared to the coupled system.

The sway motion of the AUV is shown in Figure 6. Minor overshooting is observed for both systems, especially when rotational DoFs are excited in combination with the translational DoFs. This can be observed in Figure 7. However, when the sway motion is excited, convergence to the desired set point is observed.

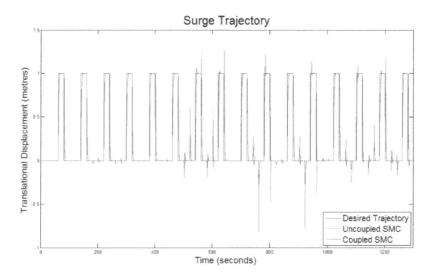

Fig. 4. Surge trajectory.

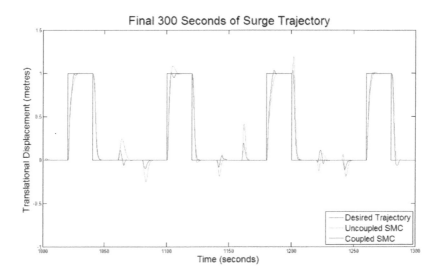

Fig. 5. Surge trajectory magnified.

Heave motion is shown in Figure 8. Here, the initial position of the vehicle is chosen to be 10 metres below the surface of the water, such that the vehicle is completely submerged for the entire trajectory. When observing the heave motion in Figure 9, it is clear that when heave is excited, convergence to the desired set point is achieved. However, it is also evident that significant coupling exists between this DoF and other DoFs as divergence from the desired set point occurs when the heave motion is not excited.

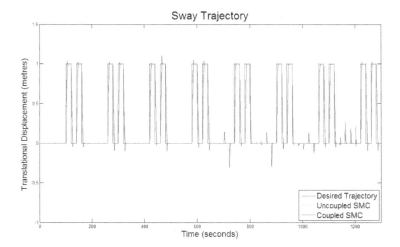

Fig. 6. Sway trajectory.

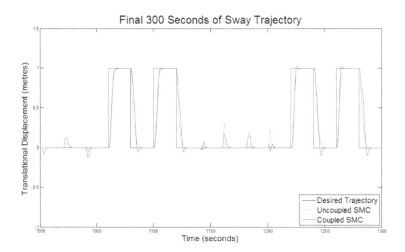

Fig. 7. Sway trajectory magnified.

Pitch motion is observed in Figure 10. The first observation that can be made from this plot is the relatively large spikes in the motion for both the coupled and uncoupled systems between 800 seconds and 1000 seconds. The second observation is that a steady state error is observed when the uncoupled system attempts a pitch of 1 radian. This is more clearly observed in Figure 11. Due to the vertical offset of the vehicle's centre of buoyancy from its centre of gravity, a restorative moment will always be applied to the vehicle for non-zero pitch angles. This plot indicates that it is difficult for the uncoupled system to compensate for this effect. However, the coupled system is able to not only eliminate this steady state error, it also has significantly smaller overshoot in general, as can be seen in Figure 10.

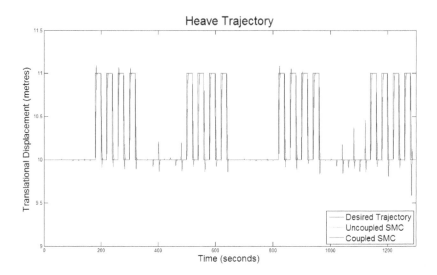

Fig. 8. Heave trajectory.

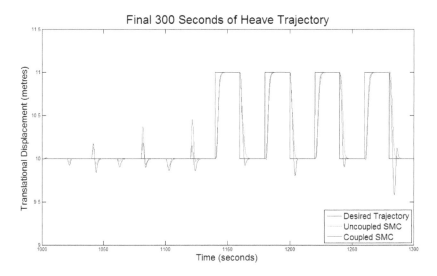

Fig. 9. Heave trajectory magnified.

Figure 12 shows the desired yaw motion of the vehicle. Here we can see significant overshooting from both systems, especially during the time period before the yaw motion is excited. By observing Figure 13, it can be seen that the coupled system achieves faster dynamics in terms of rise time, but the cost of this is overshoot. The uncoupled system has significantly less overshoot, and this is due to it taking slightly longer to react to trajectory changes.

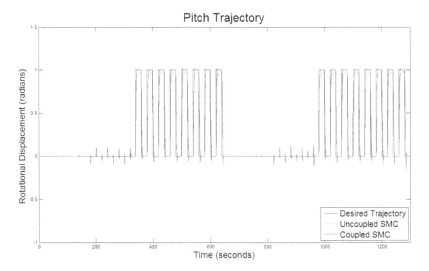

Fig. 10. Pitch trajectory.

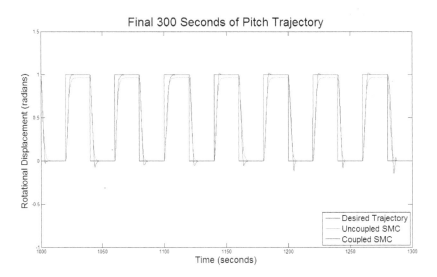

Fig. 11. Pitch trajectory magnified.

Overall, the simulation trajectory used here is extremely complex as it excites all possible combinations of DoFs without any water current present. This level of complexity allows for the coupling that exists between DoFs within an AUV model to be highlighted. For example, looking at the period of 0 to 660 seconds in Figure 12, the yaw motion is not being excited, yet motion in this DoF is observed. This coupling between multiple DoFs is the reason why controlling AUVs is a complex and challenging task.

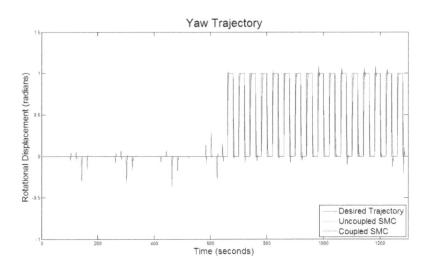

Fig. 12. Yaw trajectory.

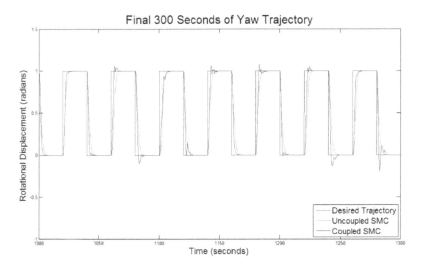

Fig. 13. Yaw trajectory magnified.

Furthermore, if the absolute error between actual and desired position is accumulated over the entire mission for each DoF, a measure of the accuracy of the control system can be obtained. A comparison can be made using Table 1. Here, we can see that by summing together the values obtained from the three translational DoFs, the coupled system improves upon the uncoupled system by 9.75%, while by summing together the values obtained from the three rotational DoFs, the coupled improves upon the uncoupled system by 23.87%. This indicates that the coupled system is superior to the uncoupled system for both translational and rotational motion, when looking at the accumulated absolute error.

Control System	Accumulated Absolute Translational Error	Accumulated Absolute Rotational Error
Uncoupled System	1.3173×10^4 (metres)	8.3713×10^3 (radians)
Coupled System	1.2003×10^4 (metres)	6.7580×10^3 (radians)

Table 1. Accumulated Absolute Translational and Rotational Errors.

5. Conclusions

Due to the increased adoption of AUVs for civilian and defence operations, accuracy and reliability are two key factors that enable an AUV to successfully complete its mission. The control system is just one of the various components within the autonomy architecture of an AUV that helps in achieving this goal. Within the control system, the control law should be robust to both external disturbances and model parameter uncertainties, while the control allocation should utilise the various actuators of the vehicle to apply the desired forces to the vehicle while minimising the power expended.

PID control has been successfully implemented on a variety of systems to effectively provide compensation. However, since PID control is better suited to linear models, the level of performance provided by PID control is not to the same standard as other, particularly nonlinear, control schemes when applied to complex nonlinear systems. Sliding mode control has proven to be a control law that is robust to parameter uncertainties, and therefore is a prime candidate for implementation within this context due to the highly complex coupled nonlinear underwater vehicle model. Active utilisation of the coupled structure of this model is what coupled SMC attempts to achieve, such that induced motion in one DoF due to motion in another DoF is adequately compensated for. This is where coupled SMC has a distinct advantage over uncoupled SMC for trajectory tracking applications when multiple DoFs are excited at once.

Various schemes exist for control allocation with the ultimate goal being to apply the desired generalised forces while minimising power consumption, both due to the actuator usage and computational demands. Non-optimal schemes exist where a generalised inverse of the force produced by all actuators is used as the allocation scheme, with the limitation being that there is no functionality to bias actuators under certain operating conditions, such as utilising control surfaces over thrusters during relatively high speed manoeuvring. Quadratic programming incorporates a weighting matrix that can bias control surface usage over tunnel thrusters, and has been implemented both online and offline, with each having advantages and disadvantages. Online optimisation allows for changes to the actuator configuration, such as failures or varied saturation limits, but is computationally demanding. Offline optimisation is less computationally demanding during mission execution, but cannot allow for altered actuator dynamics. A compromise between these schemes is the proposed 2-stage scheme where control surfaces are utilised to their full extent, and the tunnel thrusters used only when needed.

Overall, the goal of the control system is to provide adequate compensation to the vehicle, even in the presence of unknown and unmodelled uncertainties while also minimising power consumption and therefore extending mission duration. Choosing wisely both the control law and the control allocation scheme within the overall control system is fundamental to achieving both of these goals.

6. Acknowledgements

The financial support of this research from the Australian Government's Flagship Collaboration Fund through the CSIRO Wealth from Oceans Flagship Cluster on Subsea Pipelines is acknowledged and appreciated.

7. References

Fossen, T. I. (1994). *Guidance and Control of Ocean Vehicles*, John Wiley & Sons, Inc., ISBN 0-471-94113-1, Chichester, England

Fossen, T. I. (2002). *Marine Control Systems: Guidance, Navigation and Control of Ships, Rigs and Underwater Vehicles* (1), Marine Cybernetics, ISBN 82-92356-00-2, Trondheim, Norway

Fossen, T. I., Johansen, T. A. & Perez, T. (2009). A Survey of Control Allocation Methods for Underwater Vehicles, In: *Underwater Vehicles*, Inzartsev, A. V., pp. (109-128), In-Tech, Retrieved from <http://www.intechopen.com/articles/show/title/a_survey_of_control_allocatio n_methods_for_underwater_vehicles>

Healey, A. J. & Lienard, D. (1993). Multivariable Sliding Mode Control for Autonomous Diving and Steering of Unmanned Underwater Vehicles. *IEEE Journal of Oceanic Engineering*, Vol. 18, No. 3, (July), pp. (327-339), ISSN 0364-9059

Jalving, B. (1994). The NDRE-AUV Flight Control System. *IEEE Journal of Oceanic Engineering*, Vol. 19, No. 4, (October), pp. (497-501), ISSN 0364-9059

Kokegei, M., He, F. & Sammut, K. (2008). Fully Coupled 6 Degrees-of-Freedom Control of Autonomous Underwater Vehicles, *MTS/IEEE Oceans '08*, Quebec City, Canada, September 15-18

Kokegei, M., He, F. & Sammut, K. (2009), Nonlinear Fully-Coupled Control of AUVs, *Society of Underwater Technology Annual Conference*, Perth, Australia, 17-19 February

Lammas, A., Sammut, K. & He, F. (2010). 6-DoF Navigation Systems for Autonomous Underwater Vehicles, In: *Mobile Robots Navigation*, Barrera, A., pp. (457-483), In-Teh, Retrieved from <http://www.intechopen.com/articles/show/title/6-dof-navigation-systems-for-autonomous-underwater-vehicles>

Lammas, A., Sammut, K. & He, F. (2008), Improving Navigational Accuracy for AUVs using the MAPR Particle Filter, *MTS/IEEE Oceans '08*, Quebec City, Canada, 15-18 September

Marco, D. B. & Healey, A. J. (2001). Command, Control, and Navigation Experimental Results with the NPS ARIES AUV. *IEEE Journal of Oceanic Engineering*, Vol. 26, No. 4, (October), pp. (466-476), ISSN 0364-9059

Palmer, A., Hearn, G. E. & Stevenson, P. (2009). Experimental Testing of an Autonomous Underwater Vehicle with Tunnel Thrusters, *First International Symposium on Marine Propulsors*, Trondheim, Norway, 22-24 June

Prestero, T. (2001a), Development of a Six-Degree of Freedom Simulation Model for the REMUS Autonomous Underwater Vehicle, *12th International Symposium on Unmanned Untethered Submersible Technology*, University of New Hampshire, Durham, NH, 26-29 August

Prestero, T. (2001b). *Verification of a Six-Degree of Freedom Simulation Model for the REMUS Autonomous Underwater Vehicle.* Master of Science in Ocean Engineering and Master of Science in Mechanical Engineering, Massachusetts Institute of Technology and the Woods Hole Oceanographic Institution, Cambridge and Woods Hole

Yoerger, D. R. & Slotine, J.-J. E. (1985). Robust Trajectory Control of Underwater Vehicles. *IEEE Journal of Oceanic Engineering*, Vol. 10, No. 4, (October), pp. (462-470), ISSN 0364-9059

Part 3

Mission Planning and Analysis

8

Short-Range Underwater Acoustic Communication Networks

Gunilla Burrowes and Jamil Y. Khan
The University of Newcastle
Australia

1. Introduction

This chapter discusses the development of a short range acoustic communication channel model and its properties for the design and evaluation of MAC (Medium Access Control) and routing protocols, to support network enabled Autonomous Underwater Vehicles (AUV). The growth of underwater operations has required data communication between various heterogeneous underwater and surface based communication nodes. AUVs are one such node, however, in the future, AUV's will be expected to be deployed in a swarm fashion operating as an ad-hoc sensor network. In this case, the swarm network itself will be developed with homogeneous nodes, that is each being identical, as shown in Figure 1, with the swarm network then interfacing with other fixed underwater communication nodes. The focus of this chapter is on the reliable data communication between AUVs that is essential to exploit the collective behaviour of a swarm network.

A simple 2-dimensional (2D) topology, as shown in Figure 1(b), will be used to investigated swarm based operations of AUVs. The vehicles within the swarm will move together, in a decentralised, self organising, ad-hoc network with all vehicles hovering at the same depth. Figure 1(b) shows the vehicles arranged in a 2D horizontal pattern above the ocean floor

(a) AUV Swarm demonstrating stylised SeaVision©vehicles

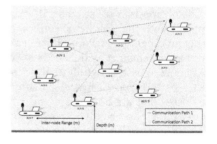

(b) 2D AUV Swarm Topology

Fig. 1. Swarm Architecture

giving the swarm the maximum coverage area at a single depth, while forming a multi-hop communication network. The coverage area will depend on application. For example, the exploration of oil and gas deposits underwater using hydrocarbon sensing would initially require a broad structure scanning a large ocean footprint before narrowing the range between vehicles as the sensing begins to target an area. Thus vehicles may need to work as closely as 10 m with inter-node communication distance extending out to 500 m. These operating distances are substantially shorter than the more traditional operations of submarines and underwater sensor to surface nodes that have generally operated at greater than 1km. Thus, the modelling and equipment development for the communication needs of these operations has focused on longer range data transmission and channel modelling. To exploit the full benefits of short range communication systems it is necessary to study the properties of short range communication channels.

Most AUV development work has concentrated on the vehicles themselves and their operations as a single unit (Dunbabin et al., 2005; Holmes et al., 2005), without giving much attention to the development of the swarm architecture which requires wireless communication networking infrastructure. To develop swarm architectures it is necessary to research effective communication and networking techniques in an underwater environment. Swarm operation has many benefits over single vehicle use. The ability to scan or 'sense' a wider area and to work collaboratively has the potential to vastly improve the efficiency and effectiveness of mission operations. Collaboration within the swarm structure will facilitate improved operations by building on the ability to operate as a team which will result in emergent behaviours that are not exhibited by individual vehicles. A swarm working collaboratively can also help to mitigate the problem of high propagation delay and lack of bandwidth available in underwater communication environments. Swarm topology will facilitate improved communication performance by utilising the inherent spatial diversity that exists in a large structure. For example, information can be transmitted more reliably within a swarm architecture by using multi-hop networking techniques. In such cases, loss of an individual AUV, which can be expected at times in the unforgiving ocean environment, will have less detrimental effect compared to a structure where multiple vehicles operate on their own. (Stojanovic, 2008).

The underwater acoustic communication channel is recognised as one of the harshest environments for data communication, with long range calculations of optimal channel capacity of less than 50kbps for SNR (Signal-to-Nosie Ratio) of 20dB (Stojanovic, 2006) with current modem capacities of less than 10kbps (Walree, 2007). Predictability of the channel is very difficult with the conditions constantly changing due to seasons, weather, and the physical surroundings of sea floor, depth, salinity and temperature. Therefore, it must be recognised that any channel model needs to be adaptable so that the model can simulate the channel dynamics to be able to fully analyse the performance of underwater networks.

In general, the performance of an acoustic communication system underwater is characterised by various losses that are both range and frequency dependent, background noise that is frequency dependent and bandwidth and transmitter power that are both range dependent. The constraints imposed on the performance of a communication system when using an acoustic channel are the high latency due to the slow speed of the acoustic signal propagation, at 0.67 ms/m (compared with RF (Radio Frequency) in air at 3.3 ns/m), and the signal fading properties due to absorption and multipath. Specific constraints on the performance due to the mobility of AUV swarms is the Doppler effect resulting from any relative motion between a transmitter and a receiver, including any natural motion present in the oceans from waves, currents and tides.

Noise in the ocean is frequency dependent. There are three major contributors to noise underwater: ambient noise which represents the noise in the far field; self noise of the vehicle (considered out of band noise); and intermittent noise sources including noises from biological sources such as snapping shrimp, ice cracking and rain. Ambient noise is therefore the component of noise taken into account in acoustic communication performance calculations. It is characterised as a Gaussian Distribution but it is not white as it does not display a constant power spectral density. For the frequencies of interest for underwater acoustic data communication, from 10 to 100 kHz, the ambient noise value decreases with increasing frequency. Therefore, using higher signal frequencies, which show potential for use in shorter range communication, will be less vulnerable to the impact of ambient noise.

Short range underwater communication systems have two key advantages over longer range operations; a lower end-to-end delay and a lower signal attenuation. End-to-end propagation at 500 m for example is approximately 0.3 sec which is considerable lower than the 2 sec at 3 km but still critical as a design parameter for shorter range underwater MAC protocols. The lower signal attenuation means potentially lower transmitter power requirements which will result in reduced energy consumption which is critical for AUVs that rely on battery power. Battery recharge or replacement during a mission is difficult and costly. The dynamics associated with attenuation also changes at short range where the spreading component dominates over the absorption component, which means less dependency on temperature, salinity and depth (pressure). This also signifies less emphasis on frequency as the frequency dependent part of attenuation is in the absorption component and thus will allow the use of higher signal frequencies and higher bandwidths at short ranges. This potential needs to be exploited to significantly improve the performance of an underwater swarm network communication system.

A significant challenge for data transmission underwater is multipath fading. The effect of multipath fading depends on channel geometry and the presence of various objects in the propagation channel. Multipath's occur due to reflections (predominately in shallow water), refractions and acoustic ducting (deep water channels), which create a number of additional propagation paths, and depending on their relative strengths and delay values can impact on the error rates at the receiver. The bit error is generated as a result of inter symbol interference (ISI) caused by these multipath signals. For very short range single transmitter-receiver systems, there could be some minimisation of multipath signals (Hajenko & Benson, 2010; Waite, 2005). For swarm operations, however, there is potentially a different mix of multipath signals that need to be taken into account, in particular, those generated due to the other vehicles in the swarm.

Careful consideration of the physical layer parameters and their appropriate design will help maximise the advantages of a short range communications system that needs to utilise the limited resources available in an underwater acoustic networking environment.

The following section will introduce the parameters associated with acoustic data transmission underwater. The underwater data transmission channel characteristics will be presented in Section 3 with a discussion of the advantages and disadvantages of the short range channel. Section 4 will show how these will impact on AUV swarm communications and the development of a short range channel model for the design and evaluation of MAC and routing protocols. This is followed in Section 5 by a discussion of the protocol techniques required for AUV swarm network design.

2. Introduction to acoustic underwater communication network

The underwater data communication link and networking environment presents a substantially different channel to the RF data communication channel in the terrestrial atmosphere. Figure 2 illustrates a typical underwater environment for data transmission using a single transmitter-receiver pair.

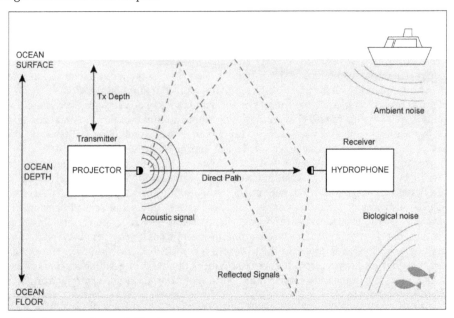

Fig. 2. Underwater Acoustic Environment

A simple schematic of the data transmission scheme involving a projector (transmitter) and a hydrophone (receiver) is presented in Figure 3. The projector takes the collected sensor and navigational data and formats it into packets at the Data Source and this is then modulated with the carrier frequency. The modulated signal is amplified to a level sufficient for signal reception at the receiver. There is an optimum amplification level as there is a trade-off between error free transmission and conservation of battery energy. The acoustic power radiated from the projector as a ratio to the electrical power supplied to it, is the efficiency η_{tx} of the projector and represented by the Electrical to Acoustic conversion block. On the receiver side, the sensitivity of the hydrophone converts the sound pressure that hits the hydrophone to electrical energy, calculated in dB/V. Signal detection, includes amplification and shaping of the input to determine a discernible signal. Here a detection threshold needs to be reached and is evaluated as the ratio of the mean signal power to mean noise power (SNR). The carrier frequency is then supplied for demodulation, before the transmitted data is available for use within the vehicle for either data storage or for input into the vehicles control and navigation requirements.

Underwater data communication links generally support low data rates mainly due to the constraints of the communication channel. The main constraints are the high propagation delay, lower effective SNR and lower bandwidth. The effects of these constraints could be reduced by using short distance links and the use of multi-hop communication techniques to cover longer transmission ranges. For an AUV swarm network, use of the above techniques could be crucial to design an effective underwater network. To develop a multi-node swarm network it is necessary to manage all point to point links using a medium access control (MAC) protocol. In a multi-access communication system like a swarm network a transmission channel is shared by many transceivers in an orderly fashion to transmit data in an interference free mode. Figure 2 shows a point to point communication link with two AUVs. When a network is scaled up to support N number of AUVs then it becomes necessary to control multiple point to point or point to multi-point links.

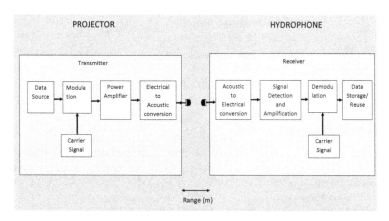

Fig. 3. Block Diagram of Projector and Hydrophone

To control the transmission of data it is necessary to design an effective MAC protocol which can control transmission of information from different AUVs. The design of a MAC protocol in a swarm network could be more complex if a multi-hop communication technique is used. The multi-hop communication technique will allow a scalable network design as well as it can support long distance transmission without the need of high power transmitter and receiver circuits. For example, using a multi-hop communication technique if AUV3 in Figure 1(b) wants to transmit packets to AUV7 then it can potentially use a number of communication paths to transmit packets. Some of the possible paths from AUV3 to AUV7 are: AUV3-AUV2-AUV1-AUV4-AUV7 or AUV3-AUV6-AUV9-AUV8-AUV7. The path selection in a network is controlled by the routing protocols. Optimum routing protocols generally select transmission paths based on a number of factors. However, the main factor used to select an optimum path in a wireless network is the SNR which indicates the quality of a link. Similarly the MAC protocol will use the transmission channel state information to develop an optimum packet access technique. To effectively design these protocols it is necessary to understand the properties of short range underwater channel characteristics. Before moving into the protocol design issues we will first evaluate the short range underwater channel characteristics in the following Sections.

3. Underwater data transmission channel characteristics

This section will focus on the parameters of the ocean channel that will affect the acoustic signal propagation from the projector to the hydrophone. There are well established underwater channel models that will be used to derive and present the data transmission characteristics for a short-range link.

3.1 Acoustic signal level

The projector source level, $SL_{tprojector}$, is generally defined in terms of the sound pressure level at a reference distance of 1 m from its acoustic centre. The source intensity at this reference range is $I = P_{tx}/Area$ (W/m^2) and measured in dB 're 1 μPa' but strictly meaning 're the intensity due to a pressure of 1 μPa'. For an omni directional projector the surface area is a sphere $(4\pi r^2 = 12.6m^2)$. Thus, $SL_{projector} = 10log((P_{tx}/12.6)/I_{ref})$ dB, where P_{tx} is the total acoustic power consumed by projector and the reference wave has an intensity: $I_{ref} = (Pa_{ref})^2/\rho * c$ (Wm^{-2}) where reference pressure level; Pa_{ref} is 1 μPa, ρ is the density of the medium and; c is the speed of sound (averages for sea water: $\rho = 1025$ kg/m^3 and c=1500 m/s) (Coates, 1989; Urick, 1967).

The equation for the transmitter acoustic signal level $(SL_{projector})$ at 1 m for an omni-directional projector can be written:

$$SL_{projector}(P) = 170.8 + 10logP_{tx} \qquad dB \qquad (1)$$

If the projector is directional, then the projector directivity index is $DI_{tx} = 10log(\frac{I_{dir}}{I_{omni}})$ where I_{omni} is the intensity if spread spherically and I_{dir} is the intensity along the axis of the beam pattern. Directivity can increase the source level by 20dB (Waite, 2005). The more general equation for the transmitter acoustic signal level $(SL_{projector})$ can be written:

$$SL_{projector}(P, \eta, DI) = 170.8 + 10logP_{tx} + 10log\eta_{tx} + DI_{tx} \qquad dB \qquad (2)$$

where the efficiency of the projector η_{tx} takes into account the losses associated with the electrical to acoustic conversion as shown in Figure 3, thus reducing the actual SL radiated by the projector. This efficiency is bandwidth dependent and can vary from 0.2 to 0.7 for a tuned projector (Waite, 2005).

3.2 Signal attenuation

Sound propagation in the ocean is influenced by the physical and chemical properties of seawater and by the geometry of the channel itself. An acoustic signal underwater experiences attenuation due to spreading and absorption. In addition, depending on channel geometry multipath fading may be experienced at the hydrophone. Path loss is the measure of the lost signal intensity from projector to hydrophone. Understanding and establishing a accurate path loss model is critical to the calculations of Signal-to-Noise ratio (SNR).

3.2.1 Spreading loss

Spreading loss is due to the expanding area that the sound signal encompasses as it geometrically spreads outward from the source.

$$PL_{spreading}(r) = k * 10log(r) \qquad dB \qquad (3)$$

where r is the range in meters and k is the spreading factor.

When the medium in which signal transmission occurs is unbounded, the spreading is spherical and the spreading factor k=2 whereas in bounded spreading, considered as cylindrical k=1. Urick (1967) suggested that spherical spreading was a rare occurrence in the ocean but recognised it may exist at short ranges. As AUV swarm operations will occur at short range it is likely that spherical spreading will need to be considered which means a higher attenuation value. Spreading loss is a logarithmic relationship with range and its impacts on the signal is most significant at very short range up to approximately 50m as seen in Figure 5(a). At these shorter ranges spreading loss plays a proportionally larger part compared with the absorption term (which has a linear relationship with range).

3.2.2 Absorption loss

The absorption loss is a representation of the energy loss in the form of heat due to the viscous friction and ionic relaxation that occurs as the wave generated by an acoustic signal propagates outwards and this loss varies linearly with range as follows:

$$PL_{absorption}(r, f) = 10log(\alpha(f)) * r \qquad dB \qquad (4)$$

where r is range in kilometres and α is the absorption coefficient.

More specifically the absorption of sound in seawater is caused by three dominant effects; viscosity (shear and volume) , ionic relaxation of boric acid and magnesium sulphate $(MgSO_4)$ molecules and the relaxation time. The effect of viscosity is significant at high frequencies above 100 kHz, whereas the ionic relaxation effects of magnesium affect the mid frequency range from 10 kHz up to 100 kHz and boric acid at low frequencies up to a few kHz. In general, the absorption coefficient, α, increases with increasing frequency and decreases as depth increases (Domingo, 2008; Sehgal et al., 2009) and is significantly higher in the sea compared with fresh water due predominately to the ionic relaxation factor.

Extensive measurements of absorption losses over the last half century has lead to several empirical formulae which take into account frequency, salinity, temperature, pH, depth and speed of sound. A popular version is Thorp's expression (Thorp, 1965), Equation 5, which is based on his initial investigations in the 60's and has since been converted into metric units (shown here). It is valid for frequencies from 100Hz to 1MHz and is based on seawater with salinity of 35% ppt, pH of 8, temp of 4°C and depth of 0 m (atmospheric pressure) which is assumed but not stated by Thorp.

$$\alpha(f) = \frac{0.11f^2}{1 + f^2} + \frac{44f^2}{4100 + f^2} + 2.75 \times 10^{-4}f^2 + 0.0033 \qquad dB/km \qquad (5)$$

Fisher and Simmons (1977) and others (Francois & Garrison, 1982) have since proposed other variations of α. In particular, Fisher and Simmons in the late 70's found the effect associated with the relaxation of boric acid on absorption and provided a more detailed form of absorption coefficient α in dB/km which varies with frequency, pressure (depth) and temperature (also valid for 100 Hz to 1 MHz with salinity 35% ppt and acidity 8 pH)(Fisher & Simmons, 1977; Sehgal et al., 2009), given in Equation 6.

$$\alpha(f, d, t) = \frac{A_1 f_1 f^2}{f_1^2 + f^2} + \frac{A_2 P_2 f_2 f^2}{f_2^2 + f^2} + A_3 P_3 f^2 \qquad dB/km \qquad (6)$$

where d is depth in meters and t is temperature in °C. The 'A' coefficients represent the effects of temperature, while the 'P' coefficients represent ocean depth (pressure) and f_1, f_2 represent the relaxation frequencies of Boric acid and $(MgSO_4)$ molecules. These terms were developed by Fisher and Simmons (1977) and presented more recently by (Domingo, 2008; Sehgal et al., 2009).

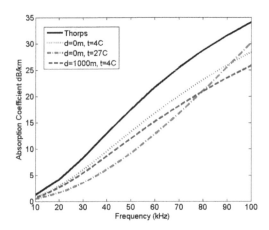

Fig. 4. Absorption Coefficient vs Frequency

Figure 4 shows the absorption coefficients in dB/km vs signal frequency for both Thorp and Fisher and Simmons coefficients and shows that in general α increases with increasing frequency at any fixed temperature and depth. Up until around 80kHz temperature change has a more significant affect on α than depth (Waite, 2005), but above these frequencies depth begins to dominate (Domingo, 2008; Sehgal et al., 2009). In any case, Thorps 'approximation' is quite close to Fisher and Simmons and is clearly more conservative at the frequencies shown. Sehgal (2009) shows that at higher frequencies above 300kHz, Thorps model predicts lower losses as it does not take into account the relaxation frequencies found by Fisher and Simmons. If depth and frequency are fixed and temperature varied from 0 to 27 °C, there is a decrease in α of approximately 4 dB/km for frequencies in the range of 30 to 60kHz which correlates to work presented by Urick (Urick, 1967)(Fig5.3 pg 89). If we consider where AUV swarms are most likely to operate, in the 'mixed surface layer', where temperature varies considerable due to latitude (but has an average temp of 17°C (Johnson, 2011)), temperature may be an important factor. It should be noted that if operating in lower temperatures α is higher and thus using 0°C will be a conservative alternative. At shorter ranges, the significance of α is expected to be less due to the linear relationship with range which will be discussed further in this chapter.

As mentioned, depth (pressure) has less of an effect on α than temperature at these lower frequencies. Domingo (Domingo, 2008) investigates the effect of depth (pressure) on absorption and confirms that for lower frequencies of less than 100kHz there is less change in α. More specifically Urick (1967) defined the variation by: $\alpha_d = \alpha * 10^{-3}(1 - 5.9 * 10^{-6}) * d$ dB/m (where d = depth in meters) but has also suggested an approximation of a 2% decrease for every 300 m depth. Thus, depth (pressure) variations are not expected to play a significant role in short range AUV swarm operations especially those that use a 2D horizontal topology as described in this chapter.

3.3 Path loss
Total path loss is the combined contribution of both the spreading and absorption losses. Urick (1967) established that this formula of spreading plus absorption yields a reasonable

agreement with long range observations.

$$PathLoss(r, f, d, t) = k * 10log(r) + \alpha(f, d, t) * r * 10^{-3} \qquad (7)$$

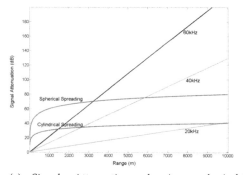
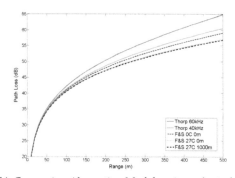

(a) Signal Attenuation showing spreading and absorption factors

(b) Comparing Absorption Models using spherical spreading. Frequency change shown using Thorp Model and Temperature °'C' and Depth 'm' changes shown in Fisher and Simmons Model.

Fig. 5. Path Loss vs Range

For very short range communication (below 50 m), see Figure 5(a), the contribution of the absorption term is less significant than the spreading term. It can be seen in Figure 5(b) that the Thorp model shows a conservative or worst case value for the ranges of interest up to 500 m. The Fisher and Simmons model for a particular frequency however provides some insight into the variations due to depth and temperature. However, the spreading factor k has the most significant affect on Path Loss, seen in Figure 5(a), at these shorter ranges according to these models.

As range increases and the absorption term begins to dominate, any variations in α also becomes more significant. For data communication, the changes in the attenuation due to signal frequency are particularly important as the use of higher frequencies will potentially provide higher data rates.

In summary using the two models, Thorp and Fisher and Simmons, the two important characteristics that can be drawn from Path Loss at the short ranges of interest for AUV swarm operation are:

- spreading loss dominates over absorption loss, and thus the 'k' term has a significant impact on the attenuation of the signal at shorter ranges as illustrated in Figure 5 (a). For AUV swarm operations and while the range between vehicles is much less than depth spherical spreading can be assumed, and

- at the ranges below 500 m the frequency component of absorption loss is most significant compared with the possible temperature and pressure (depth) changes as seen in Figure 5(b) and as range increases the difference also increases, effectively meaning that the communication channel is band-limited and available bandwidth is a decreasing function of range.

3.4 Underwater multipath characteristics

Multipath signals, in general, represent acoustic energy loss, however, for communication systems it is the Inter Symbol Interference (ISI) that will also be detrimental at the receiver as it can significantly increase the error rate of the received signal. Multipath signals are created underwater through various mechanisms described in this section, so that, at the receiver many components of the original signal will arrive at different times due to the different length of propagation paths the multipath signals have taken. It is this delay spread of the signal component arrivals that can cause ISI to occur if they overlap with previous or future signal arrivals which will cause symbol corruption or loss and therefore bit errors. As the speed of sound propagation is very slow in an acoustic channel this delay spread can be significant.

There are two major mechanisms responsible for creating multi-path signals and these are: reverberation, which refers to the reflections and scattering of the sound signal; and ray bending, which is a result of the unique sound speed structure in the oceans which create temperature gradient channels that trap acoustic signals. Multi-path signal formation is therefore determined by the geometry of the channel in which transmission is taking place, the location of the transmitter and receiver, and importantly the depth at which it is occurring. In shallow water, multi-path is due predominately to reverberation whereas in deep water it is dominated by ray bending, although reverberation will occur in deep water if the transmitter and receiver are located near the surface or bottom (Coates, 1989; Domingo, 2008; Etter, 2003; Urick, 1967).

There are several physical effects which create reverberation underwater;

- Multi-path propagation caused by boundary reflections at the sea-floor or sea-surface, seen in Figure 2.

- Multi-path propagation caused by reflection from objects suspended in the water, marine animals or plants or bubbles in the path of the transmitted signal

- Surface scattering caused by sea-surface (waves) or sea-floor roughness or surface absorption, particularly on the sea bottom depending on material

- Volume scattering caused by refractive off objects suspended in the signal path

Ray bending, causes various propagation path loss mechanisms in deep water depending on the placement of transmitter and receivers. The propagating acoustic signal bends according to Snell's Law, to lower signal speed zones. Figure 6 shows a typical ocean Sound Speed Profile, although variations occur with location and seasons. The profile is depth dependent, where sound speed is influenced more by temperature in the surface layers and by pressure at greater depths.

The various path loss mechanisms include; (Domingo, 2008)

- Surface duct, Figure 7(a) occurs when the surface layer has a positive temperature gradient, the acoustic signals can bend back towards the surface, then reflect back into the layer off the surface

- Deep Sound Channel, sometimes referred to as the SOFAR (Sound Fixing and Ranging) channel, where acoustic propagation occurs above and below the level of minimum sound speed, when the sound rays continually are bent towards the depth of minimum speed, shown in Figure 7(a)

- Convergence zone, in deep water areas when the transmitter is located quite close to the surface and the sound rays bend downwards as a result of decreasing temperatures until the increase in pressure forces the rays back towards the surface, as shown in Figure 7(b)

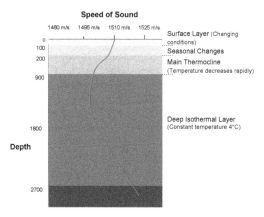

Fig. 6. Typical Sound Speed Profile in the Ocean Cox (1974)

- Reliable acoustic path, which occurs when the transmitter is located in very deep water and receiver in shallow water. Referred to as reliable as it is not generally affected by bottom or surface reflections, as shown in Figure 7(b) and

- Shadow zones that are considered a special case, as these 'zones' are void from any signal propagation. This means that in Shadow zones a hydrophone may not receive any signal at all.

Thus the geometry of the channel being used is a major determinate of the number of significant propagation paths and their relative strengths and delays. Apart from the Shadow Zones where no signal or multipath components of the signal can reach the hydrophone, the hydrophone may receive the direct signal and a combination of various multipath signals that have been reflected, scattered or bent. It is these multiple components of the signal that are delayed in time due to the various path lengths that may create ISI and errors in symbol detection.

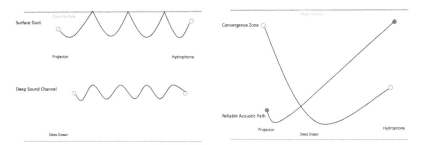

(a) Surface Duct and Deep Sound Channel   (b) Convergence Zone and Reliable Acoustic Path

Fig. 7. Ray Bending Path Loss Mechanisms

For very short range channels that will be used in AUV swarm operations, multipath will be influenced also by the range-depth ratio, which is expected to produce fewer multipath signals at the hydrophone (Hajenko & Benson, 2010; Parrish et al., 2007). In addition some improvement can be gained through directing the beam of the transmitted signal and the directional properties of the receiver (Essebbar et al., 1994), however this will require an additional level of complexity for mobile AUV's due to the need for vehicle positioning before sending or when receiving a signal.

Most of the discussion so far has focused on time-invariant acoustic channel multipath where deterministic propagation path models have been developed for the various reflective and ray bending path options. These are significant in themselves with multipath spreads in the order of 10 to 100 ms. Take Figure 2, where projector and hydrophone are separated by 100m and are at a depth of 100m, the delay spread between the direct path and the first surface reflection is $\approx$ 28 ms. Multipath in an underwater channel, however, also has time-varying components caused by the surface or volume scattering or by internal waves in deep water that are responsible for random signal fluctuations. Unlike in radio channels, the statistical characterisation of these random processes in the underwater channel are in their early development stages. Experimental results have shown that depending on the day, the location and the depth of communication link, the results of multipath can follow one of the deterministic models discussed here to worst case coherence times in the order of seconds(Stojanovic, 2006). Another source of time variability in an underwater communication channel occurs when there is relative motion between the transmitter and receiver as will be briefly discussed in the following sub-section.

3.5 The doppler effect

The motion of AUV's relative to each other will cause two possible forms of Doppler distortion in the received signal, Doppler Shifting caused by an apparent shift in frequency as the vehicles move towards or away from each other and Doppler Spreading or its time domain dual coherence time, which is the measure of the time varying nature of the frequency dispersiveness in the doppler spectrum (Rappaport, 1996). The doppler shift (Δf) of a received signal is $f_c \frac{\Delta v}{c}$ where f_c is the original signal frequency and Δv is the relative velocity between the moving vehicles. As an example, if the vehicles were moving at a moderately slow speed of 1 m/s (2 knots) relative to each other and $f_c = 40kHz$ the $\Delta f \approx 27Hz$. Doppler spread or coherence time measurements as mentioned above can be as long as 1 s. Thus doppler shifting and spreading cause complications for the receiver to track the time varying changes in the channel which need to be designed into the channel estimation algorithms and explicit delay synchronisation approach within communication protocols. As swarm operations for exploration require rigid topology where there is minimal relative speed differences between vehicles, the impact of doppler effects diminished somewhat in this context and thus will not be considered further.

3.6 Noise

There are three major contributors to noise underwater: ambient or background noise of the ocean; self noise of the vehicle; and intermittent noise including biological noises such as snapping shrimp, ice cracking and rain. An accurate noise model is critical to the evaluate the SNR at the hydrophone so that the bit error rates (BER) can be establish to evaluate protocol performance.

3.6.1 Ambient noise

Ambient Noise in the ocean has been well defined (Urick, 1967). It can be represented as Gaussian and having a continuous power spectrum density (psd). It is made up of four components (outlined below), each having a dominating influence in different portions of the frequency spectrum.

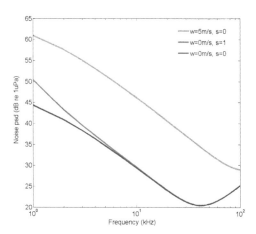

Fig. 8. Power Spectral density of the Ambient Noise, W - wind, S - shipping

For the frequency region of interest for AUV swarm communication systems (10 kHz to 100 kHz), the ambient noise psd decreases with increasing frequency, refer to Figure 8. At a frequency over 100kHz the ambient thermal noise component begins to dominate and the overall noise psd begins to increase, but this point moves further away from the frequencies of interest for AUV communication particularly as the wind speed increases.

- Turbulence noise influences only the very low frequency regions $f < 10Hz$
  $10logN_{turb}(f) = 17 - 30log(f)$
- Shipping noise dominates the 10 - 100Hz region and has defined a shipping activity factor of s whose value ranges from 0 to 1 for low to high activity respectively:
  $10logN_{ship}(f) = 40 + 20(s - 0.5) + 26log(f) - 60log(f + 0.03)$
- Wave and other surface motion caused by wind and rain is a major factor in the mid frequency region of 100Hz - 100kHz where wind speed is given by w in m/s:
  $10logN_{wind}(f) = 50 + 7.5w^{1/2} + 20log(f) - 40log(f + 0.4)$
- Thermal noise becomes dominate over 100kHz:
  $10logN_{th}(f) = -15 + 20log(f)$

where wind speed is given by w in m/s (1m/s is approximately 2 knots) and f is in kHz. Ambient Noise power also decreases with increasing depth as the distance from the surface and therefore shipping and wind noise becomes more distant. Ambient noise has been shown to be 9dB higher in shallow water than deep water (Caruthers, 1977). Swarm operations, as well as other underwater networking operations will mean that communication nodes including AUV's will be working in relatively close proximity to other nodes which will add an additional level of ambient noise to their operations due to the noise of the other vehicles in the swarm, irrespective of the operating depth. As will be discussed in the next section on Self Noise, the expectation is that this additional 'ambient noise' which relates to the 'Shipping Noise' component of ambient noise will have limited affect on the acoustic communication which generally uses frequencies above 10kHz.

3.6.2 Self noise

Self noise is defined as the noise generated by the vehicle itself as the platform for receiving signals. This noise can reach the hydrophone mounted on the AUV either through the mechanical structure or through the water passing over the hydrophone. The degree to which turbulent flows cause transducer self noise depends on the location (mounting) of the transducer and its directivity characteristics (Sullivan & Taroudakis, 2008). Self Noise can also be seen as an equivalent isotropic noise spectrum as presented by Urick from work done during WWII on submarines. In general, as with ambient noise, there is decreasing levels of self noise with increases in frequency however self noise is also significantly affected by speed with decreasing noise spectra when the vessels are travelling at slower speeds or are stationary (Eckart, 1952; Kinsler et al., 1982; Urick, 1967).

Kinsler (1982) notes that at low frequencies (<1kHz) and slow speeds machinery noise dominates and at very slow speeds self noise is usually less important than ambient noise. Whereas at higher frequencies (10kHz) propeller and flow noise begins to dominate and as speed increases the hydrodynamic noise around the hydrophone increases strongly and becomes more significant than the machinery noise. This is due to the cavitation from the propeller due to the entrainment of air bubbles under or on the blade tip of the propeller. At higher speeds, self noise can be much more significant than ambient noise and can become the limiting factor.

The self noise of different size and types of vehicles are as varied as there are vehicle designs and there is little recent published values. Each vehicle itself produces large variations in self noise with speed and operating conditions (Eckart, 1952). Self noise can be controlled by selection of motor type, configuration, mounting and motor drivers. The trend for most AUV's will be the use of small brush-less DC electric motors which have been used on the development of the SeaVision vehicle (Mare, 2010). Preliminary testing of self noise on these vehicles shows an increase in noise due to increases in speed, as has been predicted, but there was no way to distinguish between machinery and hydrodynamic affects. Higher frequency components (up to 20kHz) were present as the speed increased due to the increased work from the thrusters. When the SeaVision vehicle hovered in a stationary position the frequency of the noise psd centred around 2kHz, which is out of band noise.

Holmes (Holmes et al., 2005) at WHOI recently investigated the self noise of REMUS, their torpedo shape AUV, used as a towed array. At the maximum RPM of the AUV, the 1/3rd octave noise level, when converted to source level by the calibrated transmission factor, was 130 dB re $1\mu Pa$ at 1m directly aft of the vehicle for a centre frequency of 1000 Hz. This would represent the radiated noise source level for a vehicle moving at 3 knots (1.5m/s). Vehicles typically radiate less noise in free operating conditions than in tethered conditions, so the second trail on the REMUS was measuring the radiated noise of the vehicle to examine the power spectral density of the noise as recorded on the hydrophone array as it was towed behind the vehicle. The results showed the RPM dependent radiated noise in the aft direction at a distance of 14.6 m behind the vehicle looking at frequency range up to 2500Hz which is again out of band noise.

As the operating frequencies of the communication system is likely to be higher than most self noise, and the vehicles will operate relatively slowly, the expected contribution of self noise on the hydrophone reception will be low.

3.6.3 Intermittent sources of noise

The sources of intermittent noise can become very significant in locations or times that they occur close to operating AUV swarms. The two major areas where research has been undertaken are in the marine bio-acoustic fields and also the effect of rain and water bubbles created by raindrops.

Major contributors to underwater bio-acoustic noise include;

- Shellfish - Crustacea - most important here are the snapping scrimp who produce a broad spectrum of noise between 500Hz and 20kHz

- Fish - toadfish 10 to 50Hz

- Marine mammals - Cetacea - porpoises 20 to 120Hz

Rain creates different noise spectrum to wind and needs to be dealt with separately as it is not a constant source of noise. Urick (1967) showed examples of increases of almost 30dB in the 5 to 10kHz portion of spectrum in heavy rain, with steady rain increasing noise by 10dB or sea state equivalent increase from 2 to 6. Eckart (1952) presented average value of rain at the surface from 100Hz to 10kHz of -17 to 9dB.

These main contributors to intermittent sources predominate in the lower frequency ranges up to 20kHz. Thus, interference in the operating frequencies of communication data signals is considered low.

4. Short range channel modelling

Utilising the full capacity of the underwater acoustic channel is extremely important as the channel exhibits such challenging and limited resources as has been discussed. For short range data transmission operations there are a number of benefits that can be gained over current longer range underwater acoustic transmission. These will now be explored further in terms of data communication protocol design and development. In particular, finding the optimum signal frequency and bandwidth at different ranges and under various channel conditions will be evaluated on the basis of using the best Signal-to-Noise ratio (SNR) possible at the hydrophone. Investigation into channel capacity and BER for various possible modulation schemes will also be analysed to set up the background to the challenges for MAC and routing protocol design.

4.1 Frequency dependent component of SNR

The narrowband Signal-to-Noise-Ratio (SNR) observed at the receiver, assuming no multi-path or doppler losses is given by:

$$SNR(r, f, d, t, w, s, P_{tx}) = \frac{SL_{projector}(P_{tx}, \eta, DI)}{PathLoss(r, f, d, t) \sum Noise(f, w, s) \times B} \qquad (8)$$

where B is the receiver bandwidth and the Signal Level (SL), PathLoss and Noise terms have been previously developed.

Taking the frequency dependent portion of the SNR from Equation 8, as developed by Stojanovic (2006), is the $PathLoss(r, f, d, t) \sum Noise(f, w, s)$ product. Since SNR is inversely proportional to $PathLoss(r,f,d,t)\sum Noise(f,w,s)$ factor, the $\frac{1}{PathLoss(r,f,d,t)Noise(f,w,s)}$ term is illustrated in the Figures 9 (a) for longer ranges and Figure 9 (b) for shorter ranges. The first of these figures, shows various ranges up to 10km using Thorps absorption model (spherical spreading) and has been presented by several authors (Chen & Mitra, 2007; Nasri et al., 2008; Sehgal et al., 2009; Stojanovic, 2006). Figure 9(b) highlights shorter ranges of 500 m and 100 m and also illustrates the variation between the Thorp and Fisher and Simmons absorption loss models developed in Section 3.2.2.

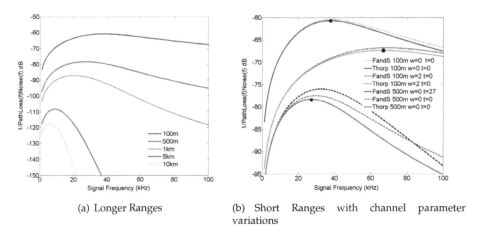

(a) Longer Ranges   (b) Short Ranges with channel parameter variations

Fig. 9. Frequency Dependent component of narrowband SNR

These figures show that there is a signal frequency where the frequency dependent component of SNR is optimised assuming that the projector parameters, including transmitter power and projector efficiency behave uniformly over the frequency band. The black dot at the apex of each of the three curves based on the Thorp absorption model indicate this optimum point. The two absorption models present similar responses and optimum frequencies. There is a minor variation in optimum frequency at 100 m where the absorption coefficient has a significantly lower contribution. The impact that the Ambient Noise component has on optimum signal frequency is seen most dramatically at the 100 m range when the wind state is changed from 0 to 2 m/s, the optimal signal frequency changes from 38 kHz to 68 kHz. From a communication perspective, if two vehicles were operating at 100 m and 38kHz and the wind state changed from 0 m/s to 2 m/s, there is a reduction of 9dB in the frequency dependent component of SNR. This is not an absolute reduction in SNR as the projector parameters and in particular the transmitter power level has not been considered here. It does however indicate the significant impact that wind and wave action can play with data communication underwater, and in addition this reduced SNR value does not include any increased losses associated with the increase scattering that wave action can generate. The impact of shipping, found in the Ambient Noise term, is not included here as its effect is minor on signal frequencies above 10kHz. In Section 3.2.2 Figure 4, temperature variations were seen to have the most impact on the signal frequency associated with the ranges of interest. Figure 9(b) illustrates this difference in terms of the frequency dependent component of SNR. This is further explored in Figure 10(a) in terms of the signal frequency variations over range.

The impact of changes in range can be seen if the vehicles moved from 100 m to 500 m (at wind state 0 m/s), the optimum signal frequency to maintain highest SNR decreases from 38 kHz to $\approx$ 28 kHz, Figure 9(b). Reduction in signal frequency implies a potential reduction in absolute bandwidth and with that a reduction in data rate which needs to be managed. This will be investigated further in the next sub sections. Figure 10 (a) and (b) show the optimum signal frequency verses range up to 500 m for the various parameters; temperature and depth, within the Thorp and Fisher and Simmons Absorption Loss models as well as the wind in the Ambient Noise model. The optimum frequency, decreases with increasing range due to the dominating characteristic of the absorption loss. It can be seen in Figure 10(a) that as the range increases there is an increasing deviation between the two models and between the parameters within the Fisher and Simmons model. There is approximately a 2.5 kHz difference between the models themselves at 500 m and up to 6 kHz when temperature increases are included. When wind is included, Figure 10(b), there is a dramatic change in optimum signal frequency at very short ranges and this difference reduces substantially over the range shown. This is due to the increasing significance of the Absorption Loss term relative to the constant Ambient Noise term (as it is not range dependent), which reduces the affect of the Noise term and therefore the wind parameter. In both Figure 10(a) and (b), the Fisher and Simmons model provides higher optimum frequencies due to the more accurate inclusion of the relaxation frequencies of boric acid and magnesium sulphate.

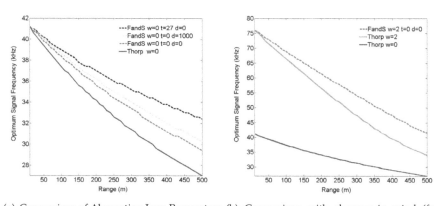

(a) Comparison of Absorption Loss Parameters (b) Comparison with changes in wind (from Ambient Noise Characteristics)

Fig. 10. Optimum Frequency determined from frequency dependent component of narrowband SNR

4.2 Channel bandwidth

Having established that at different ranges there is an optimum signal frequency that provides a maximum SNR, assuming constant transmitter power and projector efficiency, there is therefore an associated channel bandwidth with these conditions for different ranges. To determine this bandwidth a heuristic of 3dB around the optimum frequency is used. Following a similar approach to Stojanovic (2006) the bandwidth is calculated according to the frequency range using $\pm$3dB around the optimum signal frequency $f_o(r)$ which has been chosen as the centre frequency. Therefore, the $f_{min}(r)$ is the frequency when

$PathLoss(r, d, t, f_0(r))N(f_0(r)) - PathLoss(r, d, t, f))N(f) \geq 3dB$ holds true and similarly for $f_{max}(r)$ when $PathLoss(r, d, t, f))N(f) - PathLoss(r, d, t, f_0(r))N(f_0(r)) \geq 3dB$ is true. The system bandwidth B(r,d,t) is therefore determined by:

$$B(r, d, t) = f_{max}(r) - f_{min}(r) \tag{9}$$

Thus, for a given range, there exists an optimal frequency from which a range dependent 3dB bandwidth can be determined as illustrated in Figure 11. The changes discussed in Section 4.1, related to changes in the optimum signal frequency with changes in range and channel conditions such as temperature, depth and wind. These variations are reflected in a similar manner to the changes seen here in channel bandwidth and in turn will reflect in the potential data transmission rates. Figure 11 demonstrates that both the optimal signal frequency and the 3dB channel bandwidth decrease as range increases. The impact of changing wind conditions on channel bandwidth is significant, however as discussed wind and wave action will also include time variant complexities and losses not included here. Temperature increases show an increase in channel bandwidth, at ranges of interest, due to the reduction in absorption loss as temperature increases, which means some benefits in working in the surface layers. The discussion here highlights that the underwater acoustic channel is severely band-limited and bandwidth efficient modulation will be essential to maximise data throughput and essentially that major benefits can be gained when performing data transmission at shorter ranges or in multi-hop arrangements.

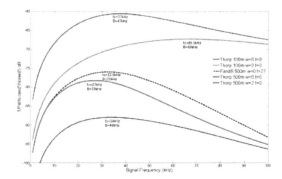

Fig. 11. Range dependent 3dB Channel Bandwidth shown as dashed lines. Where Y-axis is the frequency dependent component of the narrowband SNR

4.3 Channel capacity

Prior to evaluating the more realistic performance of the underwater data communication channel, the maximum achievable error-free bit rate C for various ranges of interest will be determined using the Shannon-Hartley expression, Equation 10. In these channel capacity calculations, all the transmitted power P_{tx} is assumed to be transferred to the hydrophone except for the losses associated with the deterministic Path Loss Models developed earlier. The Shannon-Hartley expression using the Signal-to-Noise ratio, SNR(r), defined in Equation 8, is:

$$C = Blog_2(1 + SNR(r)) \tag{10}$$

where C is the channel capacity in bps and B is the channel bandwidth in Hz

Thus using the optimum signal frequency and bandwidths at 100 m and 500 m found in the Section 4.1 and 4.2, the maximum achievable error free channel capacities against range are shown in Figure 12. The signal frequency and channel bandwidth values for 100 m were $f_o = 37\text{kHz}$ and B=47kHz and for 500 m were $f_o = 27\text{kHz}$ and B=33kHz. These are significantly higher than values currently available in underwater operations(Walree, 2007), however they provide an insight into the theoretical limits. Two different transmitter power levels are used, 150dB re 1μPa which is approximately 10mW (Equation 1) and 140dB re 1μPa is 1mW. Looking at the values associated with the same power level in Figure 12, the higher channel capacities are those associated with the determined optimum frequency and bandwidth for that range as would be expected. The change in transmitter power, however, by a factor of 10, does not produces a linear change in channel capacity across the range. These variations are important to consider as minimising energy consumption will be critical for AUV operations. In general, current modem specifications indicate possible data rate capacities of less than 10kbps (LinkQuest, 2008) for modem operations under 500 m, well short of these theoretical limits. This illustrates the incredibly severe data communication environment found underwater and that commercial modems are generally not yet designed to be able to adapt to specific channel conditions and varying ranges. The discussion here is to understand the variations associated with the various channel parameters at short range that may support adaptability and improved data transmission capacities.

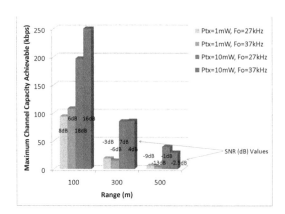

Fig. 12. Theoretical limit of Channel Capacity (kbps) verse Range

4.4 BER in short range underwater acoustic communication

Achieving close to the maximum channel capacities as calculated in the previous section is still a significant challenge in underwater acoustic communication. The underwater acoustic channel presents significant multipaths with rapid time-variations and severe fading that lead to complex dynamics at the hydrophone causing ISI and bit errors. The probability of bit error, BER, therefore provides a measure of the data transmission link performance. In underwater systems, the use of FSK (Frequency Shift Keying) and PSK (Phase Shift Keying) have occupied researchers approaches to symbol modulation for several decades. One approach is using the simpler low rate incoherent modulation frequency hopping FSK signalling with strong error correction coding that provides some resilience to the rapidly varying multipath. Alternatively, the use of a higher rate coherent method of QPSK signalling that incorporates a Doppler tolerant multi-channel adaptive equalizer has gained in appeal over that time (Johnson et al., 1999).

The BER formulae are well known for FSK and QPSK modulation techniques (Rappaport, 1996), which require the Energy per Bit to Noise psd, $\frac{E_b}{N_o}$, that can be found from the SNR (Equation 8) by:

$$\frac{E_b}{N_o} = SNR(r) \times \frac{B_c}{R_b} \tag{11}$$

where R_b is the data rate in bps and B_c is the channel bandwidth. Equation 12 and 13 are the uncoded BER for BPSK/QPSK and FSK respectively:

$$QPSK: \quad BER = \frac{1}{2}erfc[\frac{E_b}{N_o}]^{1/2} \tag{12}$$

$$FSK: \quad BER = \frac{1}{2}erfc[\frac{1}{2}\frac{E_b}{N_o}]^{1/2} \tag{13}$$

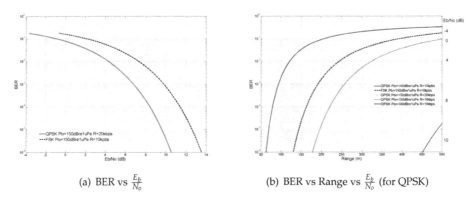

(a) BER vs $\frac{E_b}{N_o}$ (b) BER vs Range vs $\frac{E_b}{N_o}$ (for QPSK)

Fig. 13. Probability of Bit Error for Short Range Acoustic Data Transmission Underwater

The data rates R_b used are 10 and 20 kbps to reflect the current maximum commercial achievable levels. Figure 13 (a) and (b) show the BER for $\frac{E_b}{N_o}$ and Range respectively. Taking a BER of 10^{-4} or 1 bit error in every 10, 000 bits, the $\frac{E_b}{N_o}$ required for QPSK is 8dB for a transmitter power of 10mW and a data rate of 20kbps. This increases to 12dB if using FSK with half the data rate (10 kbps) and same Transmitter Power. From Figure 13 (b), these settings will provide only a 150 m range. The range can be increased to 250 m using QPSK if the data rate was halved to 10 kbps or out to 500 m if the transmitter power was increased to 100mW in addition to the reduced data rate. Transmitter power plays a critical role, as illustrated here, by the comparison of ranges achieved from $\approx$ 75 m to 500 m with a change of transmitter power needed from 1mW to 100mW for this BER.

5. Swarm network protocol design techniques

A short range underwater network, as shown in Figure 1(b) is essentially a multi-node sensor network. To develop a functional sensor network it is necessary to design a number of protocols which includes MAC, DLC (Data Link Control) and routing protocols. A typical protocol stack of a sensor network is presented in Figure 14. The lowest layer is the physical layer which is responsible for implementing all electrical/acoustic signal conditioning techniques such as amplifications, signal detection, modulation and demodulation, signal conversions, etc.. The second layer is the data link layer which accommodates the MAC and DLC protocols. The MAC is an important component of a sensor networks protocol stack, as it allows interference free transmission of information in a shared channel. The DLC protocol includes the ARQ (Automatic Repeat reQuest) and flow control functionalities necessary for error free data transmission in a non zero BER transmission environment. Design of the DLC functionalities are very closely linked to the transmission channel conditions. The network layers main operational control is the routing protocol; responsible for directing packets from the source to the destination over a multi-hop network. Routing protocols keep state information of all links to direct packets through high SNR links in order to minimise the end to end packet delay. The transport layer is responsible for end to end error control procedures which replicates the DLC functions but on an end to end basis rather than hop to hop basis as implemented by the DLL. The transport layer could use standard protocols such as TCP (Transmission Control Protocol) or UDP (User Datagram Protocol). The application layer hosts different operational applications which either transmit or receive data using the lower layers. To develop efficient network architectures, it is necessary to develop network and/or application specific DLL and network layers. The following subsections will present MAC and routing protocol design characteristics required for underwater swarm networking.

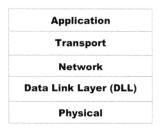

Fig. 14. A typical protocol stack for a sensor network

5.1 MAC protocol

Medium access protocols are used to coordinate the transmission of information from multiple transmitters using a shared communication channel. MAC protocols are designed to maximise channel usage by exploiting the key properties of transmission channels. MAC protocols can be designed to allocate transmission resources either in a fixed or in a dynamic manner. Fixed channel allocation techniques such as Frequency Division Multiplexing (FDM) or Time Division Multiplexing (TDM) are commonly used in many communication systems where ample channel capacity is available to transmit information (Karl & Willig, 2006). For low data rate and variable channel conditions, dynamic channel allocation techniques are generally used to maximise the transmission channel utilisation where the physical transmission channel condition could be highly variable. Based on the dynamic channel allocation technique it is possible to develop two classes of MAC protocols known as random access and scheduled access protocol. The most commonly used random access protocols is the CSMA (Carrier Sense Multiple Access) widely used in many networks including sensor network designs. Most commonly used scheduled access protocol is the polling protocol. Both the CSMA and polling protocols have flexible structures which can be adopted for different application environments. As discussed in this chapter, the underwater communication channel is a relatively difficult transmission medium due to the variability of link quality depending on location and applications. Also, the use of an acoustic signal as a carrier will generate a significant delay which is a major challenge when developing a MAC protocol. In the following subsection we discuss the basic design characteristics of the standard CSMA/CA protocol and its applicability for underwater applications.

5.1.1 CSMA/CA protocol

Carrier Sense Multiple Access with Collision Detection protocol is a distributed control protocol which does not require any central coordinator. The principle of this protocol is that a transmitter that wants to initiate a transmission, checks the transmission channel by checking the presence of a carrier signal. If no carrier signal is present which indicates the channel is free and the transmitter can initiate a transmission. For a high propagation delay network such a solution does not offer very high throughput due to the delay.

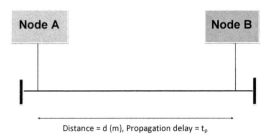

Fig. 15. CSMA/CA protocol based packet transmission example

Consider Figure 15, where two nodes are using CSMA/CA protocol, are spaced apart by 100 meters. In this case, if at t=0, Node A senses the channel then it will find the channel to be free and can go ahead with the transmission. If Node A starts transmission of a packet immediately then it can assume that the packet will be successfully transmitted. However, if Node B starts sensing the channel before the propagation delay time t_p then it will also find the channel is free and could start transmission. In this case both packet will collide and the transmission channel capacity will be wasted for a period of L+t_p where L is the packet transmission time. On the other hand, if Node B checks the channel after time t_p from the commencement of A's packet transmission, then it will find the channel is busy and will not transmit any packets. Now this simple example shows how the performance of random access protocol is dependent on the propagation delay. If propagation delay is small then there is much lower probability that a packet will be transmitted before the packet from A arrives at B. As the propagation delay increases the collision probability will also increase.
The CSMA/CA protocol is generally used in RF (Radio Frequency) networks where 100 m link delay will incur a propagation delay of 0.333 μsec whereas an underwater acoustic link

of same distance will generate a propagation delay of 0.29 sec which is about 875,000 times longer than the RF delay. One can easily see why an acoustic link will produce much lower throughput than is predicted by the Shannon-Hartley theorem as discussed in Section 4.3. If we assume that we are transmitting a 100 byte packet, then the packet will take about 0.08 sec to transmit on a 10 kbps RF link. The same packet will take 0.3713 sec on a 10 kbps acoustic link offering a net throughput of 2.154 kbps. This calculation is based on the assumption that the transmission channel is ideal i.e. BER=0. If the BER of the channel is non zero then the throughput will be further reduced.

Previous sections have shown that the BER of a transmission link is dependent on the link parameters, geometry of the application environment, modulation techniques, and presence of various noise sources. Non zero BER conditions introduce a finite packet error rate (PER) on a link which is described by Equation 14, where K represents the packet length. The PER will depend on the BER and the length of the transmitted packet. For a BER of 10^{-3} using a packet size of 100 bytes, the link will generate a PER value of 0.55 which means that almost every second packet will be corrupted and require some sort of error protection scheme to reduce the effective packet error rate.

There are generally two types of packet error correction techniques used in communication systems, one is forward error correction (FEC) scheme which uses a number of redundant bits added with information bits to offer some degree of protection against the channel error. The second technique involves the use of packet retransmission techniques using the DLC function known as the ARQ. The ARQ protocol will introduce retransmissions when a receiver is unable to correct a packet using the FEC bits. The retransmission procedure could effectively reduce the throughput of a link further because the same information is transmitted multiple times. From this brief discussion one can see that standard CSMA/CA protocols used in sensor networks are almost unworkable in the underwater networking environment unless the standard protocol is further enhanced. This is a major research issue which is currently followed up by many researchers and authors. Readers can find some of the current research work on the MAC protocol in the following references (Chirdchoo et al., 2008; Guo et al., 2009; Pompili & Akyildiz, 2009; Syed et al., 2008).

$$PER = 1 - (1 - BER)^K \qquad (14)$$

5.2 Packet routing

Packet routing is another challenging task in the underwater networking environment. Packet routing protocols are very important for a multi-hop network because the receivers and the transmitters are distributed in a geographical area where nodes can also change their positions over time. Each node maintains a routing table to forward packets through multi-hop links. Routing tables are created by selecting the best cost paths from transmitters to receivers. The cost of a path can be expressed in terms of delay, packet loss, BER, real monetary cost $, etc.. For underwater networks, the link delay could be used as a cost metric, to transmit packets with a minimum delay. Routing protocols are generally classified into two classes: distance vector and link state routing protocols (LeonGarcia & Widjaja, 2004). The distance vector algorithms generally select a path from a transmitter to receiver based on shortest path through neighbouring networks. When the status of a link changes, for example, if the delay or SNR of a link is increased then the node next to the link will detect and inform its neighbour about the change and suggest a new link. This process will continue until all the nodes in the network have updated their routing table. The link state routing protocols work in a different manner. In this case all the link state information is periodically transmitted to all nodes in the network. In case of any change of state of a link, all nodes get notification and modify their routing table. In a swarm network link qualities will be variable which will require regular reconfiguration of routing tables. The performance of routing algorithms is generally determined by a number of factors including the convergence delay. In the case of a swarm network the convergence delay will be a critical factor because of high link delays. For underwater swarm applications, each update within a network will take considerably longer time than a RF network, causing additional packet transmission delays. Hence, it is necessary to develop the network structure in different ways than a conventional sensor network. For example, it may be necessary to develop smaller size clustered networks where cluster heads form a second tier network. Within this topology, local information will flow within the cluster and inter-cluster information will flow through the cluster head network. Cluster based communication architectures are also being used in Zigbee based and wireless personal communication networks (Karl & Willig, 2006). Further research is necessary to develop appropriate routing algorithms to minimise packet transmission delay in swarm networks. Readers can consult the following references to follow some of the recent progress in the area (Aldawibio, 2008; Guangzhong & Zhibin, 2010; LeonGarcia & Widjaja, 2004; Zorzi et al., 2008).

Discussion in this section clearly shows that the MAC and routing protocol designs require transmission channel state information in order to optimise their performance. Due to the high propagation delay of an underwater channel, any change of link quality such as SNR will significantly affect the performance of the network. Hence, it is necessary to develop a new class of protocols which can adapt themselves with the varying channel conditions and offer reasonable high throughput in swarm networks.

6. Conclusion

The increasing potential of Autonomous Underwater Vehicle (AUV) swarm operations and the opportunity to use multi-hop networking underwater has led to a growing need to work with a short-range acoustic communication channel. Understanding the channel characteristics for data transmission is essential for the development and evaluation of new MAC and Routing Level protocols that can better utilise the limited resources within this harsh and unpredictable channel.

The constraints imposed on the performance of a communication system when using an acoustic channel are the high latency due to the slow speed of the acoustic signal (compared with RF), and the signal fading properties due to absorption and multipath signals, particularly due to reflections off the surface, sea floor and objects in the signal path. The shorter range acoustic channel has been shown here to be able to take advantage of comparatively lower latency and transmitter power as well as higher received SNR and signal frequencies and bandwidths (albeit still only in kHz range). Each of these factors influence the approach needed for developing appropriate protocol designs and error control techniques while maintaining the required network throughput and autonomous operation of each of the nodes in the swarm.

Significant benefits will be seen when AUVs can operate as an intelligent swarm of collaborating nodes and this will only occur when they are able to communicate quickly and clearly between each other in a underwater short range ad-hoc mobile sensor network.

7. References

Aldawibio, O. (2008). A review of current routing protocols for ad hoc underwater acoustic networks, *First International Conference on the Applications of Digital Information and Web Technologies. ICADIWT*, pp. 431 –434.

Caruthers, J. (1977). *Fundamentals of Marine Acoustics*, Elsevier Scientific Publishing.

Chen, W. & Mitra, U. (2007). Packet scheduling for multihopped underwater acoustic communication networks, *IEEE OCEANS'07*, pp. 1–6.

Chirdchoo, N., Soh, W. & Chua, K. (2008). Ript: A receiver-initiated reservation-based protocol for underwater acoustic networks, *IEEE Journal on Selected Areas in Communications* 26(9): 1744 –1753.

Coates, R. (1989). *Underwater Acoustic Systems*, John Wiley and Sons.

Cox, A. W. (1974). *Sonar and Underwater Sound*, Lexington Books.

Domingo, M. (2008). Overview of channel models for underwater wireless communication networks, *Physical Communication* pp. 163 – 182.

Dunbabin, M., Roberts, J., Usher, K., Winstanley, G. & Corke, P. (2005). A hybrid auv design for shallow water reef navigation, *Proc. International Conference on Robotics and Automation (ICRA)*, pp. 2117–2122.

Eckart, C. (1952). *Principles of Underwater Sound*, Research Analysis Group, National Research Council, California University.

Essebbar, A., Loubet, G. & Vial, F. (1994). Underwater acoustic channel simulations for communication, *IEEE OCEANS '94. 'Oceans Engineering for Today's Technology and Tomorrow's Preservation.'*, Vol. 3, pp. III/495 –III/500 vol.3.

Etter, P. (2003). *Underwater Acoustic Modeling and SImulation*, third edn, Spon Press.

Fisher, F. & Simmons, V. (1977). Sound absorption in sea water, *Journal of the Acoustical Society of America* 62(3).

Francois, R. & Garrison, G. (1982). Sound absorption based on ocean measurements: Part 1 and 2, *Journal of the Acoustical Society of America* 72(3,6): 896–907, 1879 – 1890.

Guangzhong, L. & Zhibin, L. (2010). Depth-based multi-hop routing protocol for underwater sensor network, *2nd International Conference on Industrial Mechatronics and Automation (ICIMA)*, Vol. 2, pp. 268 –270.

Guo, X., Frater, M. & Ryan, M. (2009). Design of a propagation-delay-tolerant mac protocol for underwater acoustic sensor networks, *IEEE Journal of Oceanic Engineering* 34(2): 170 –180.

Hajenko, T. & Benson, C. (2010). The high frequency underwater acoustic channel, *IEEE OCEANS 2010, Sydney*, pp. 1 –3.

Holmes, J., Carey, W., Lynch, J., Newhall, A. & Kukulya, A. (2005). An autonomous underwater vehicle towed array for ocean acoustic measurements and inversions, *IEEE Oceans 2005 - Europe*, Vol. 2, pp. 1058 – 1061 Vol. 2.

Johnson, M., Preisig, J., Freitag, L.& Stojanovic, M. (1999). FSK and PSK performance of the utility acoustic modem, *IEEE OCEANS '99 MTS. Riding the Crest into the 21st Century*, Vol. 3, pp. 1512 Vol. 3.

Johnson, R. (2011). *University Corporation of Atmospheric Research, Window to the Universe*, University Corporation of Atmospheric Research, http://www.windows. ucar. edu/ tour/link/earth/Water/overview.html.

Karl, H. & Willig, A. (2006). *Protocols and Architecures for Wireless Sensor Networks*, John Wiley and Sons, Ltd.

Kinsler, L., Frey, A., Coppens, A. & Sanders, J. (1982). *Fundementals of Acoustics*, John Wiley and Sons.

LeonGarcia, A. & Widjaja, I. (2004). *Communication Networks: Fundamental Concepts and Key Architecture*, second edn, McGraw Hill.

LinkQuest (2008). *SoundLink Underwater Acoustic Modems, High Speed, Power Efficient, Highly Robust*, LinkQuest Inc., http://www.link-quest.com/.

Mare, J. (2010). Design considerations for wireless underwater communication transceiver, *OCEANS10, Sydney*.

Nasri, N., Kachouri, A., Andrieux, L. & Samet, M. (2008). Design considerations for wireless underwater communication transceiver, *International Conference on Signals, Circuits and Systems*.

Parrish, N., Roy, S., Fox, W. & Arabshahi, P. (2007). Rate-range for an fh-fsk acoustic modem, *Proceedings of the second workshop on Underwater networks, WuWNet '07*, pp. 93–96.

Pompili, D. & Akyildiz, I. (2009). Overview of networking protocols for underwater wireless communications, *IEEE Communications Magazine* 47(1): 97 –102.

Rappaport, T. (1996). *Wireless Communications, Principles and Practice*, Prentice Hall.

Sehgal, A., Tumar, I. & Schonwalder, J. (2009). Variability of available capacity due to the effects of depth and temperature in the underwater acoustic communication channel, *IEEE OCEANS 2009, EUROPE*, pp. 1–6.

Stojanovic, M. (2006). On the relationship between capcity and distance in an underwater acoustic communication channel, *International Workshop on Underwater Networks, WUWNet'06*.

Stojanovic, M. (2008). Underwater acoustic communications: Design considerations on the physical layer, *5th Annual Conference on Wireless on Demand Network Systems and Services, WONS*, pp. 1–10.

Sullivan, E. & Taroudakis, M. (2008). *Handbook of Signal Processing in Acoustics Volume 2*, RSpringer.

Syed, A., Ye, W. & Heidemann, J. (2008). Comparison and evaluation of the t-lohi mac for underwater acoustic sensor networks, *IEEE Journal on Selected Areas in Communications* 26(9): 1731 –1743.

Thorp, W. H. (1965). Deep-ocean sound attenuation in the sub- and low-kilocycle-per-second region, *Journal of the Acoustical Society of America* 38(4): 648–654.

Urick, R. (1967). *Principles of Underwater Sound for Engineers*, McGraw-Hill.

Waite, A. (2005). *Sonar for Practicing Engineers*, third edn, Wiley.

Walree, P. (2007). Acoustic modems: Product survey, *Hydro International Magazine* 11(6): 36–39.

Zorzi, M., Casari, P., Baldo, N. & Harris, A. (2008). Energy-efficient routing schemes for underwater acoustic networks, *IEEE Journal on Selected Areas in Communications* 26(9): 1754 –1766.

9

Deep-Sea Fish Behavioral Responses to Underwater Vehicles: Differences Among Vehicles, Habitats and Species

Franz Uiblein

Institute of Marine Research, Bergen, Norway and South African Institute of Aquatic Biodiversity, Grahamstown, South Africa

1. Introduction

Fishes have a wide range of perceptual capabilities allowing them to behaviorally respond to various environmental stimuli such as visual, acoustic, mechanical, chemical, and electromagnetic signals. In our "noisy" world of today many artificially evoked signals pass through aquatic habitats, where fishes perceive them and respond to in often unpredictable manner. Proper distinction between natural and artificially evoked (="disturbed") behavior is of utmost importance in ecological studies that try to identify the prevailing factors and mechanisms influencing fish abundance, distribution and diversity.

As we know today, the need to consider human-induced behavioral disturbance as an important factor in ecological studies (Beale 2007) applies even to inhabitants of remote aquatic habitats such as the deep sea. *In situ* studies using various types of underwater vehicles (UV's) have significantly changed the conception that the inhabitants of the deep, dark and mostly cold ocean are less behaviorally active and hence less susceptible to anthropogenic disturbance. While direct observation of deep-sea animals goes back to the time of William Beebe in the 1930s, *in situ* studies of deep ocean organisms and their habitats have become increasingly more common during the last 50 years.

After initial use for exploration and discovery of yet unknown habitats and organisms, UV's were adopted to systematically investigate the ecology of deep-sea organisms, especially the larger and easier observable fauna in the open water and close to the bottoms. In analogy to census studies conducted by divers in shallow waters, vertical or horizontal transects with underwater vehicles were used to obtain density or distributional data of fishes (e.g., Yoklavich et al. 2007, Uiblein et al. 2010). Distinct fish species or closely related taxonomic groups were found to occur at relatively high densities during such transects allowing quantitative behavioral investigations.

Early *in situ* exploration encountered first evidence of pelagic and bottom-associated (demersal) fishes living at depths well below 200 m being behaviorally active similar to shallow-water species (Beebe 1930, Heezen & Hollister 1971). These preliminary behavioral observations were followed by detailed studies of locomotion behavior and habitat utilization based mainly on video equipment employed during bottom transects with manned submersibles (e.g., Lorance et al. 2002, Uiblein et al. 2002, 2003) and later with

ROV's (e.g., Trenkel et al. 2004a, Lorance et al. 2006). Quantitative behavioral comparisons conducted with the submersible Nautile clearly showed that fish species differ among each other in the way they swim and in their vertical positioning above the bottom (Uiblein et al. 2003). Moreover, distinct responses to the approaching vehicle were identified which needed to be analyzed in detail so to be able to distinguish natural behavior from responses to anthropogenic disturbance. That underwater vehicles have a disturbance effect on fish behavior has also important consequences for fish density calculations from *in situ* transects, as the data may not reflect natural conditions when disturbance responses are intense and/or occur frequently (Trenkel et al. 2004b, Stone et al. 2008).

Disturbance responses in deep-sea fishes may be caused by a number of factors like noise produced by motors and thrusters, light used for illumination purposes, motion, electromagnetic fields, or odor plumes deriving from the vehicle. Detailed investigations regarding the actual source(s) of disturbance are generally lacking. Here, a description and categorization of disturbance responses is provided and differences between vehicles, habitats, and species are elaborated. These data suggest that disturbance responses are manifold and can – by themselves – reveal interesting insights into the life modes of deep-sea fishes. In addition, when disturbance responses are identified, natural behavior (e.g., locomotion and vertical positioning above bottom) can be filtered out and studied independently of artificial evocation.

Here, nine case studies based on manned submersible and ROV video transects in the deep North Atlantic are presented dealing subsequently with differences in disturbance responses between underwater vehicles, dive transects (habitats), and co-occurring species/species groups. In addition, a separate section is devoted to combined analyses of natural behavior and disturbance responses, to show the full picture. These results are discussed referring to (1) novel insights about deep-sea fish artificially and naturally aroused behavior, (2) the need for consideration and integration of all influential factors in the behavioral analysis and interpretation, and (3) future technological possibilities and challenges towards optimizing *in-situ* investigations on the behavior and ecology of deep-sea fishes.

2. Material and methods

Video recordings from the areas of the Bay of Biscay and the northern Mid-Atlantic Ridge made during six dives with four different underwater vehicles were studied. The underwater vehicles were as follows (Table 1): the manned submersible Nautile and the ROV Victor 6000 (both at IFREMER, www.ifremer.fr), the ROV Aglantha (IMR, www.imr.no), and the ROV Bathysaurus (ARGUS, www.argus-rs.no). Each dive consisted of one to three horizontal transects close to the bottom which lasted between 10 and 174 minutes and covered various depth ranges between 812 and 1465 m (Table 1). During transects the respective vehicle moved slowly (ca. 0.5 to 1.0 knots on average) above the bottom, mostly in straight lines, sometimes interrupted by short stops.

Each of the 10 total transects crossed a distinct habitat within canyons or deep-sea terraces of the Bay of Biscay (Nautile, Victor 6000) and slopes or valleys of the northern Mid-Atlantic Ridge (Aglantha, Bathysaurus) (Table 1). The Mid-Atlantic Ridge study area was divided in a southern investigation box, close to the Azores, and a northern box situated in the area south and north of the Charlie Gibbs Fracture Zone (Table 1).

Cruise	OBSERVHAL	VITAL					MAR-ECO				
Region	Bay of Biscay						Northern Mid-Atlantic Ridge				
Vehicle	Submersible Nautile	ROV Victor 6000					ROV Bathysaurus	ROV Aglantha			
Dive	OB22	VT1			VT3		ME4-2 (SS 44)	ME10 (SS 60)			ME16 (SS 70)
Transect	OB22-1	VT1-1	VT1-2	VT3-1		VT3-2	ME4-2-1	ME10-1	ME10-2	ME10-3	ME16-1
Area	Mériadzek Terrace			Belle Isle Canyon			Southern Box	Northern Box, south			Northern Box, north
Date	17.05.98	23.8.02	24.8.02	31.8.02		1.9.02	12.7.04	19.7.04			26.7.04
Duration (min)	126	174	111	91		85	50	21	10	19	105
Depth	1250-990	1454 - 1392	1228 – 1208	1193 - 1159		1465 - 1446	1228-1272	954-848	847-812	842-879	1373-1270
Roundnose grenadier	29	1	1	19		20	3	7	33	20	20
Orange roughy	5	2	0	3		0	0	0	73	43	0
Codling	32	29	37	32		33	3	29	11	31	0
False boarfish	0	33	0	0		0	41	4	3	1	0

SS ... MAR-ECO super-station

Table 1. Overview of dives, vehicles and video transects, with numbers of encountered fish per transect. Samples analyzed are highlighted. For further explanations see text.

Behaviour	Category			
Disturbance response	No response	Close distance	Far distance	Arriving disturbed
Vertical position in water column	Close to bottom	Well above bottom	Far above bottom	
Locomotion behaviour	Inactive	Drifting	Station holding	Forward moving

Table 2. Overview of the behavioral categories studied

The four species/species groups selected for detailed analysis were the roundnose grenadier *Coryphaenoides rupestris* (family Macrouridae; Fig.1), the orange roughy *Hoplostethus atlanticus* (family Trachichthyidae; Fig.1) the false boarfish *Neocyttus helgae* (family Oreosomatidae; Fig.1) and codling (family Moridae). The term "codling" includes the most common *Lepidion eques* (North Atlantic codling; Fig. 1), its congeners *L. guentheri* and *L. schmidti*, and the slender codling *Halagyreus johnssonii*. Identification of species/species groups was based on the size and form of the body, head and fins, and color patterns and distributional data from the respective area deriving from collected material.

The recording of all behaviors started immediately after a fish appeared on the video screen. Four main behaviors, overall activity level, disturbance response, locomotion, and vertical positioning above the bottom, each consisting of two or more categories, were recorded for subsequent statistical analysis (Table 2). Fishes visualized on video with high or increasing swimming speed indicating burst swimming in response to prior disturbance by the submersible ("arriving disturbed") were excluded from further-going behavioral analyses. During the subsequent behavioral recordings, the UV frequently got closer to the fishes, with increasing illumination intensity caused by the front lights. If a disturbance response was observed during this process (i.e. a marked change in activity level and/or locomotion behavior), the recordings of locomotion or vertical body positioning were stopped immediately before the occurrence of this behavioral change. The disturbance response during UV approach was split into two separate categories, depending if it happened still at far distance or at close distance to the UV and mostly within the highest illumination radius.

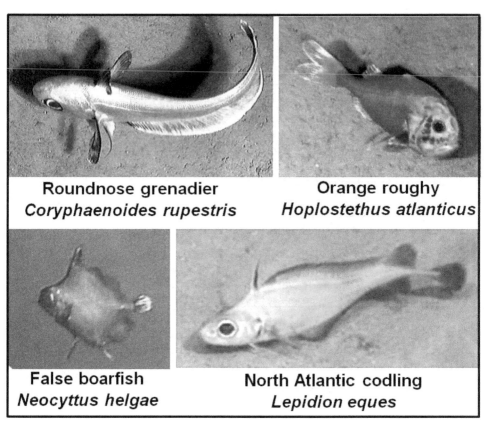

Roundnose grenadier
Coryphaenoides rupestris

Orange roughy
Hoplostethus atlanticus

False boarfish
Neocyttus helgae

North Atlantic codling
Lepidion eques

Fig. 1. Photographs of studied fish species (North Atlantic codling was the most common species of the codling group)

For the analysis of undisturbed natural behavior, four locomotion activity categories were identified: "inactive" (Table 2) (= without any movement), "station holding" (= body stationary with active swimming against current), "drifting" (= movement in lateral or backward direction with or without swimming activity), and "forward movement" (= clear active forward swimming movements). Three categories for vertical body positioning in relation to the bottom surface were determined: "close to bottom" (= positioned at the bottom or at distances of less than one body length above the bottom), "well above bottom" (= distance from bottom exceeds one body length), and "far above bottom" (= distance from bottom exceeds three body lengths).

In order to reduce the number of influential factors comparisons between underwater vehicles and species/species groups were mostly restricted to the same transect or area and comparisons among habitats were restricted to single species. Only samples with 19 or more individuals per species/species group encountered per transect were analyzed to allow statistical comparisons in all instances. For statistical comparisons of categorical data among species/species groups and habitats, G-tests of independency were carried out (Sokal & Rohlf 1981).

3. Results

The behavioral data of 501 fishes from the four selected species/species groups were analyzed. Apart from a single exception (codling in dive transect OB22-1) disturbance responses occurred during all transects and in all species/species groups. On average 44 % of all fishes showed disturbance and in 7 of the 15 total observational sets (= species-transect combinations) that were analyzed, more than 50 % of the fish displayed disturbance responses. While pre-arrival disturbance was relatively rare (14 % of all disturbed behavior registered), disturbance responses at far distance occurred most frequently (59 %). The disturbance responses were only rarely directed towards any of the four UV's used. No clear signs of attraction or aggressive responses triggered by the UV's could be observed in any of the four species/species groups.

Differences between underwater vehicles (Fig.2)

The codling showed a significant difference (p<0.005) in disturbance responses between two dive transects performed in the same area at the Mériadzek terrace, Bay of Biscay, one with the manned submersible Nautile (transect OB22-1, Table 1) and the other with the ROV Victor 6000 (transect VT-1, Table 1). While no disturbance response was registered during the dive with Nautile, 35 % of the individuals encountered during the ROV transect showed clear signs of disturbance. Among the disturbed fish 23 % showed pre-arrival disturbance, while 54 % responded at far distance and 23 % responded at short distance to the approaching vehicle. Regarding undisturbed natural behavior, no significant differences in both vertical positioning and locomotion behavior were found between the two transects.

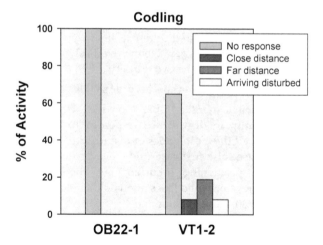

Fig. 2. Disturbance responses of codling during a manned submersible transect (left) and a ROV transect (right) in the area of Mériadzek Terrace, Bay of Biscay

Differences between dive transects and habitats (Fig. 3)

Orange roughy showed significant differences in disturbance responses (p<0.01; Fig. 3a) between two transects that crossed adjacent habitats at similar depths (812-879 m) during dive ME10 (Table 1) with the ROV Aglantha on the northern Mid-Atlantic Ridge, just south of the Charlie Gibbs Fracture zone. Each of the three categories of disturbance responses decreased in frequency between the first and the second transect thus indicating less responsiveness. Both vertical positioning and locomotion behavior did not differ significantly between transects.

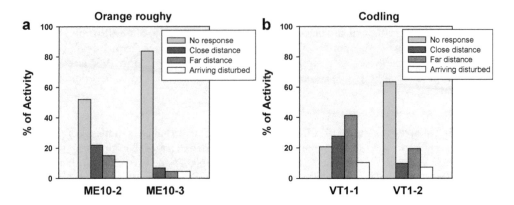

Fig. 3. Disturbance responses of (a) orange roughy during two subsequent ROV transects in the area of the northern Mid-Atlantic Ridge and (b) codling during two subsequent ROV transects in the area of Mériadzek Terrace, Bay of Biscay

The codling showed a significant decrease in disturbance responses (p<0,005; Fig. 3b) between the first and second transect of ROV dive VT1 (Table 1) on the Mériadzek Terrace, Bay of Biscay. These two transects covered different depth zones (1392-1454 vs. 1208-1228 m), the first (VT1-1) being clearly deeper. Neither vertical positioning nor locomotion behavior differed significantly between the two transects.

Differences between co-occurring species/species groups (Fig. 4)

During the manned submersible transect OB22-1 on the Mériadzek terrace, roundnose grenadier differed significantly in disturbance responses (p<0.0001; Fig. 4a) from the codling. The former showed all three categories of disturbance, while the latter showed no disturbance responses at all (see also first case study; Fig. 2). Regarding natural behaviour, no differences in vertical positioning occurred, but roundnose grenadier showed significantly more forward movement and less station holding than codling (p<0.01).

During ROV dive transect VT1-1 the codling and the boarfish differed significantly from each other in disturbance responses (p<0.005; Fig. 4b) with the boarfish showing clearly less disturbed arrival and close-distance responses to the approaching vehicle. At far distance from the ROV, the frequency of disturbance responses was similar in both taxa. In addition, significant differences occurred both in vertical positioning and locomotion behavior which are dealt with at the end of the next section.

Variation in natural behavior and disturbance responses

Four different comparative data sets were selected (1) to exemplify situations with disturbance responses occurring at constant or variable rates between transects/habitats or between species/species groups and (2) to analyze in detail the undisturbed, natural vertical

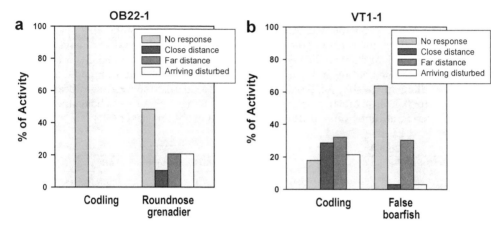

Fig. 4. Disturbance responses of (a) codling and roundnose grenadier during a manned submersible transect and (b) codling and false boarfish during a ROV transect in the area of Mériadzek Terrace, Bay of Biscay

positioning and locomotion behavior of these species/species groups during the same dive transects.

a. Differences only in locomotion behavior (Fig. 5)

The disturbance responses of codling did not differ significantly between two transects (ME10-1, ME10-3, Table 1) of a dive with the ROV Bathysaurus on the Mid Atlantic Ridge (Fig. 5a). There was however a significant difference in locomotion behavior (p<0.05). During the first transect all individuals encountered were active and mostly station holding, while several were inactively sitting on the bottom during the third transect, with less fish station holding. Drifting and forward moving occurred in both transects at rather similar rates. No significant differences in vertical positioning occurred.

During the ROV transect VT3-2 in the Bay of Biscay roundnose grenadier and codling did not differ significantly from each other in disturbance response and vertical positioning (Fig. 5b). However, they clearly differed in locomotion behavior (p<0.005), with roundnose grenadier showing less frequently station holding and more often drifting and forward movement than the codling.

b. Differences only in locomotion and vertical positioning (Fig. 6a)

In roundnose grenadier disturbance responses did not vary significantly between the ROV transects VT3-1 in the Bay of Biscay and ME16-1 on the Mid Atlantic Ridge (Table 1). In both cases only few individuals were recorded as being entirely undisturbed (10-21 %) and 67-87 % of all the disturbed individuals encountered responded to the vehicles at far distance. Both the locomotion behavior and the vertical positioning registered prior to disturbance responses differed significantly between the two habitats (locomotion: p=0.0005; vertical positioning: p<0.025). Roundnose grenadier occurred much higher above the bottom and showed a much higher rate of drifting on the ridge site. Station holding was frequently registered in the Bay of Biscay habitat, whereas it did not occur on the ridge site.

c. Differences in disturbance response and natural behavior between habitats and species (Fig. 6b)

The false boarfish showed significantly more disturbance responses (p<0.05), reacting more frequently at far distances during ROV transect VT1-1 in the Bay of Biscay compared to ROV transect ME4-2-1 on the Mid-Atlantic Ridge. In the latter habitat this species was positioned slightly higher above the bottom (p=0.08 for well- and far-above bottom categories combined) and showed a significant difference in locomotion behavior (p<0.005) with much less station holding and a higher rate of forward movement. Compared to the co-occurring codling in the Bay of Biscay transect, false boarfish showed significantly less disturbance (p<0.005), a much more frequent positioning well or far above the bottom (p<0.0001) and significantly more drifting and less station holding (p<0.0001)

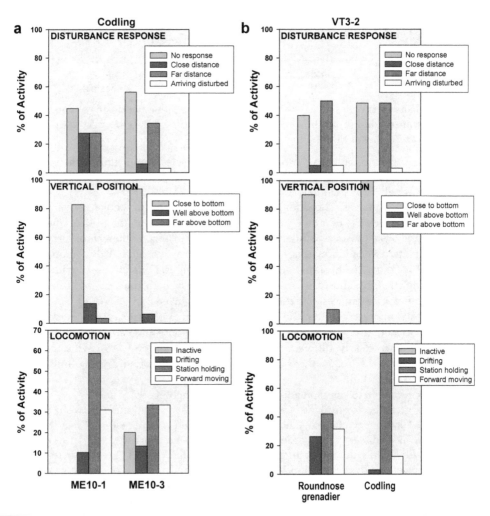

Fig. 5. Disturbance responses, vertical positioning above bottom and locomotion behavior in (a) codling during two ROV transects on the northern Mid-Atlantic Ridge and (b) roundnose grenadier and codling during a ROV transect in Belle Isle Canyon, Bay of Biscay

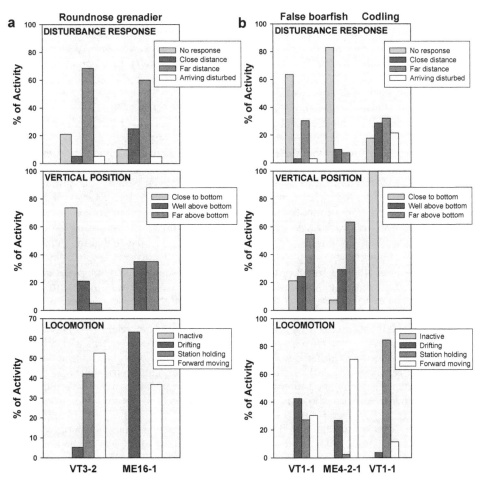

Fig. 6. Disturbance responses, vertical positioning above bottom and locomotion behavior in (a) roundnose grenadier during two ROV transects in Belle Isle Canyon, Bay of Biscay (left), and on the northern Mid-Atlantic Ridge (right) and (b) false boarfish (left and middle) and codling (right) during two transects, one on Mériadzek Terrace, Bay of Biscay (VT1-1) and the other one on the Mid-Atlantic Ridge (ME4-2-1)

4. Discussion and conclusions

Deep-sea fish disturbance responses

The underwater vehicles involved in this study elicited disturbance responses in deep-sea fishes encountered during bottom transects that can be best interpreted as avoidance or flight behavior. Clear signs of attraction to the UV's as they have been reported elsewhere (e.g., Stoner et al. 2007; Moore et al. 2008) were not observed. No longer vehicle stops and no point or selective long-term observations (e.g., by following individual fish) were conducted during the dive transects. In the studies presented here behavioral recordings were only made during the fishes' appearance on the forward directing video screen during transects. It is well possible that additional disturbance responses occurred at larger distances before appearance or after the fish disappeared on the screen, but those were not recorded. Apart from these obvious restrictions, the registration and subsequent quantitative comparison of disturbance responses recorded during UV video transects is a solid method to investigate the influences of various factors such as different vehicles, habitats, or species on the frequency and intensity of evoked reactions (see also, Lorance et al. 2002, Uiblein et al. 2002, 2003).

While the manned submersible did not evoke any response in codling (first case study), they responded considerably disturbed when encountered in the same area with an ROV. A large portion of the disturbance responses happened at far distance or even before encounter indicating early detection, before the main illumination focus reached the fish. Sound may therefore be seen as a main source of disturbance. No exact comparative measurements are however available of the light and sound intensity produced by the two vehicles during those dives. Also, the possibilities cannot be ruled out that the signals acted in combination and that other disturbance sources such as, e.g., pressure waves produced by the moving vehicle body were involved, too. The present findings provide however no evidence that the much larger-bodied manned submersible elicited a comparatively higher disturbance response than any of the four ROV's used, whereas an opposite effect was demonstrated in the first case study.

In orange roughy, light may play an important role in addition to sound in eliciting disturbance responses, because a considerable portion of the reactions occurred at short distances only. Interestingly, the responsiveness to the ROV Bathysaurus decreased between the two adjacent habitats on the ridge. No differences in natural behavior (vertical positioning and locomotion) were observed. One additional difference, however, was a much higher density of orange roughy during the first transect, indicating aggregation formation. Does orange roughy remain particularly vigilant when residing in dense conspecific aggregations? During transects with the manned submersible Nautile an aggregation of orange roughy in the central St. Nazaire canyon did not differ in disturbance responses from conspecifics encountered in the peripheral area (Lorance et al. 2002, Uiblein et al. 2003). Aggregation formation in this species may be related to rather different activities such as resting, spawning or feeding (Lorance et al. 2002). More detailed studies of this ROV dive in the area of the northern Mid-Atlantic Ridge are planned that shall also include comparisons with roundnose grenadier and associated habitat conditions encountered during these transects.

Depth may be an important factor influencing disturbance responses, as can be concluded from the behavior of codling during ROV transects in the Bay of Biscay. These results corroborate with behavioral observations of the northern cutthroat eel *Synaphobranchus kaupii* which also showed more frequent disturbance responses at a deeper located dive in the Bay of Biscay (Uiblein et al. 2002, 2003). The latter species shows a deeper-bigger pattern, hence larger fish living at greater depth have a larger sensory surface that should facilitate signal perception. Also, as food becomes scarcer with larger depths, fish need to pay more attention to environmental stimuli. Both these argumentations may also apply to codling, however, more field and biological data would be necessary to test these assumptions.

Species differences in disturbance responses during single dive transects provide the best evidence for the importance of intrinsic, organism-dependent factors that need to be considered when studying anthropogenic disturbance. Codling showed no response during the manned submersible dive in the Bay of Biscay (OB22), while roundnose grenadier responded considerably and hence may be more sensitive to the signals emitted from this vehicle. It reacted mainly at far distance or immediately before encounter what points towards the perception of rather far-ranging signals (e.g., rather noise than light). On the other hand, codling showed considerable disturbance responses when confronted with an ROV. In the same situation, false boarfish responded to a lesser extent. These three taxa differ fundamentally from each other in their biology: the codling typically holds station close to soft bottoms, the false boarfish prefers to swim or drift closely to shelter provided by hard bottom structures and corals, while the roundnose grenadier is more flexible showing different locomotion behavior and vertical positioning depending on habitat context. Among these three species/species groups, false boarfish appears least prone to predation risk, also given their rather high body (see also Moore et al. 2008). Probably the response to UV's reveals also something about a species' vigilance and assessment of predation risk.

Deep-sea fish disturbance responses and natural behavior: the full picture

When disturbance responses are properly identified, recorded and analyzed, natural behaviour can be studied separately thus allowing to gain insights into the ecology of deep-sea fishes even in the presence of anthropogenic influences. To illustrate this, four case studies were conducted, three elaborating different aspects of natural behavior (locomotion, vertical positioning) with disturbance effects remaining constant and one with all three behaviors varying. In the first two instances only locomotion varied for codling between two separated transects during an ROV dive on the Mid-Atlantic Ridge and for roundnose grenadier and codling during a single ROV transect in the Bay of Biscay. These data indicate that while species clearly differ among each other ("species-specific" behavior), it is also of high importance to understand their behavioral flexibility in adaptation to different habitats. Behavioral flexibility or plasticity allows a choice among different locomotion modes and to select those that fit best to the prevailing conditions in the respective habitat. For instance, less station holding and increased inactivity ("sit and wait") as exemplified by codling in one of two ridge habitats (Fig. 5a) should allow efficient, energy-saving foraging when currents are weak or absent and food abundance is relatively high.

As deep-sea fishes are behaviorally flexible, one can expect to find considerable differences among contrasting habitats, as demonstrated for the roundnose grenadier by ROV dives in the Bay of Biscay and the Mid-Atlantic Ridge. While disturbance responses remained rather similar in both areas, the fish displayed more drifting and no station holding and were positioned significantly higher in the water column on the ridge. This reflects obviously behavioral adjustment to typical ridge conditions (see also, Zaferman 1992) with food particles arriving at the bottom mainly through the water column, while food input deriving from the productive shelf areas is lacking.

A rather complex picture of deep-sea fish behavioral ecology is obtained when all behaviors differ and different habitats are contrasted with different species or species groups, like in the last case study. False boarfish from habitats in the Bay of Biscay and the Mid-Atlantic Ridge were compared showing less disturbance responses, a slightly higher vertical position, less station holding, and more forward movement on the ridge site. The boarfish's behavior in the Bay of Biscay clearly contrasts with codling during the same transect, the latter showing a higher disturbance response, a position on or very close to the bottom, and more station holding. Interpretations are however complicated through one (or several) additional factor(s) that need to be considered in this as well as in the anterior case study featuring roundnose grenadier, because two different UV's were used.

Towards optimizing *in situ* behavioral ecology of deep-sea fishes and related research

A promising approach towards reaching best possible interpretations of what deep-sea fishes do, why they do it, and how they respond to human-induced environmental changes is to consider all influential external and internal factors in the data analysis and in the interpretation of the results. The central method to approach this goal is to analyze video-recordings made during UV transects based on detailed description, categorization and registration of the entire behavior observed with special emphasis to separate human-induced responses from natural behavior, followed by statistical comparisons. Additional data on the biology and ecology of the target species, the physical and biological environment, and the effects and possible impacts of anthropogenic disturbance need to be integrated, too.

To reduce complexity, the number of influential variables should be minimized whenever possible. Optimally, the same design models of UV's should be used during all dives that need to be compared. Dive transects, video recordings, and data analysis should follow standardized protocols. During each transect representative size measurements combined with estimates of absolute swimming speed should be obtained from each studied fish species. Visually well identifiable species should be preferably selected for study so to minimize possible informational noise introduced by species differences within composed groups. Groups of closely related species should be used only exceptionally, when *in situ* species identification is impossible and the species have a very similar body structure, hence similar behavior can be expected. Short video or photographic close-ups of each individual fish from problematic species groups should be taken (preferably by a second camera) to visualize diagnostic details helpful for species identification. Advice and assistance from taxonomists specialized in problematic fish groups should be gathered.

Use of different UV's in comparative studies cannot be recommended, because it may turn out to be difficult, if not impossible, to adjust for disturbance effects. Most certainly more than a single signal source of disturbance needs to be considered. Experimental manipulation of light, sound, and vehicle velocity, but possibly also the magnetic field, singly or in combination, might certainly assist to better understand the relative importance of these potential sources of disturbance (see also Stoner et al. 2007). However, for full control of disturbance effects from UV's, one would also need to investigate the receiver bias and in particular the sensory equipment (Popper & Hastings 2009) and reaction norms (Tuomainen & Candolin 2010) which may differ considerably among fish species, populations, size classes, and ontogenetic stages.

The longer the encounter with an UV the more increases the chance of interactions and evocation of disturbance responses. During longer UV stops, odor plumes deriving from collected organisms or bait brought along may be formed and scavengers may be attracted (Trenkel & Lorance 2011). If point observation are made during longer stops of an UV, these data should be treated separately from transect data. Also, when stationary, the vehicle itself may be perceived in quite different ways than when transitionally encountered during transects and disturbance responses may change and in some cases shift from avoidance to attraction or even to aggression (see for instance, Moore et al. 2008). Observations of deep-sea fishes deriving from longer-term interactions with UV's are certainly interesting *per se*, but may not always contribute to properly understand natural behavior. To reduce interactions it may be of advantage to position the vehicle firmly on the ground and switch off the motors for behavioral observations close to the bottom. During point observations in the open water as well as close to the bottom switching off the illumination and use of infrared light combined with infrared-sensitive cameras should be considered (Widder et al. 2005).

As stated initially, investigations of the effects of UV's on deep-sea fish behavior have important implications for many other studies of deep-sea fishes, as for instance, *in situ* assessments of abundances, populations dynamics, habitat associations, community structure, and patterns of biological diversity (Stoner et al. 2008). Hence the suggestions and recommendations towards optimization of *in situ* behavioral ecology may prove useful also for broader applications in deep-sea fish research and management.

5. Summary

An important prerequisite for *in situ* ecological investigations of deep-sea fishes using underwater vehicles (UV's) is to distinguish between disturbance responses elicited by the vehicles and undisturbed natural behavior. Nine case studies deriving from ten video transects along deep bottoms of the North Atlantic (Bay of Biscay, Mid-Atlantic Ridge) with a manned submersible and three remotely operated vehicles (ROV's) are presented to demonstrate differences in behavioral disturbance between vehicles, habitats, and species. Three species, roundnose grenadier (*Coryphaenoides rupestris*), orange roughy (*Hoplostethus atlanticus*) and false boarfish (*Neocyttus helgae*), and codling, a group of closely related species (North Atlantic codling, *Lepidion eques*, being the most common), were studied. During each UV transect recordings of disturbance responses and two activity patterns shown by undisturbed fishes, vertical positioning in the water column and locomotion mode, were made. Each behavior was subdivided into several categories and analyzed quantitatively using sample sizes larger than 18 individuals per species/species group and transect. Codling showed no disturbance responses to a manned submersible, while reacting intensely to a ROV during two transects performed in the same area. When the same UV was used, clear differences in disturbance responses were found between both adjacent dive transects and species/species groups indicating habitat- and species-specific responsiveness to signals emitted by the vehicle, in particular sound and light, but possibly also other sources. In three additional case studies, disturbance responses remained rather constant between transects or species, but natural behavior differed. The final study provides the fullest picture with all three behaviors differing, the interpretations being however complicated by the fact that different vehicles were used in different habitats. The findings are discussed emphasizing the significance of *in situ* quantitative behavioral studies of UV-based video transects in deep-sea fish ecology and related research fields. Detailed suggestions and recommendations towards optimization of vehicle-disturbance control and observation techniques are provided.

6. Acknowledgements

Travel support provided by a bilateral Amadeus project (between Austria and France,) no. V13, made comparative studies of video footage from the Bay of Biscay UV dives possible. Special thanks to Daniel Latrouite, Pascal Lorance and Verena Trenkel, IFREMER, who provided copies of relevant video footage from OBSERVHAL and VITAL dives and assistance in behavioral data recordings. Thanks to Jan Bryn, Reidar Johannesen, and Asgeir Steinsland for assistance during MAR-ECO ROV dives. Thanks to Mirjam Bachler for valuable comments on the manuscript.

7. References

Beale, C.M. 2007. The Behavioral Ecology of Disturbance Responses. International Journal of Comparative Psychology 20: 111-120

Beebe, W. 1933. Preliminary Account of Deep Sea Dives in the Bathysphere with Especial Reference to One of 2200 Feet. Proceedings of the National Academy of Sciences of the USA. 1933 January; 19(1): 178–188.

Heezen, B.C. & C.D. Hollister. 1971. The Face of the Deep. Oxford UP, New York.

Lorance, P., Trenkel, V.M. & F. Uiblein, 2005. Characterizing natural and reaction behaviour of large mid-slope species: consequences for catchability. In: Shotton, R. (Ed). 2005. Deep Sea 2003: Conference on the Governance and Management of Deep-sea Fisheries. Part1: Conference Reports. FAO Fisheries Proceedings 3/1: 162-164

Lorance, P. & V. Trenkel, 2006. Variability in natural behaviour, and observed reactions to an ROV, by mid-slope fish species. Journal of Experimental Marine Biology and Ecology 332 : 106-119.

Lorance, P., Uiblein, F., & D. Latrouite, 2002: Habitat, behaviour and colour patterns of orange roughy *Hoplostethus atlanticus* (Pisces: Trachichthyidae) in the Bay of Biscay. Journal of the Marine Biological Association of the UK 82: 321-331

Moore, J.A., Auster, P.J., Calini, D., Heinonen, K., Barber, K. & B. Hecker, 2008. False boarfish *Neocyttus helgae* in the Western North Atlantic. Bulletin of the Peabody Museum of Natural History 49: 31-41.

Popper, A.N. & M.C. Hastings 2009; The effects of human-generated sound on fish. Integrative Zoology 4: 43-52

Sokal, R.R. & F.J. Rohlf, 1981. Biometry. WH Freeman, New York.

Stoner, A.W., Ryer, C.H., Parker, S.J., Auster, P.J. & W.W. Wakefield, 2008. Evaluating the role of fish behavior in surveys conducted with underwater vehicles. Canadian Journal of Fisheries and Aquatic Sciences 65: 1230-1241

Trenkel V.M., Francis, R.I.C.C., Lorance, P. Mahévas, S., Rochet M-J. & D.M. Tracey. 2004a. Availability of deep-water fish to trawling and visual observation from a remotely operated vehicle (ROV). Marine Ecology Progress Series 284:293–303.

Trenkel V.M., Lorance P. & S. Mahévas 2004b. Do visual transects provide true population density estimates for deep-water fish? ICES Journal of Marine Science 62:1050–1056.

Trenkel V.M., Lorance P., 2011. Estimating *Synaphobranchus kaupii* densities; Contribution of fish behavior to differences between bait experiments and visual strip transects. Deep-Sea Research I 58:63-71.

Tuomainen, U. & U. Candolin, 2011. Behavioural responses to human-induced environmental change. Biological Reviews, 86:640-657

Uiblein, F., Lorance, P. & D. Latrouite, 2002: Variation in locomotion behaviour in northern cutthroat eel (*Synaphobranchus kaupi*) on the Bay of Biscay continental slope. Deep-Sea Research I 49: 1689-1703

Uiblein, F., Lorance, P. & D. Latrouite 2003: Behaviour and habitat utilization of seven demersal fish species on the Bay of Biscay continental slope, NE Atlantic. Marine Ecology Progress Series 257: 223–232

Uiblein, F., Bordes, F., Lorance, P., Nielsen, J.G., Shale, D., Youngbluth, M. & R. Wienerroither, 2010: Behavior and habitat selection of deep-sea fishes: a methodological perspective. In: S. Uchida (ed.): Proceedings of an International Symposium Into the Unknown, researching mysterious deep-sea animals 2007, Okinawa Churaumi Aquarium, Okinawa, Japan, 5-21.

Widder, E.A., Robison, B.H., Reisenbichler, K.R. & S.H.D. Haddock, 2005. Using red light for in situ observations of deep-sea fishes. Deep-Sea Research I 52: 2077-2085.

Yoklavich, M.M., Love, M.S. & K.A. Forney, 2007. A fishery-independent assessment of an overfished rockfish stock, cowcod (*Sebastes levis*), using direct observations from an occupied submersible. Canadian Journal of Fisheries and Aquatic Sciences 64: 1795-1804.

10

Embedded Knowledge and Autonomous Planning: The Path Towards Permanent Presence of Underwater Networks

Pedro Patrón, Emilio Miguelañez and Yvan R. Petillot
Ocean Systems Laboratory, Heriot-Watt University
United Kingdom

1. Introduction

Oceanographic observatories, year-round energy industry subsea field inspections and continuous homeland security coast patrolling now all require the routine and permanent presence of underwater sensing tools.

These applications require underwater networks of fixed sensors that collaborate with fleets of unmanned underwater vehicles (UUVs). Technological challenges related to the underwater domain, such as power source limitations, communication and perception noise, navigation uncertainties and lack of user delegation, are limiting their current development and establishment. In order to overcome these problems, more evolved embedded tools are needed that can raise the platform's autonomy levels while maintaining the trust of the operator.

Embedded decision making agents that contain reasoning and planning algorithms can optimize the long term management of heterogeneous assets and provide fast dynamic response to events by autonomously coupling global mission requirements and resource capabilities in real time. The problem, however, is that, at present, applications are mono-domain: Mission targets are simply mono-platform, and missions are generally static procedural list of commands described *a-priori* by the operator. All this, leaves the platforms in isolation and limits the potential of multiple coordinated actions between adaptive collaborative agents.

In a standard mission flow, operators describe the mission to each specific platform, data is collected during mission and then post-processed off-line. Consequently, the main use for underwater platforms is to gather information from sensor data on missions that are static and incapable to cope with the long term environmental challenges or resource changes.

In order for embedded service agents to make decisions and interoperate, it is necessary that they have the capability of dealing with and understanding the highly dynamic and complex environments where these networks are going to operate. These decision making tools are constrained to the quality and scope of the available information.

Shared knowledge representation between embedded service-oriented agents is therefore necessary to provide them with the required common situation awareness. Two sources can provide this type of information: the domain knowledge extracted from the expert (orientation) and the inferred knowledge from the processed sensor data (observation). In both cases, it will be necessary for the information to be stored, accessed and shared efficiently by the deliberative agents while performing a mission. These agents, providing different capabilities and working in collaboration, might even be distributed among the different platforms or sharing some limited resources.

1.1 Contribution

In this chapter, we first provide a review to the different approaches solving the decision making process for UUV missions. Then, we propose a semantic framework that provides a solution for hierarchical distributed representation of knowledge for multidisciplinary agent interaction. This framework uses a pool of hierarchical ontologies for representation of the knowledge extracted from the expert and the processed sensor data. It provides a common machine understanding between embedded agents that is generic and extendable. It also includes a reasoning interface for inferring new knowledge from the observed data and guarantee knowledge stability by checking for inconsistencies. This framework improves local (machine level) and global (system level) situation awareness at all levels of service capabilities, from adaptive mission planning and autonomous target recognition to deliberative collision avoidance and escape. It acts as as an enabler for on-board decision making.

Based on their capabilities, service-oriented agents can then gain access to the different levels of information and contribute to the enrichment of the knowledge. If the required information is unavailable, the framework provides the facility to request other agents with the necessary capabilities to generate the required information, i.e. an target classification algorithm could query the correspondent agent to provide the required object detection analysis before proceeding with its classification task.

Secondly, we present an algorithm for autonomous mission adaptation. Using the knowledge made available by the semantic framework, our approach releases the operator from decision making tasks. We show how adaptation plays an important role in providing long term autonomy as it allows the platforms to react to events from the environment while at the same time requires less communication with the operator. The aim is to be effective and efficient as a plan costs time to prepare. Once the initial time has been invested preparing the initial plan, when changes occur, it might be more efficient to try to reuse previous efforts by repairing it. Also, commitments might have been made to the current plan: trajectory reported to other intelligent agents, assignment of mission plan sections to executors or assignment of resources, etc. Adapting an existing plan ensures that as few commitments as possible are invalidated. Using plan proximity metrics, we prove how similar plans are more likely to be accepted and trusted by the operator than one that is potentially completely different.

Finally, we show during a series of in-water trials how these two elements combined, a decision making algorithm and shared knowledge representation, provide the required interoperability between embedded service-oriented agents to achieve high-level mission goals, detach the operator from the routinary mission decision making and, ultimately, enable the permanent presence of dynamic sensing networks underwater.

2. Unmanned decision making loop

In this section, we describe the decision making process currently used by UUV systems and we introduce the unmanned decision loop, where observations, orientations, decisions and actions (OODA) occur in a loop enabling adaptive mission planning.

In order to describe the unmanned decision loop, we need to start by modelling the mission environment. A mission environment is defined by the tuple $\Pi = (\Sigma, \Omega)$, where:

- Σ is the mission domain model containing information about domain, i.e. the platform and the environment of execution, and

- Ω is the mission problem model containing information about the problem, i.e. mission status, requirements, and objectives.

The set of all possible mission environments for a given domain is defined as the *domain space* (e.g., the domain space of the underwater domain). It is denoted by Θ. A mission environment Π is an element of one and only one Θ.

From this model, a mission plan π that tries to accomplish the mission objectives can be produced. However, this mission environment evolves over time t as new observations of the domain model Σ_t and the problem model Ω_t continuously modify it:

$$\Pi_t \leftarrow \Pi_{t-1} \cup \Sigma_t \cup \Omega_t \tag{1}$$

The decision making process to calculate a mission plan π_t for a given mission environment Π_t occurs in a cycle of observe-orient-decide-act. This process was termed by Boyd (1992) as the OODA-loop, and it was modelled on human behaviour. Inside this loop, the *Orientation* phase contains the previously acquired knowledge and initial understanding of the situation of the mission environment (Π_{t-1}). The *Observation* phase corresponds to new perceptions of the mission domain model (Σ_t) and the mission problem model (Ω_t) that modify the mission environment. The *Decision* component represents the level of comprehension and projection, the central mechanism enabling adaptation before closing the loop with the *Action* stage.

Note that it is possible to make decisions by looking only at orientation inputs without making any use of observations. In this case, Eq. 1 becomes $\Pi_t \leftarrow \Pi_{t-1}$. In the same way, it is also possible to make decisions by looking only at the observation inputs without making use of available prior knowledge. In this case, Eq. 1 becomes $\Pi_t \leftarrow \Sigma_t \cup \Omega_t$.

In current UUVs implementations, the human operator constitutes the decision phase. See Figure 1 for a schematic representation of the control loop. When high bandwidth communication links exist, the operator remains in the OODA-loop during the mission execution taking the decisions. For each update of the mission environment Π_t received, the operator decides on the correspondent mission plan π_t to be performed. From the list of actions in this mission plan, the mission executive issues the correspondent commands to the platform. Examples of the implementation of this architecture are existing Remotely Operated Vehicles (ROVs).

However, when communication is unreliable or unavailable, the operator must attempt to include all possible if-then-else cases to cope with execution alternatives before the mission starts. This is the case of current UUVs implementations that follow an orientation-only model. Figure 2 shows this model, where the OODA-loop is broken because observations are not reported to the human operator.

We will now discuss a few recent UUV implementations which show where the state-of-the-art is currently positioned. Most implementations rely on pre-scripted mission

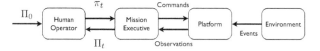

Fig. 1. Observation, Orientation, Decision and Action (OODA) loop for unmanned vehicle systems with decision making provided by the human operator.

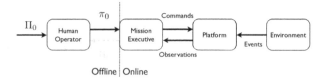

Fig. 2. Broken OODA-loop. Decision stage on the human operator based only on initial pre-mission orientation.

plan managers that are procedural and static and might not even consider conditional executions (Hagen, 2001). At this level, the mission executive follows a sequence of basic command primitives and issues them to the functional control layer of the platform. Description about how these approaches maintain control of underwater vehicles can be found in Fossen (1994), Ridao et al. (1999) and Yuh (2000). In this situation, decisions taken by the operator are made using only orientation inputs related to some previous experience and *a-priori* knowledge. This has unpredictable consequences, in which unexpected situations can cause the mission to abort and might even cause the loss of the vehicle (Griffiths, 2005; von Alt, 2010).

More modern approaches are able to mitigate this lack of adaptability by introducing sets of behaviours that are activated based on observations (Arkin, 1998). Behaviours divide the control system into a parallel set of competence-levels. They can be seen as manually scripted plans generated *a-priori* to encapsulate the decision loop for an individual task. Under this approach, the key factor is to find the right method for coordinating these competing behaviours.

The subsumption model, attributed to Brooks (1986), arbitrates behaviour priorities through the use of inhibition (one signal inhibits another) and suppression (one signal replaces other) networks. Most recent UUV control systems are a variant of the subsumption architecture.

This model was first applied to the control of UUVs by Turner (1995) during the development of the ORCA system. This system used a set of schemas in a case-based framework. However, its scalability remains unclear as trials for its validation were not conducted.

Later, Oliveira et al. (1998) developed and deployed the CORAL system based on Petri nets. The system was in charge of activating the vehicle primitives needed to carry out the mission. These primitives were chained by preconditions and effects.

The scaling problem was addressed by Bennet & Leonard (2000) using a layered control architecture. Layered control is a variant of the subsumption model that restricts of interaction between layers in order to keep it simple (Bellingham et al., 1990). The system was deployed for the application of adaptive feature mapping.

Another approach for coordinating behaviours is vector summation that averages the action between multiple behaviours. Following this principle, the DAMN system developed by Rosenblatt et al. (2002) used a voting-based coordination mechanism for arbitration implementing utility fusion with fuzzy logic.

The MOOS architecture developed by Newman (2002) was also able to guide UUVs by using a mission control system called Helm. Helm's mission plan was described by a set of prioritised primitive tasks. The most suitable action was selected using a set of prioritised mission goals. It used a state-machine for execution, a simplified version of a Petri net.

The O^2CA^2 system (Carreras et al., 2007) also used a Petri net representation of the mission plan (Palomeras et al., 2009). The system maintains the low level control (dynamics) from the

guidance control (kinematics) uncoupled (Caccia & Veruggio, 2000). Although it contained a declarative mission representation, missions were programmed manually.
A detailed survey of other behaviour-based approaches applied to mission control systems for UUVs can be found in Carreras et al. (2006).
More recently, Benjamin et al. (2009) has applied multiple objective decision theory to provide a suitable framework for formulating behaviour-based controllers that generate Pareto-optimal and satisfying behaviours. This approach was motivated by the infeasibility of optimal behaviour selection for real-world applications. This approach has been implemented and deployed as part of the IvP Helm extension to MOOS. This method seems to be a more suitable for behaviour selection, although more computationally expensive. Also, the approach is limited to the control of only the direction and velocity parameters of the host platform.
After reviewing this related work, two problems affecting the effectiveness of the decision loop become evident. Firstly, orientation and observation should be linked together because it is desirable to place the new observations in context. Secondly, decision and action should be iterating continuously. These two problems have not been addressed together by previous approaches. These are the two of the goals that we address in this chapter. In order to achieve them autonomously, two additional components are required: a status monitor and a mission plan adapter. The status monitor reports any changes detected in the mission environment during the execution of a mission. When the mission executive is unable to handle the changes detected by the status monitor, the mission planner is called to generate a new modified mission plan that agrees with the updated mission environment. Figure 3 shows the OODA-loop for autonomous decision making. Comparing it to the previous Figure 2, the addition of status monitor and mission planner removes the need for human decisions in the loop. Note that the original mission plan π_0 could also be autonomously generated as long as the high-level goals are provided by the human operator in Π_0.

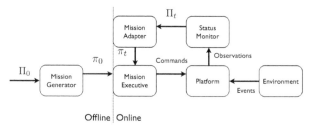

Fig. 3. Required OODA-loop for autonomous decision making in UUVs. Decision stage for adaptation takes place on-board based on initial orientation provided by the operator and observations provided by the status monitor.

Adaptive mission planning enables a true unmanned OODA-loop. This autonomous decision making loop copes with condition changes in the mission environment during the mission execution. As a consequence, it releases the operator from decision making tasks in stressful environments containing high levels of uncertainty and dynamism.
The potential benefits of adaptive mission planning capabilities for autonomous decision making in UUVs were promoted by Turner (2005), Bellingham et al. (2006) and Patrón & Petillot (2008). Possibly the most advanced autonomous decision making framework for UUVs has been developed at the Monterey Bay Aquarium Research Institute. This architecture, known as T-REX, has been deployed successfully inside the Dorado AUV (Rajan

et al., 2009). This is now providing adaptive planning capabilities to oceanographers for maximising the science return of their UUV missions (McGann et al., 2007; Rajan et al., 2007). Using deliberative reactors for the concurrent integration of execution and planning (McGann, Py, Rajan & Henthorn, 2008), live sensor data can be analysed during mission to adapt the control of the platform in order to measure dynamic and episodic phenomenon, such as chemical plumes (McGann, Py, Rajan, Henthorn & McEwen, 2008; McGann et al., 2009). Alternative approaches to adaptive plume tracing can also be found in the works of Farrell et al. (2005) and Jakuba (2007). Their research goals of all these approaches have been motivated by scientific applications and do not consider the needs of the human operators or the maritime industry.

However, autonomy cannot be achieved without humans, as it is necessary for this autonomy to be ultimately accepted by an operator. Our research is geared towards improving human access to UUVs in order to solve the maritime industry's primary requirement of improving platform operability (Patrón et al., 2007). We propose a goal-based approach to solving adaptive mission planning. The advantage of this approach is that it provides high levels of mission abstraction. This makes the human interface simple, powerful and platform independent, which greatly eases the operator's task of designing and deploying missions (Patrón, 2009). Ultimately, operators will not need any specialist training for an UUV specific platform, and instead missions will be described purely in terms of their goals. Apart from ease of use, we have also demonstrated using a novel metric (Patrón & Birch, 2009) that adaptive mission planners can produce solutions which are close to what a human planner would produce (Patrón et al., 2009a). This means that our solutions can be trusted by an operator.

Another advantage of our research over other state-of-the-art UUV implementations, is that we are industry focussed. Our service-oriented approach provides goal-based mission planning with discoverable capabilities, which meets industry's need for platform independence (Patrón et al., 2009b). Finally, our plan repair approach optimises the resources required for adaptability and maximises consistency with the original plan, which improves human acceptance of autonomy. Resource optimisation and consistency are very important properties for real world implementations, as we demonstrate in our sea trials (Patrón, Miguelanez, Petillot & Lane, 2008).

Section 3 describes how do we link together orientation and observation. Section 4 presents an approach to the continuous iteration of decision and action.

3. Semantic knowledge-based situation awareness

Unmanned vehicle situation awareness SA_V consists in enabling the vehicle to autonomously understand the '*big picture*' (Adams, 2007). This picture is composed of the experience gained from previous missions (orientation) and the information obtained from the sensors while on mission (observation). Ontologies allow the representation of knowledge of these two components.

Ontologies are models of entities and interactions, either generically or in some particular practice of knowledge (Gruber, 1995). The main components of an ontology are concepts and axioms. A concept represents a set or class of entities within a domain (e.g., a *fault* is a concept within the domain of diagnostics). Axioms are used to constrain the range and domain of the concepts (e.g., a *driver* is a *software* that has a *hardware*). The finite set of concept and axiom definitions is called the *Terminology Box* *TBox* of the ontology.

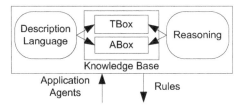

Fig. 4. Knowledge Base representation system including the *TBox*, *ABox*, the description language and the reasoning components. Its interface is made of orientation rules and agent queries.

Instances are the individual entities represented by a concept of the ontology (e.g. a *remus* is an instance of the concept *UUV*). Relations are used to describe the interactions between individuals (e.g. the relation *isComponentOf* might link the individual *SensorX* to the individual *PlatformY*). This finite set of instances and relations about individuals is called the *Assertion Box ABox*. The combination of *TBox* and *ABox* is what is known as a *Knowledge Base*. *TBox* aligns naturally to the orientation component of SA_V while *ABox* aligns to the observation component.

In the past, authors such as Matheus et al. (2003) and Kokar et al. (2009) have used ontologies for situation awareness in order to assist humans during information fusion and situation analysis processes. Our work extends these previous works by using ontologies for providing unmanned situation awareness in order to assist autonomous decision making algorithms in underwater vehicles. One of the main advantages of using a knowledge base over a classical data base schema to represent SA_V is the extended querying that it provides, even across heterogeneous data systems. The meta-knowledge within an ontology can assist an intelligent agent (e.g., status monitor, mission planner, etc.) with processing a query. Part of this intelligent processing is due to the capability of reasoning. This enables the publication of machine understandable meta-data, opening opportunities for automated information processing and analysis.

For instance, a status monitor agent using meta-data about sensor location could automatically infer the location of an event based on observations from nearby sensors (Miguelanez et al., 2008). Inferences over the ontology are made by reasoners. A reasoner enables the domain's logic to be specified with respect to the context model and applied to the corresponding knowledge i.e., the instances of the model (see Fig. 4). A detailed description of how a reasoner works is outside of the scope of this article. For the implementation of our approach, we use the open source reasoner called Pellet (Sirin et al., 2007).

3.1 Semantic knowledge-based framework

A library of knowledge bases comprise the overall knowledge framework used in our approach for building SA_V (Miguelanez et al., 2010; Patrón, Miguelanez, Cartwright & Petillot, 2008). Reasoning capabilities allow concept consistency providing reassurance that SA_V remains stable through the evolution of the mission. Also, inference of concepts and relationships allows new knowledge to be extracted or derived from the observed data. In order to provide with a design that supported maximum reusability (Gruber, 1995; van Heijst et al., 1996), we adopt a three-level segmentation structure that includes the (1) Foundation, (2) Core and (3) Application ontology levels (see Fig. 5).

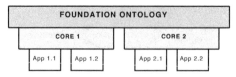

Fig. 5. Levels of generality of the library of knowledge bases for SA_V. They include the Foundation Ontology, the Core Ontology, and the Application Ontology levels.

Foundational Ontologies (FOs) represents the very basic principles and includes Upper and Utility Ontologies. Upper ontologies describe generic concepts (e.g., the Suggested Upper Merged Ontology or SUMO (Niles & Pease, 2001)) while Utility ontologies describe support concepts or properties (e.g. OGC_GML for describing geospatial information (Portele, 2007)). FOs meet the requirement that a model should have as much generality as possible, to ensure reusability across different domains.

The Core Ontology provides a global and extensible model into which data originating from distinct sources can be mapped and integrated. This layer provides a single knowledge base for cross-domain agents and services (e.g., vehicle resource / capabilities discovery, vehicle physical breakdown, and vehicle status). A single model avoids the inevitable combinatorial explosion and application complexities that results from pair-wise mappings between individual metadata formats and ontologies.

In the bottom layer, an Application Ontology provides an underlying formal model for agents that integrate source data and perform a variety of extended functions. As such, higher levels of complexity are tolerable and the design is motivated more by completeness and logical correctness than human comprehension. Target areas of these Application Ontologies are found in the status monitoring of the vehicle and its environment and the planning of the mission.

Figure 6 represents the relationship between the Foundation Ontologies (Upper and Utility), the Core Ontology and the Application Ontology for each service-oriented agent. Raw data gets parsed from sensors into assertions during the mission using a series of adapter modules for each of the sensing capabilities. It also shows that the knowledge handling by the agent during its decision making process is helped by the reasoner and the rule engine process.

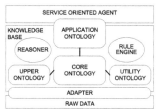

Fig. 6. SA_V representation in the Knowledge Base using Core and Application ontologies supported by Upper and Utility ontologies. Generation of instances from raw data is performed by the Adapter. Handling of knowledge is done by the Reasoner, Rule Engine and the Service-Oriented Agent.

3.2 Foundation and core ontology

To lay the foundation for the knowledge representation of unmanned vehicles, consideration was placed on the Joint Architecture for Unmanned Systems (JAUS) (SAE, 2008a). This standard was originally developed for the Unmanned Ground Vehicles (UGVs) environment only but has recently been extended to all other environments, such as air and water, trying to provide a common set of architecture elements and concepts. The JAUS model classifies four different sets of Knowledge Stores: Status, World map, Library and Log. Our experience has shown that overlap exists between these different sets of knowledge stores. The approach proposed in this paper provides more flexibility in the way the information can be accessed and stored, while still providing JAUS 'Message Interoperability' (SAE, 2008b) between agents.

Within the proposed framework, JAUS concepts are considered as the Foundation Ontology for the knowledge representation. The Core Ontology developed in this work extends these concepts while remaining focused in the domain of unmanned systems. Some of the knowledge concepts identified related with this domain are:

- Platform: Static or mobile (ground, air, underwater vehicles),

- Payload: Hardware with particular properties, sensors or modules,

- Agent: Software with specific capabilities,

- Sensor: A device that receives and responds to a signal or stimulus,

- Driver: Module for interaction with a specific sensor / actuator,

Additionally, the Standard Ontology for Ubiquitous and Pervasive Applications (SOUPA) (Chen et al., 2004) is used as an Utility Ontology. By providing generic context-aware concepts, it enables the spatio-temporal representation of concepts in the Core Ontology.

3.3 Application Ontology

Each service-oriented agent has its own Application Ontology. It represents the agent's awareness of the situation by including concepts that are specific to the expertise of the agent. In the case study presented in this chapter, these agents are the status monitor and the mission planner. Together, they provide the *status monitor* and *mission adapter* components described in Fig. 3 required for closing the OODA-loop and provide on-board decision making adaptation.

3.3.1 Status Monitoring Application Ontology

The Status Monitoring Application Ontology is used to express the SA_V of the status monitor agent. To model the behaviour of all components and subsystems considering from sensor data to possible model outputs, the Status Monitoring Application Ontology is designed and built based on ontology design patterns (Blomqvist & Sandkuhl, 2005). Ontology patterns facilitate the construction of the ontology and promote re-use and consistency if it is applied to different environments. In this work, the representation of the monitoring concepts are based on a system observation design pattern. Some of the most important concepts identified for status monitoring are:

- Data: all internal and external variables (gain levels, water current speed),

- Observation: patterns of data (sequences, outliers, residuals,...),

- Symptom: individuals related to interesting patterns of observations (e.g., low gain levels, high average speed),

- Event: represents a series of correlated symptoms (low power consumption, position drift), Two subclasses of Events are defined: *CriticalEvent* for high priority events and IncipientEvent for the remaining ones.

- Status: links the latest and most updated event information to the systems being monitored (e.g. sidescan transducer),

Please note how some of these concepts are related to concepts of the Core Ontology (e.g. an *observation* comes from a *sensor*). These Core Ontology elements are the enablers for the knowledge exchange between service-oriented agents. This will be shown in the demonstration scenario of Section 5.3.

3.3.2 Planning Application Ontology

The Plan Application Ontology is used to express the SA_V of the mission planner agent. It uses concepts originally defined by the *Planning Domain Definition Language* (PDDL). The PDDL language was originally created by Ghallab et al. (1998) to standardise plan representation. Concepts are extracted from the language vocabulary and the language grammar is used for describing the relationships and constraints between these concepts.

For the adaptation mission planning process, the Planning Application Ontology also requires concepts capable of representing the diagnosis of incidents or problems occurring in some parts of the mission plan (van der Krogt, 2005). Some of the most important concepts identified for mission plan adaptability are:

- Resource: state of an object (physical or abstract) in the environment (vehicle, position, sensor, etc.),

- Action: Modification of the state of resources (calibrate, classify, explore, etc.),

- Gap: A non-executable action,

- Execution: When an action is executed successfully,

- Failure: An unsuccessful execution of an action,

Please note how some of these concepts are also related to concepts of the Core Ontology (e.g. a list of *capability* concepts is required to perform a mission *action*).

4. Adaptive mission planning

The adaptive mission planning process involves the detection of events, the effects that these events have on the mission plan and the response phase. The detection of events is performed by the status monitoring agent. The mission plan diagnosis and repair is undertaken by the adaptive mission planner agent.

4.1 Status Monitor Agent

The Status Monitor Agent considers all symptoms and observations from environmental and internal data in order to identify and classify events according to their priority and their nature (critical or incipient). Based on internal events and context information, this agent is able to infer new knowledge about the current condition of the vehicle with regard to the availability for operation of its components (i.e. status). In a similar way, environmental data is also considered for detecting and classifying external events in order to keep the situation awareness of the vehicle updated.

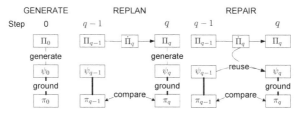

Fig. 7. Schematic representation of the autonomous mission generation, replan and repair processes using partial plan representation of the mission plans.

4.2 Mission plan adaptation agent

This section describes the mission environment model used by the mission plan repair techniques presented in this chapter. We continue using the mission environment model previously described. We generalise this model to any step q in the mission execution timeline. An instance of an UUV mission environment at a given step q can be simply defined as $\Pi_q = \{\Sigma_q, \Omega_q\}$. The mission domain model Σ_q contains the set of propositions defining the available resources in the system P_V and the set of actions or capabilities A_V. The mission problem model Ω_q contains the current platform state x_q and the mission requirements Q_O.

Based on this model, we analyse how to calculate mission plans on the plan space (Sacerdoti, 1975). A plan space is an implicit directed graph whose vertices are partially specified plans and whose edges correspond to refinement operations. In a real environment where optimality can be sacrificed by operability, partial plans are seen as a suitable representation because they are a flexible constrained-based structure capable of being adapted.

A partial plan ψ is a tuple containing a set of partially instantiated actions and a set of constraints over these partially grounded actions. Constraints can be of the form of ordering constraints, interval preservation constraints, point truth constraints and binding constraints. Ordering constraints indicate the ordering in which the actions should be executed. Interval preservation constraints link preconditions and effects over actions already ordered. Point truth constraints assure the existence of precondition facts at certain points of the plan. Binding constraints on the variables of actions are used to ground the actions to variables of the domain. Figure 8 shows a partial plan representation of an UUV mission. Partial plans are flexible to modification. They provide an open approach for handling extensions such as temporal and resource constraints. Due to nature of the constraints, it is easy to explain a partial plan to a user. Additionally, it is easily extensible to distributed multi-agent mission planning.

Figure 7 explains the processes of mission plan repair and mission plan replan for mission plan adaptation for UUVs using a partial plan representation of the mission plans. At the initial step, a partial ordered plan ψ_0 is generated satisfying the original mission environment Π_0. The ψ_0 is then grounded into the minimal mission plan π_0 including all constraints in ψ_0. At step q, the semantic knowledge-based framework is updated by the diagnosis information $\dot{\Pi}_q$ providing a modified awareness of the mission environment Π_q. From here, two mission adaptation processes are possible: *Mission replan* generates a new partial plan ψ_q, as done at the first stage, based only on the knowledge of Π_q. On the other hand, *mission plan repair* re-validates the original plan by ensuring minimal perturbation of it. Given the partial plan at the previous step ψ_{q-1} and the diagnosis information $\dot{\Pi}_q$, the mission repair problem produces a solution partial plan ψ_q that satisfies the updated mission problem Π_q, by modifying ψ_{q-1}. The final step for both approaches is to ground ψ_q to its minimal mission plan

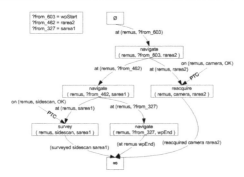

Fig. 8. Example of a partial ordered plan representation of an autonomously generated UUV mission. The ordering constraints are represented using the graph depth, interval preservation constraints are represented with black arrows, point truth constraints are represented with PTC-labelled arrows, and binding constraints are shown in the top left box.

π_q. It can be seen that mission repair better exploits the orientation capabilities for decision making: instead of taking the new mission environment as a given, it uses the diagnosis information about the changes occurred to guide the adaptation process.

We have now identified the benefits of mission plan repair over mission replan. Mission plan repair modifies the partial plan ψ_q, so that it uses a different composition, though it still maintains some of the actions and the constraints between actions from the previous partial plan. However, mission plan adaptation can also be achieved by mission execution repair by looking directly at the mission plan instantiation π_q. Execution repair modifies the instantiation of the mission plan π_q such that a ground action $g_q^{a_h}$ that was previously instantiated by some execution e_q is newly bound by another action execution instance e'_q.

Executive repair is less expensive and it is expected to be handled directly by the mission executive agent. Plan repair, however, is computationally more expensive and requires action of the mission planner agent.

The objective is to maximise the number of execution repairs over plan repairs and, at the plan repair level, maximise the number of decisions reused from the previous mission instantiation. The information provided by the semantic-base knowledge base during the plan diagnosis phase is critical.

Executive repair fixes plan failures identified in the mission plan during the diagnosis stage. Our approach uses ontology reasoning in combination with an action execution template to adapt the mission plan at the executive level.

Once a mission plan π_q is calculated by the mission planner, its list of ground actions is transferred to the executive layer. In this layer, each ground action $g_q^{a_h}$ of π_q gets instantiated into an action execution instance e_q^t using the action template for the action a_h available in the Core Ontology of the knowledge base. At the end of this phase, each e_q^t contains the script of commands required to perform its correspondent ground action. Flexibility in the execution of an action instance is critical in real environments. This is provided by a timer, an execution counter, a time-out register and a register of the maximum number of executions in the action execution instance. Additionally, three different outputs control the success, failure or time-out of its execution. These elements handle the uncertainty during the execution phase and enable the executive repair process. This minimise the number of calls to the adaptive mission planner agent and therefore the response time for adaptation.

Plan repair uses a strategy to repair with new partial plans the plan gaps identified during the plan diagnosis stage. Our approach uses an iteration of unrefinement and refinement strategies on a partial-ordered planning framework to adapt the mission plan.

Planning in the plan space is slower than in the state space because the nodes are more complex. Refinement operations are intended to achieve an open goal from the list of mission requirements or to remove a possible inconsistency in the current partial plan. These techniques are based on the least commitment principle, and they avoid adding to the partial plan any constraint that is not strictly needed. A refinement operation consists of one or more of the following steps: adding an action, an ordering constraint, a variable binding constraint or a causal link.

A partial plan is a solution to the planning problem if has no flaw and if the sets of constraints are consistent. Flaws are either subgoals or threats. Subgoals are open preconditions of actions that have not been linked to the effects of previous actions. Threats are actions that could introduce inconsistencies with other actions or constraints. We implemented a recursive non-deterministic approach based on the Partial ordered Planning (PoP) framework (Penberthy & Weld, 1992). This framework is sound, complete, and systematic. Unlike other Plan space planners that handle both types of flaws (goals and threats) similarly, each PoP recursive step first refines a subgoal and then the associated threats (Ghallab et al., 2004).

In our implementation, we introduce a previous step capable of performing an unrefinement of the partial plan when necessary. During the unrefinement strategy we remove refinements from the partial plan that are reported by the plan diagnosis phase to be affecting the consistency of the mission plan with the mission environment, i.e. to remove constraints and finally the actions if necessary.

In simple terms, when changes on the $ABox$ Planning Application Ontology are sensed (Π_q) that affect the consistency of the current partial plan ψ_{q-1}, the plan repair process is initiated. The plan repair stage starts an unrefinement process that relaxes the constraints in the partial plan ψ_{q-1} that are causing the mission plan to fail.

The remaining temporal mission partial plan ψ'_{q-1} is now relaxed to be able to cope with the new mission environment. However, this relaxation could open some subgoals and introduce threats in the partial plan that need to be addressed. The plan repair stage then executes a refinement process searching for a new mission plan ψ_q that is consistent with the new mission environment Π_q and removing these possible flaws. By doing this, it can be seen that the new mission plan ψ_q is not generated again from Π_q (re-planned) but recycled from ψ_{q-1} (repaired). This allows re-use of the parts of the plan ψ_{q-1} that were still consistent with Π_q.

5. Results

5.1 Architecture

The combination of the status monitor agent, the adaptive mission planner, the mission executive and the semantic knowledge-based framework is termed as the Semantic-based Adaptive Mission Planning system (SAMP). The SAMP system implements the four stages of the OODA-loop. Figure 9 represents the customised version of Figure 3 for SAMP.

The status monitor agent reports to the knowledge base the different changes occurring in the environment and the modifications of the internal status of the platform. The knowledge base stores the ontology-based knowledge containing the expert orientation provided *a priori* and the observations reported by the status monitor. A mission planner agent generates and

Fig. 9. Architecture of the SAMP system. The embedded agents are the planner, executive, monitor, and knowledge base. These agents interconnect via set of messages. The system integrates to the functional layer of a generic host platform by an abstract layer interface (ALI).

adapts mission plans based on the situation awareness stored in the knowledge base. The mission executive agent executes mission commands in the functional layer based on the sequence of ground actions received from the mission planner. An Abstract Layer Interface (ALI) based on JAUS-like messages (SAE, 2008a) over UDP/IP packages implemented using the OceanSHELL protocol (Oce, 2005) provides independence from the platform's functional layer making the system generic and platform independent.

5.2 Simulation results

A set of synthetic simulated scenarios have been implemented to test the performance of the SAMP system. The tests are based on the mine counter measure (MCM) operation, where UUVs support and provide solutions for mine-hunting and neutralisation.

A set of 15 selected MCM scenarios were simulated covering the variability of missions described by the concepts of operations for unmanned underwater vehicles presented in the UUV (2004) and the JRP (2005). For each scenario, the detection of a failure in one of the components of the system was simulated. The mission plan was adapted to the new constraints using replanning methods and the mission plan repair approach based on partial plans introduced in Section 4.

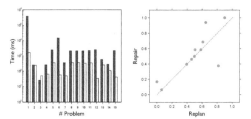

Fig. 10. Left: A semi-log plot displaying the computational time in miliseconds for replan (dark grey bars) and repair approaches (light grey bars). Right: Comparison of Plan Proximity ($PP_{0.5}$) of the replan and repair approaches to the original plan.

The performance of the two approaches was compared by looking at the computation time and the Plan Proximity (Patrón & Birch, 2009) of the adaptive mission plan provided to the original reference mission plan. Figure 10 left shows the computation time in milliseconds required for adapting the mission to the new constraints for replan (dark grey bars) and repair approaches (light grey bars). Note that a logarithmic scale is used for the time values. Figure 10 right displays the Plan Proximity to the original plan of the replan strategy result versus the repair strategy result. It can be seen that plans adapted using the mission repair strategy tend to be closer to the original plan than using the mission replan strategy. In these results, 14 out of 15 scenarios were computed faster by using mission plan repair. This computation was on average 9.1x times faster. Also, 14 out of 15 scenarios showed that mission plan repair had greater or equal Plan Proximity values as compared to mission replan. In general, our mission repair approach improves performance and time response while at the same time finds a solution that is closer to the original mission plan available before adaptation.

5.3 Experimental results

This section shows the performance of the SAMP system inside the real MCM application using a REMUS 100 UUV platform (see Fig. 11.left) in a set of integrated in-water field trial demonstration days at Loch Earn, Scotland ($56^o23.1$N,$4^o12.0$W). The REMUS UUV had a resident guest PC/104 1.4GHz payload computer where the SAMP system was installed. SAMP was capable of communicating with the vehicle's control and status modules and taking control of it by using an interface module that translated the ALI protocol into the manufacturer's *Remote Control* protocol (REM, 2008).

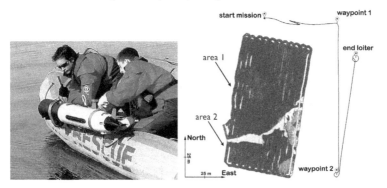

Fig. 11. Left: REMUS UUV deployment before starting one of its missions. Right: Procedural mission uploaded to the vehicle control module and *a priori* seabed classification information stored in the knowledge base. The two dark grey areas correspond to the classified seabed regions.

Figure 11.right shows the procedural waypoint-based mission as it was described to the vehicle's control module. This was known as the baseline mission. It was only used to start the vehicle's control module with a mission in the area of operation before taking control of it using the SAMP system. The baseline mission plan consisted on a start waypoint and two waypoints describing a North to South mission leg at an approximate constant Longitude ($4^o16.2$W). This leg was approximately 250 meters long and it was followed by a loiter pattern at the recovery location. The track obtained after executing this baseline mission using the vehicle control module is shown in Fig. 11 with a dark line. A small adjustment of the vehicle's location can be observed on the top trajectory after the aided navigation module corrects its solution to the fixes received from the Long Baseline (LBL) transponders previously deployed in the area of operations.

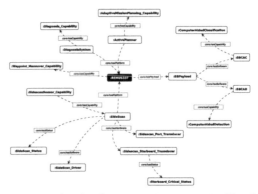

Fig. 12. Core Ontology instances for the demonstration scenario. The diagram represents the main platform, its components and their capabilities.

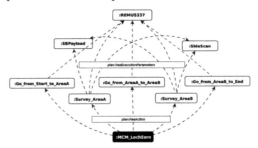

Fig. 13. Plan Application Ontology concepts representing the mission planning actions and their execution parameters and relationships.

On the payload side, the SAMP system was oriented (in the OODA-loop sense) using *a priori* information about the environment and the platform and a declarative description of the goals of the mission. The *a priori* knowledge and the platform configuration capabilities was represented using Core Ontology concepts (see Fig. 12). Knowledge about the environment was provided based on automatic computer-aided seabed classification information generated from previous existent data (Reed et al., 2006). The two classified seabed areas are shown in Fig. 11. The declarative description of the mission requirements was represented using concepts from the Planning Application Ontology. They could be resumed as 'survey all known areas maximizing efficiency'.

5.3.1 Pre-mission reasoning
Please note that the previously described separation between Core knowledge and Planning knowledge gracefully aligns with the separation between platform engineers and mission scientists on current UUV operations. If the platform capabilities were described in Core Ontology terms by the engineers that manufactured the platform, it can be seen how, by using the SAMP approach, a scientific operator that only cares about the data should be able to describe the mission to the platform without knowing anything about the custom properties of the platform. It is, therefore, important to assist the operator in knowing if the platform capabilities can match the mission requirements before starting the mission:

- *Is this platform configuration suitable to successfully perform this mission?*

In order to answer this question, new knowledge could be inferred from the initial Core Ontology orientation. The Core Ontology rule engine was executed providing with additional knowledge. A set of predefined rules helped orienting the knowledge base into inferring new relationships between instances. An example of a rule dealing with the transfer of payload capabilities to the platform is represented in Eq. 2.

$$
\begin{aligned}
core : isCapabilityOf(?Capability, ?Payload) \wedge \\
core : isPayloadOf(?Payload, ?Platform) \\
\rightarrow core : isCapabilityOf(?Capability, ?Platform)
\end{aligned}
\tag{2}
$$

Once all the possible knowledge was extracted, it was possible to query the knowledge base in order to extract the list of capabilities of the platform (see Eq. 3) and the list of requirements of the mission (see Eq. 4).

$$
SELECT\ ?Platform\ ?Cap\ WHERE\ \{\ rdf{:}type(\ ?Platform,\ core{:}Platform)\ \wedge \\
core{:}hasCapability(?Platform,?Cap)\ \}
\tag{3}
$$

$$
SELECT\ ?Mission\ ?Req\ WHERE\ \{\ plan{:}hasAction(\ ?Mission,\ ?Action)\ \wedge \\
plan{:}hasRequirement(\ ?Action,?Req\ )\ \}
\tag{4}
$$

This way, it was possible to autonomously extract that the requirements of the mission of the experiment were[1]:

- *core:WaypointManeuver_Capability* $\in$ *jaus:Maneuver_Capability*
- *core:ComputerAidedClassification_Capability* $\in$ *jaus:Autonomous_RSTA-I_Capability*
- *core:ComputerAidedDetection_Capability* $\in$ *jaus:Autonomous_RSTA-I_Capability*
- *core:SidescanSensor_Capability* $\in$ *jaus:Environmental_Sensing_Capability*

which were a subset of the platform capabilities. Therefore, for this particular case, the platform configuration suited the mission requirements.

5.3.2 In mission adaptation

For these experiments, SAMP was given a static location in which to take control of the host vehicle. At this point, the mission planner agent generated a mission plan based on the knowledge available and the mission requirements. The instantiation of this mission plan is described in Fig. 13 using Planning Application Ontology concepts. The mission was then passed to the executive agent that took control of the vehicle for its execution.

While the mission was executed the status monitor agent maintained the knowledge base updated (in the OODA-loop sense) by reporting changes in the status of hardware components, such as batteries and sensors, and external parameters, such as water currents.

When observations indicated that some of these changes were affecting the mission under execution, the mission planner was activated in order to adapt the mission to the changes. This indication was detected by the planner agent by querying the knowledge base with the following question:

---

[1] RSTA-I i.e., Reconnaissance/Surveillance/Target Acquisition & Identification capability concepts inherited from JAUS (SAE, 2006)

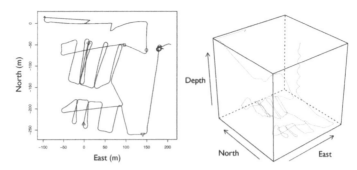

Fig. 14. Left: Vehicle's track during mission in a North-East coordinate frame projection with the origin at the starting point of the mission. Right: Three-dimensional display of the vehicle's track during the mission (Note that depth coordinates are not to scale).

- *Are the observations coming from the environment affecting the mission currently under execution?*

In order to explain the reasoning process involved during the event detection, diagnosis and response phases of the mission adaptation process, a component fault as an internal event was temporarily simulated in the host vehicle. The fault simulated the gains of the starboard transducer of the sidescan sonar dropping to their minimum levels half way through the lawn mower survey of the first area.

For the detection phase, the low gain signals from the transducer triggered a symptom instance, which had an associated event level. This event level, represented in the Status Monitoring Application Ontology using a value partition pattern, plays a key role in the classification of the instances in the *Event* concept between being critical or incipient. This classification is represented axiomatically in the Eqs. 5 and 6.

$$
\begin{aligned}
&\text{status:CriticalEvent} \sqsubseteq \text{status:Event} \sqcap \exists\text{status:causedBySymptom} \ldots \\
&(\text{status:Symptom} \sqcap \exists\text{status:hasEventLevel} \ldots \\
&(\text{status:Level} \sqcap \exists\text{status:High}))
\end{aligned} \tag{5}
$$

$$
\begin{aligned}
&\text{status:IncipientEvent} \sqsubseteq \ldots \\
&\text{status:Event} \sqcap \exists\text{status:causedBySymptom} \ldots \\
&(\text{status:Symptom} \sqcap \exists\text{status:hasEventLevel} \ldots \\
&(\text{status:Level} \sqcap \exists\text{status:Med}))
\end{aligned} \tag{6}
$$

After the *Event* individuals were re-classified, the *Status* property of the related component in the Core Ontology was updated.

During the diagnosis event phase, a critical status of a component is only considered to be caused by a critical event. Therefore, due to the fact that the sidescan sonar component is composed of two transducers, port and starboard, one malfunctioned transducer was only diagnosed as an incipient *Status* of the overall sidescan sonar component.

During the response phase, the *Status* property of the Core Ontology components were used by the mission planner to perform the plan diagnosis of the mission under execution. The query to the knowledge base shown in Eq. 7 reported that the two survey actions in the mission plan were affected by the incipient status of the sidescan sonar.

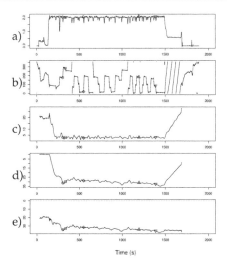

Fig. 15. Vehicle telemetry (top to bottom): a) vehicle velocity (m/s), b) compass heading (degrees), c) altitude (m), d) depth (m) and e) reconstructed profile of the seabed bathymetry (m) during the mission, all plotted against mission time (s).

$$SELECT \quad ?Mission \: ?Action \: ?Param \: ?Status$$
$$WHERE \: \{ \: plan{:}hasAction(\: ?Mission, \: ?Action) \: \wedge$$
$$plan{:}hasExecParam(\: ?Action, ?Param) \: \wedge$$
$$plan{:}hasStatus(\: ?Param, \: ?Status) \: \}$$

(7)

An incipient *Status* of the action parameters indicates that the action can still be performed by adapting the way it is being executed, an execution repair. If both transducers were down, a critical status of the sidescan sensor is diagnosed and a plan repair adaptation of the mission plan would have been required instead. In that case, the adaptive mission planner would have looked for redundant components or similar capabilities to perform the action or to drop the action from the plan.

The same procedure was used after the transducer recovery was reported to adapt the survey action to the normal pattern during the second lawn mower survey. In a similar process, SAMP adapted the lawnmower pattern survey of the areas to the detected water current *Status* at the moment of initialising the survey of the areas.

The timeline of the mission executed using the SAMP approach is described in the following figures: Figure 14 represents the final trajectory of the vehicle in 2D and 3D using a North-East coordinate frame projection with the origin at the starting point of the mission. Figure 15 displays the vehicle's telemetry recorded during the mission. It includes vehicle's velocity, compass heading, altitude, depth measurements and processed bathymetry over time. Figure 16 shows subset of the variables being monitored by the status monitor agent that were relevant to this experiment. These variables include direction of water current, remaining battery power, the availability of the transducers in the sidescan sensor and the mission execution status. Figure 17 represents the system activity of the payload computer recorded during the mission. The system activity logs show percentage of processor usage, memory usage, network activity and disk usage.

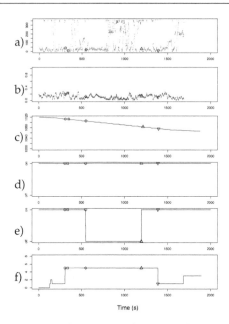

Fig. 16. Status monitoring (top to bottom): a) direction of water current (degrees), b) speed of water current (m/s), c) battery power (Wh), d) sidescan sensor port and e) starboard transducers availability (on/off) and f) mission status binary flag, all plotted against mission time (s).

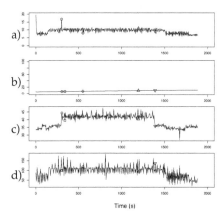

Fig. 17. System activity (top to bottom): a) % processor usage, b) % memory usage, c) network activity (packets/s) and d) disk usage (I/O sectors/s), all plotted against mission time (s).

Each of the symbols ◯, □, ◇, △ and ▽ on the aforementioned figures represents a point during the mission where an event occurred. Symbol ◯ represents the point where SAMP takes control of the vehicle. Note a change on the host platform mission status binary flag that becomes 0x05, i.e. the mission is active (0x01) and the payload is in control (0x04) (Figure 16.e).

Also, a peak on the CPU usage can be noted as this is the point where the mission partial plan gets generated (Figure 17.a).

Symbol □ represents the point where the vehicle arrives to perform the survey of the area. At this point, the action survey gets instantiated based on the properties of the internal elements and external factors. Although the Loch waters where the trials were performed were very still (see Figure 16.b), note how the vehicle heading during the lawnmower pattern performed to survey the areas follows the water current direction sensed at the arrival (approx. 12°, Symbol □ - Figure 16.a) in order to minimize drag and maximise battery efficiency. The heading of the vehicle during the survey can be observed in Figure 14 and Figure 15.b. The link between the vehicle heading in relation to the water current direction and its effect on the battery consumption was expert orientation knowledge captured by a relationship property between the two concepts in the Core Ontology.

Symbol ◇ represents the point when the status monitor agent detects and reports a critical status in the starboard transducer of the sidescan sonar (Figure 16.d). It can be seen how the lawnmower pattern was adapted to cope with the change and to use the port transducer to cover the odd and even spacing of the survey. This pattern avoids gaps in the sidescan data under the degraded component configuration and maximises sensor coverage for the survey while the transducer is down.

Symbol △ indicates the point where the starboard transducer recovery is diagnosed. It can be observed how the commands executing the action are modified in order to optimise the survey pattern and minimise distance travelled. Although also being monitored, the power status does not report any critical status during the mission that requires modification of the actions (Figure 16.c).

Symbol ▽ shows the location where all the mission goals are considered achieved and the control is given back to the mission control of the host vehicle (see Symbol ▽ - Figure 16.e shows the mission is still active but the payload is not longer in control (0x01)). From this point the host vehicle's control module takes the control back and drives the vehicle to the loiter at the recovery location.

6. Conclusion and future work

The underwater domain is a challenging environment in which to maintain the operability of an UUV. Operability can be improved with the embedded adaptation of the mission plan. We implement a system capable of adapting mission plans autonomously in the face of events while during a mission. We do this by using a combination of ontological hierarchical representation of knowledge and adaptive mission plan repair techniques. The advantage of this approach is that it maximises robustness, system performance and response time. The system performance has been demonstrated in simulation. Additionally, the mission adaptation capability is shown during an in-water field trial demonstration.

In our fully integrated experiments we achieved the following:

- *Knowledge based framework:* We have presented a semantic-based framework that provides the core architecture for knowledge representation for service oriented agents in UUVs. The framework combines the initial expert orientation and the observations acquired during mission in order to improve the situation awareness in the vehicle. This is currently unavailable in UUVs.

- *Goal-oriented plan vs. waypoint-based plan:* The system uses a goal-oriented approach in which the mission is described in terms of 'what to do' instead of a 'how to do' it. The mission is parametrised and executed based on the available knowledge and vehicle capabilities. This is the first time that an approach to goal-based planning is applied to the adaptation of an underwater mission in order to maintain platform's operability.

- *Adaptation to environmental parameters and internal issues:* The approach shows adaptability to environmental elements, such as water current flows in order to improve mission performance. The approach is also capable of dealing with the critical status of certain components in the platform and can react accordingly.

- *Platform agnostic:* The approach is platform independent making it readily applicable to other domains, such as ground or air vehicles.

SAMP is open to event detections coming from other embedded service-oriented agents. We are planning to apply the approach to more complex scenarios involving other embedded agents, such as agents for automatic target recognition. We are also planning to extend it to a team of vehicles performing a collaborative mission. In this scenario, agents are distributed across the different platforms. We are currently working towards a shared situation awareness for a team of vehicles to which every team member possess the awarenessrequired for its responsibilities. The main advantage of our semantic-based approach is the low bandwidth required to share id-coded ontological concepts and, therefore, to cope with the underwater acoustic communication limitations.

7. References

Adams, J. A. (2007). Unmanned vehicle situation awareness: A path forward, *Proceedings of the 2007 Human Systems Integration Symposium*, Annapolis, Maryland, USA.

Arkin, R. C. (1998). *Behavior-based Robotics*, Massachusetts Institute of Technology Press, ISBN: 978-0262011655.

Bellingham, J., Consi, T., Beaton, R. & Hall, W. (1990). Keeping layered control simple, *Proceedings of the (1990) Symposium on Autonomous Underwater Vehicle Technology (AUV'90)*, pp. 3–8.

Bellingham, J., Kirkwood, B. & Rajan, K. (2006). Tutorial on issues in underwater robotic applications, *16th International Conference on Automated Planning and Scheduling (ICAPS'06)*.

Benjamin, M. R., Leonard, J. J., Schmidt, H. & Newman, P. M. (2009). An overview of MOOS-IvP and a brief users guide to the IvP Helm autonomy software, *Technical Report MIT-CSAIL-TR-2009-028*, Computer Science and Artificial Intelligence Lab, MIT.

Bennet, A. A. & Leonard, J. J. (2000). A behavior-based approach to adaptive feature detection and following with autonomous underwater vehicles, *IEEE Journal of Oceanic Engineering* 25(2): 213–226.

Blomqvist, E. & Sandkuhl, K. (2005). Patterns in ontology engineering âĂŞ classification of ontology patterns, *7th International Conference on Enterprise Information Systems*, Miami, USA.

Boyd, J. (1992). A discourse on winning and losing - organic design for command and control, *Technical report*, http://www.d-n-i.net/dni/john-r-boyd/.

Brooks, R. (1986). A robust layered control system for a mobile robot, *Journal of Robotics and Automation* RA-2(1): 14–23.

Caccia, M. & Veruggio, G. (2000). Guidance and control of a reconfigurable unmanned underwater vehicle, *Control Engineering Practice* 8(1): 21 – 37.

Carreras, M., Palomeras, N., Ridao, P. & Ribas, D. (2007). Design of a mission control system for an AUV, *International Journal of Control* 80(7): 993–1007.

Carreras, M., Ridao, P., Garcia, R. & Batlle, J. (2006). *Behaviour Control of UUVs*, G. Roberts and R. Sutton, Eds Advances in Unmanned Maritime Vehicles (IEE Control Engineering), ISBN: 978-0863414503, chapter 4, pp. 67–86.

Chen, H., Perich, F., Finin, T. & Joshi, A. (2004). Soupa: Standard ontology for ubiquitous and pervasive applications, *International Conference on Mobile and Ubiquitous Systems: Networking and Services, Boston, MA.*

Farrell, J. A., Pang, S. & Li, W. (2005). Chemical plume tracing via an autonomous underwater vehicle, *IEEE Journal of Oceanic Engineering* 30, 2: 428–442.

Fossen, T. I. (1994). *Guidance and Control of Ocean Vehicles*, John Wiley and Sons, ISBN: 0-471-94113-1.

Ghallab, M., Howe, A., Knoblock, C., McDermott, D., Ram, A., Veloso, M., Weld, D. & Wilkins, D. (1998). PDDL: The planning domain definition language, *Technical report*, Yale Center for Computational Vision and Control.

Ghallab, M., Nau, D. & Traverso, P. (2004). *Automated Planning: Theory and Practice*, Morgan Kaufmann, ISBN: 1-55860-856-7.

Griffiths, G. (2005). AUTOSUB under ice, *Ingenia, ISSN: 1472-9768* 22: 30–32.

Gruber, T. R. (1995). Towards principles for the design of ontologies used for knowledge sharing, *International Journal Human-Computer Studies* 43: 907–928.

Hagen, P. E. (2001). AUV/UUV mission planning and real time control with the HUGIN operator systems, *Proceedings of the IEEE International Conference Oceans (Oceans'01)*, Vol. 1, Honolulu, HI, USA, pp. 468–473.

Jakuba, M. (2007). *Stochastic Mapping for Chemical Plume Source Localization with Application to Autonomous Hydrothermal Vent Discovery*, Doctor of philosophy in mechanical engineering, MIT/WHOI Joint Program in Applied Ocean Physics and Engineering.

JRP (2005). Joint robotics program master plan FY2005, *Technical report*, US Department of Defense.

Kokar, M. M., Matheus, C. J. & Baclawski, K. (2009). Ontology-based situation awareness, *Information Fusion* 10(1): 83–98.

Matheus, C., Kokar, M. & Baclawski, K. (2003). A core ontology for situation awareness, *Proceedings of the Sixth International Conference of Information Fusion* 1: 545 – 552.

McGann, C., Py, F., Rajan, K. & Henthorn, R. (2008). Adaptive control for autonomous underwater vehicles, *Association for the Advancement of Artificial Intelligence (AAAI'08)*, Chicago, IL, USA.

McGann, C., Py, F., Rajan, K., Henthorn, R. & McEwen, R. (2008). A deliberative architecture for AUV control, *International Conference on Robotic and Autonomation (ICRA'08)*, Pasadena, CA, USA.

McGann, C., Py, F., Rajan, K. & Olaya, A. G. (2009). Integrated planning and execution for robotic exploration, *International Workshop on Hybrid Control of Autonomous Systems*, Pasadena, CA, USA.

McGann, C., Py, F., Rajan, K., Thomas, H., Henthorn, R. & McEwen, R. (2007). T-REX: A model-based architecture for AUV control, *Workshop in Planning and Plan Execution for Real-World Systems: Principles and Practices for Planning in Execution, International Conference of Autonomous Planning and Scheduling (ICAPS'07)*.

Miguelanez, E., Lewis, R., K.Brown, Roberts, C. & Lane, D. (2008). Fault diagnosis of a train door system based on the semantic knowledge representation, *The 4th IET International Conference on Railway Condition Monitoring RCM 2008*, Derby Conference Centre, Derby, UK, pp. 1 –6.

Miguelanez, E., Patrón, P., Brown, K., Petillot, Y. R. & Lane, D. M. (2010). Semantic knowledge-based framework to improve the situation awareness of autonomous underwater vehicles, *IEEE Transactions on Knowledge and Data Engineering (In Press)* PP(99).

Newman, P. (2002). MOOS – a mission oriented operating suite, *Technical report*, Department of Ocean Engineering, Massachusetts Institute of Technology.

Niles, I. & Pease, A. (2001). Towards a standard upper ontology, *Proceedings of the 2nd International Conference on Formal Ontology in Information Systems (FOIS-2001)* .

Oce (2005). OceanSHELL: An embedded library for distributed applications and communications, *Technical report*, Ocean Systems Laboratory, Heriot-Watt University.

Oliveira, P., Pascoal, A., Silva, V. & Silvestre, C. (1998). Mission control of the MARIUS AUV: System design, implementation, and sea trials, *International Journal of Systems Science* 29(10): 1065–1085.

Palomeras, N., Ridao, P., Carreras, M. & Silvestre, C. (2009). Using petri nets to specify and execute missions for autonomous underwater vehicles, *Proceedings of the IEEE/RSJ International Conference on Intelligent Robots and Systems (IROS'09)*.

Patrón, P. (2009). Embedded knowledge and autonomous planning, *Sea Technology Magazine, ISSN. 0093-3651* 50(4): 101.

Patrón, P. & Birch, A. (2009). Plan proximity: an enhanced metric for plan stability, *Workshop on Verification and Validation of Planning and Scheduling Systems, 19th International Conference on Automated Planning and Scheduling (ICAPS'09)*, Thessaloniki, Greece, pp. 74–75.

Patrón, P., Evans, J. & Lane, D. M. (2007). Mission plan recovery for increasing vehicle autonomy, *Proceedings of the Conference of Systems Engineering for Autonomous Systems from the Defence Technology Centre (SEAS-DTC'07)*, Edinburgh, UK.

Patrón, P., Lane, D. M. & Petillot, Y. R. (2009a). Continuous mission plan adaptation for autonomous vehicles: balancing effort and reward, *4th Workshop on Planning and Plan Execution for Real-World Systems, 19th International Conference on Automated Planning and Scheduling (ICAPS'09)*, Thessaloniki, Greece, pp. 50–57.

Patrón, P., Lane, D. M. & Petillot, Y. R. (2009b). Interoperability of agent capabilities for autonomous knowledge acquisition and decision making in unmanned platforms, *Proceedings of the IEEE International Conference Oceans Europe (Oceans Europe'09)*, Bremen, Germany.

Patrón, P., Miguelanez, E., Cartwright, J. & Petillot, Y. R. (2008). Semantic knowledge-based representation for improving situation awareness in service oriented agents of autonomous underwater vehicles, *Proceedings of the IEEE International Conference Oceans (Oceans'08)*, Quebec, Canada.

Patrón, P., Miguelanez, E., Petillot, Y. R. & Lane, D. M. (2008). Fault tolerant adaptive mission planning with semantic knowledge representation for autonomous underwater vehicles, *Proceedings of the IEEE/RSJ International Conference on Intelligent Robots and Systems (IROS'08)*, Nice, France, pp. 2593–2598.

Patrón, P. & Petillot, Y. R. (2008). The underwater environment: A challenge for planning, *Proceedings of the Conference of the UK Planning Special Interest Group (PlanSIG'08)*, Edinburgh, UK.

Penberthy, J. & Weld, D. S. (1992). UCPOP: A sound, complete, partial order planner for ADL, *Proceedings of the National Conference on Knowledge Representation and Reasoning (KR)*, pp. 103–114.

Portele, C. (2007). OpenGISÂő geography markup language (GML) encoding standard v.3.2.1, *Technical report*, Open Geospatial Consortium Inc.

Rajan, K., McGann, C., Py, F. & Thomas, H. (2007). Robust mission planning using deliberative autonomy for autonomous underwater vehicles, *Workshop in Robotics in challenging and hazardous environments, International Conference on Robotics and Automation (ICRA'07)*.

Rajan, K., Py, F., McGann, C., Ryan, J., O'Reilly, T., Maughan, T. & Roman, B. (2009). Onboard adaptive control of AUVs using automated planning and execution, *International Symposium on Unmanned Untethered Submersible Technology (UUST'09)*, Durham, NH, USA.

Reed, S., Ruiz, I., Capus, C. & Petillot, Y. (2006). The fusion of large scale classified side-scan sonar image mosaics, *IEEE Transactions on Image Processing* 15(7): 2049–2060.

REM (2008). REMUS remote control protocol v1.12, *Technical report*, Hydroid Inc.

Ridao, P., Batlle, J., Amat, J. & Roberts, G. (1999). Recent trends in control architectures for autonomous underwater vehicles, *International Journal of Systems Science* 30(9): 1033–1056.

Rosenblatt, J. K., Williams, S. B. & Durrant-Whyte, H. (2002). Behavior-based control for autonomous underwater exploration, *International Journal of Information Sciences*, 145(1–2): 69–87.

Sacerdoti, E. (1975). The nonlinear nature of plans, *International Joint Conference on Artificial Intelligence (IJCAI'75)*, pp. 206–214.

SAE (2006). Society of automotive engineers AS-4 AIR5664 JAUS history and domain model, *Technical report*, SAE International Group.

SAE (2008a). Society of automotive engineers AS-4 AIR5665 JAUS architecture framework for unmanned systems, *Technical report*, SAE International Group.

SAE (2008b). Society of automotive engineers AS-4 ARP6012 JAUS compliance and interoperability policy, *Technical report*, SAE International Group.

Sirin, E., Parsia, B., Grau, B. C., Kalyanpur, A. & Katz, Y. (2007). Pellet: A practical owl-dl reasoner, *Web Semantics: Science, Services and Agents on the World Wide Web* 5(2): 51–53.

Turner, R. (2005). Intelligent mission planning and control of autonomous underwater vehicles, *Workshop on Planning under uncertainty for autonomous systems, 15th International Conference on Automated Planning and Scheduling (ICAPS'05)*.

Turner, R. M. (1995). Context-sensitive, adaptive reasoning for intelligent auv control: Orca project update, *In Proceedings of the Ninth International Symposium on Unmanned, Untethered Submersible Technology (UUST'95)*, Durham, NH, USA, pp. 426–435.

UUV (2004). The navy unmanned undersea vehicle (UUV) master plan, *Technical report*, US Department of the Navy.

van der Krogt, R. (2005). *Plan repair in single-agent and multi-agent systems*, PhD thesis, Netherlands TRAIL Research School.

van Heijst, G., Schreiber, A. & Wielinga, B. (1996). Using explicit ontologies in kbs development, *International Journal of Human-Computer Studies* 46(2-3).

von Alt, C. (2010). Remus technology 2020 and beyond, *NATO Naval Armaments Group – Maritime Capability Group 3 on Mines, Mine Countermeasures and Harbour Protection – Industry Day*, Porton Down, UK.

Yuh, J. (2000). Design and control of autonomous underwater robots: A survey, *Journal of Autonomous Robots* 8(1): 7–24.

11

Mapping and Dilution Estimation of Wastewater Discharges Based on Geostatistics Using an Autonomous Underwater Vehicle

Patrícia Ramos[1] and Nuno Abreu[2]
[1]*Institute for Systems and Computer Engineering of Porto (INESC Porto)*
School of Accounting and Administration of Porto – Polytechnic Institute of Porto
[2]*Institute for Systems and Computer Engineering of Porto (INESC Porto)*
Faculty of Engineering of University of Porto
Portugal

1. Introduction

Wastewaters are often discharged into coastal waters through outfall diffusers that efficiently dilute effluent and usually restrict any environmental impact within a small area. However, predicting this impact is difficult because of the complexity of the hydrodynamic processes that mix the wastewater and also because of the variability in oceanic conditions. Despite great improvements over the years in the understanding of these mixing processes, since models are now available that can make reasonable predictions under steady-state conditions (Hunt et al., 2010), many aspects remain unknown and unpredictable. For this reason, much effort has been recently devoted to improve ways of monitoring and characterizing sewage plumes under a variety of oceanographic conditions.

1.1 MARES AUV

Autonomous Underwater Vehicles (AUVs) have been used efficiently in a wide range of applications. They were first developed with military applications in mind, for example for mine hunting missions. Later on, scientists realized their true potential and started to use them as mobile sensors, taking measurements in difficult scenarios and at a reasonable cost (Bellingham, 1997; Bellingham et al., 1992; Fernandes et al., 2000; Nadis, 1997; Robinson et al., 1999; Yu et al., 2002). MARES (Modular Autonomous Robot for Environment Sampling) AUV has been successfully used to monitor sea outfalls discharges (Abreu et al., 2010; Abreu & Ramos, 2010; Ramos & Abreu, 2010; 2011a;b;c) (see Fig. 1). MARES is 1.5 m long, has a diameter of 8-inch and weighs about 40 kg in air. It features a plastic hull with a dry mid body (for electronics and batteries) and additional rings to accommodate sensors and actuators. Its modular structure simplifies the system's development (the case of adding sensors, for example). It is propelled by two horizontal thrusters located at the rear and two vertical thrusters, one at the front and the other at the rear. This configuration allows for small operational speeds and high maneuverability, including pure vertical motions. It is equipped with an omnidirectional acoustic transducer and an electronic system that allows for long baseline navigation. The vehicle can be programmed to follow predefined trajectories while collecting relevant data using the onboard sensors. A Sea-Bird Electronics 49 FastCAT CTD had already been installed onboard the MARES AUV to measure conductivity, temperature and depth. MARES' missions for environmental monitoring of wastewater discharges are conducted using a GUI software that fully automates the operational procedures of the campaign (Abreu et al., 2010). By providing visual and audio information, this software guides the user through a series of steps which include: (1) real time data acquisition from CTD and ADCP sensors, (2) effluent plume parameter modeling using the CTD and ADCP data collected, (3) automatic path creation using the plume model parameters, (4) acoustic buoys and vehicle deployment, (5) automatic acoustic network setup and (6) real time tracking of the AUV mission.

Fig. 1. Autonomous Underwater Vehicle MARES.

1.2 Data processing

Data processing is the last step of a sewage outfall discharge monitoring campaign. This processing involves the ability to extrapolate from monitoring samples to unsampled locations. Although very chaotic due to turbulent diffusion, the effluent's dispersion process tends to a natural variability mode when the plume stops rising and the intensity of turbulent fluctuations approaches to zero (Hunt et al., 2010). It is likely that after this point the pollutant substances are spatially correlated. In this case, geostatistics appears to be an appropriate technique to model the spatial distribution of the effluent. In fact, geostatistics has been used with success to analyze and characterize the spatial variability of soil properties, to obtain information for assessing water and wind resources, to design sampling strategies for monitoring estuarine sediments, to study the thickness of effluent-affected sediment in the vicinity of wastewater discharges, to obtain information about the spatial distribution of sewage pollution in coastal sediments, among others. As well as giving the estimated values, geostatistics provides a measure of the accuracy of the estimate in the form of the kriging variance. This is one of the advantages of geostatistics over traditional methods of assessing pollution. In this work ordinary block kriging is used to model and map the spatial distribution of temperature and salinity measurements gathered by an AUV on a Portuguese sea outfall monitoring campaign. The aim is to distinguish the effluent plume from the receiving waters, characterize its spatial variability in the vicinity of the discharge and estimate dilution.

2. Geostatistical analysis

2.1 Stationary random function models

The most widely used geostatistical estimation procedures use stationary random function models. A random function is a set of random variables that have some spatial locations and whose dependence on each other is specified by some probabilistic mechanism. A random function is stationary if all the random variables have the same probability distribution and if any pair of random variables has a joint probability distribution that depends only on the separation between the two points and not on their locations. If the random function is stationary, then the expected value and the variance can be used to summarize the univariate behavior of the set of random variables. The parameter that is commonly used to summarize the bivariate behavior of a stationary random function is its covariance function, its correlogram, and its variogram. The complete definition of the probabilistic generating mechanism of a random function is usually difficult even in one dimension. Fortunately, for many of the problems we typically encounter, we do not need to know the probabilistic generating mechanism. We usually adopt a stationary random function as our model and specify only its covariance or variogram (Isaaks & Srivastava, 1989; Kitanidis, 1997; Wackernagel, 2003).

2.2 Ordinary kriging

Ordinary kriging method is often referred with the acronym BLUE which stands for "Best Linear Unbiased Estimator". "Linear" because its estimates are weighted linear combinations of the available data; "Unbiased" since it tries to have the mean error equal to 0; and "Best" because it aims at minimizing the variance of the errors. Let us then see how the concept of a random function model can be used to decide how to weight the nearby samples so that our estimates are unbiased. For any point at which we want to estimate the unknown value, our model is a stationary random function that consists of n random variables, one for the value at each of the n sample locations, $Z(\mathbf{x}_1), Z(\mathbf{x}_2), \ldots, Z(\mathbf{x}_n)$, and one for the unknown value at the point we are trying to estimate $Z(\mathbf{x}_0)$. Each of these random variables has the same probability law; at all locations, the expected value of the random variable is m and the variance is σ^2. Every value in this model is seen as an outcome (or realization) of the random variable. Our estimate is also a random variable since it is a weighted linear combination of the random variables at the n sampled locations (Cressie, 1993; Goovaerts, 1997; Isaaks & Srivastava, 1989; Kitanidis, 1997; Stein, 1999; Wackernagel, 2003; Webster & Oliver, 2007):

$$\hat{Z}(\mathbf{x}_0) = \sum_{i=1}^{n} w_i \cdot Z(\mathbf{x}_i). \tag{1}$$

The estimation error is defined as the difference between the random variable modeling the true value and the estimate:

$$\varepsilon(\mathbf{x}_0) = Z(\mathbf{x}_0) - \hat{Z}(\mathbf{x}_0). \tag{2}$$

The estimation error is also a random variable. Its expected value, often referred to as the bias, is

$$E\left[\varepsilon(\mathbf{x}_0)\right] = m \left(1 - \sum_{i=1}^{n} w_i \right). \tag{3}$$

Setting this expected value to 0, to ensure an unbiasedness estimate results in:

$$\sum_{i=1}^{n} w_i = 1. \tag{4}$$

This is known as the condition of unbiasedness (Isaaks & Srivastava, 1989; Kitanidis, 1997; Wackernagel, 2003). The expression of the variance of the modeled error is

$$\sigma^2_{\varepsilon(\mathbf{x}_0)} = var\,[Z(\mathbf{x}_0)] - 2\,cov\,[\hat{Z}(\mathbf{x}_0), Z(\mathbf{x}_0)] + var\,[\hat{Z}(\mathbf{x}_0)]\,. \tag{5}$$

Since we have already assumed that all of the random variables have the same variance σ^2, then $var\,[Z(\mathbf{x}_0)] = \sigma^2$. The second term in Equation 5 can be written as

$$- 2\,cov\,[\hat{Z}(\mathbf{x}_0), Z(\mathbf{x}_0)] = -2 \sum_{i=1}^{n} w_i \cdot cov\,[Z(\mathbf{x}_i), Z(\mathbf{x}_0)]$$

$$= -2 \sum_{i=1}^{n} w_i\, C_{i0}. \tag{6}$$

The third term of Equation 5 can be expressed as

$$var\,[\hat{Z}(\mathbf{x}_0)] = \sum_{i=1}^{n} \sum_{j=1}^{n} w_i \cdot w_j \cdot cov\,\left[Z(\mathbf{x}_i), Z(\mathbf{x}_j)\right]$$

$$= \sum_{i=1}^{n} \sum_{j=1}^{n} w_i\, w_j\, C_{ij}. \tag{7}$$

Then, the expression of the error variance comes in the following way:

$$\sigma^2_{\varepsilon(\mathbf{x}_0)} = \sigma^2 - 2 \sum_{i=1}^{n} w_i\, C_{i0} + \sum_{i=1}^{n} \sum_{j=1}^{n} w_i\, w_j\, C_{ij}. \tag{8}$$

Equation 8 expresses the error variance as function of the n weights, once chosen the random model function parameters, namely the variance σ^2 and all the covariances C_{ij}. The minimization of $\sigma^2_{\varepsilon(\mathbf{x}_0)}$ is constrained by the unbiasedness condition imposed earlier, which can be solved using the method of Lagrange multipliers. We start by introducing a new parameter μ, called the Lagrange multiplier, in Equation 8 in the following way:

$$\sigma^2_{\varepsilon(\mathbf{x}_0)} = \sigma^2 - 2 \sum_{i=1}^{n} w_i\, C_{i0} + \sum_{i=1}^{n} \sum_{j=1}^{n} w_i\, w_j\, C_{ij} + 2\,\mu \underbrace{\left(\sum_{i=1}^{n} w_i - 1 \right)}_{=0}. \tag{9}$$

Then we minimize $\sigma^2_{\varepsilon(\mathbf{x}_0)}$ by calculating the $n+1$ partial first derivatives of Equation 9 with respect to the n weights and the Lagrange multiplier, and setting each one to 0, which produces the following system of equations:

$$\frac{\partial(\sigma^2_{\varepsilon(x_0)})}{\partial(w_1)} = -2\,C_{10} + 2\sum_{j=1}^{n} w_j C_{1j} + 2\,\mu = 0 \quad \Rightarrow \quad \sum_{j=1}^{n} w_j C_{1j} + \mu = C_{10}$$

$$\vdots \qquad\qquad\qquad\qquad \vdots$$

$$\frac{\partial(\sigma^2_{\varepsilon(x_0)})}{\partial(w_i)} = -2\,C_{i0} + 2\sum_{j=1}^{n} w_j C_{ij} + 2\,\mu = 0 \quad \Rightarrow \quad \sum_{j=1}^{n} w_j C_{ij} + \mu = C_{i0}$$

$$\vdots \qquad\qquad\qquad\qquad \vdots$$

$$\frac{\partial(\sigma^2_{\varepsilon(x_0)})}{\partial(w_n)} = -2\,C_{n0} + 2\sum_{j=1}^{n} w_j C_{nj} + 2\,\mu = 0 \quad \Rightarrow \quad \sum_{j=1}^{n} w_j C_{nj} + \mu = C_{n0}$$

$$\frac{\partial(\sigma^2_{\varepsilon(x_0)})}{\partial(\mu)} = 2\left(\sum_{i=1}^{n} w_i - 1\right) = 0 \quad \Rightarrow \quad \sum_{i=1}^{n} w_i = 1$$

which can also be written in a compact way as

$$\sum_{j=1}^{n} w_j C_{ij} + \mu = C_{i0}, \quad \forall i = 1, 2, \ldots, n; \qquad \sum_{i=1}^{n} w_i = 1. \tag{10}$$

This system of equations, often referred to as the *ordinary kriging system*, can be written in matrix notation as

$$\mathbf{C} \qquad\quad \cdot \quad \mathbf{W} \quad = \quad \mathbf{D}$$

$$\begin{bmatrix} C_{11} & \cdots & C_{1n} & 1 \\ \vdots & \vdots & \ddots & \vdots \\ C_{n1} & \cdots & C_{nn} & 1 \\ 1 & \cdots & 1 & 0 \end{bmatrix} \cdot \begin{bmatrix} w_1 \\ \vdots \\ w_n \\ \mu \end{bmatrix} = \begin{bmatrix} C_{10} \\ \vdots \\ C_{n0} \\ 1 \end{bmatrix} \tag{11}$$

$$\underbrace{\qquad\qquad}_{(n+1)\times(n+1)} \quad \underbrace{\quad}_{(n+1)\times 1} \quad \underbrace{\quad}_{(n+1)\times 1}$$

The set of weights and the Lagrange multiplier that will produce an unbiased estimate of $Z(x_0)$ with the minimum error variance are then given by

$$\mathbf{W} = \mathbf{C}^{-1} \cdot \mathbf{D}. \tag{12}$$

The value of $\sigma^2_{\varepsilon(x_0)}$ can be obtained in a quicker way using an alternative expression to Eq. 8. Multiplying each of the n equations given in Eq. 10 by w_i and summing these n equations leads to the following:

$$\sum_{i=1}^{n}\sum_{j=1}^{n} w_i w_j C_{ij} = \sum_{i=1}^{n} w_i C_{i0} - \mu.$$

Substituting this into Equation 8 the minimized error variance comes as follows:

$$\sigma^2_{\varepsilon(x_0)} = \sigma^2 - \left(\sum_{i=1}^{n} w_i C_{i0} + \mu\right), \tag{13}$$

or, in terms of matrices as

$$\sigma^2_{\varepsilon(\mathbf{x}_0)} = \sigma^2 - \mathbf{W}^T \cdot \mathbf{D}. \tag{14}$$

The minimized error variance is usually called the *ordinary kriging variance*.

2.3 Block kriging

A consideration in many environmental applications has been that ordinary kriging usually exhibits large prediction errors (Bivand et al., 2008). This is due to the larger variability in the observations. When predicting averages over larger areas, i.e. within blocks, much of the variability averages out and consequently block mean values have lower prediction errors. If the blocks are not too large the spatial patterns do not disappear. The block kriging system is similar to the point kriging system given by Equation 11. The matrix $\mathbf{C}$ is the same since it is independent of the location at which the block estimate is required. The covariances for the vector $\mathbf{D}$ are point-to-block covariances. Supposing that the mean value over a block V is approximated by the arithmetic average of the N point variables contained within that block (Goovaerts, 1997; Isaaks & Srivastava, 1989), i.e.

$$Z_V \approx \frac{1}{N} \sum_{j=1}^{N} Z(\mathbf{x}_j), \tag{15}$$

the point-to-block covariances required for vector $\mathbf{D}$ are

$$\overline{C}_{iV} = cov\,[Z(\mathbf{x}_i), Z_V] = \frac{1}{N} \sum_{j=1}^{N} C_{ij}, \quad \forall i = 1, 2, \dots, n. \tag{16}$$

The block kriging variance is

$$\sigma^2_V = \overline{C}_{VV} - \left(\sum_{i=1}^{n} w_i \overline{C}_{iV} + \mu \right), \tag{17}$$

where $\overline{C}_{VV}$ is the average covariance between pairs of points within V:

$$\overline{C}_{VV} = \frac{1}{N^2} \sum_{i=1}^{N} \sum_{j=1}^{N} C_{ij}. \tag{18}$$

An equivalent procedure, that can be computationally more expensive than block kriging, is to obtain the block estimate by averaging the N kriged point estimates within the block (Goovaerts, 1997; Isaaks & Srivastava, 1989).

2.4 Spatial continuity

Spatial continuity exists in most earth science data sets. When we look at a contour map, or anything similar, the values do not appear to be randomly located, but rather, low values tend to be near other low values and high values tend to be near other high values. I.e. two measurements close to each other are most likely to have similar values than two measurements far apart (Isaaks & Srivastava, 1989). To compute the set of weights and the Lagrange multiplier, that will produce each estimate and the resulting minimized error variance, we need to know the covariances of $\mathbf{C}$ and $\mathbf{D}$ matrices. As we said before, since our random function is stationary, all pairs of random variables separated by a distance and direction $\mathbf{h}$ (known as lag) have the same joint probability distribution. The covariance function, $C(\mathbf{h})$ is the covariance between random variables separated by a lag $\mathbf{h}$ (Isaaks & Srivastava, 1989; Kitanidis, 1997; Wackernagel, 2003). For a stationary random function, the covariance function $C(\mathbf{h})$ is:

$$C(\mathbf{h}) = E\left[Z(\mathbf{x})\, Z(\mathbf{x}+\mathbf{h})\right] - \{E\left[Z(\mathbf{x})\right]\}^2. \tag{19}$$

The covariance between random variables at identical locations is the variance of the random function:

$$C(\mathbf{0}) = E\left[\{Z(\mathbf{x})\}^2\right] - \{E\left[Z(\mathbf{x})\right]\}^2 = var\left[Z(\mathbf{x})\right] = \sigma^2. \tag{20}$$

The semivariogram, or simply variogram, is half the expected squared difference between random variables separated by a lag $\mathbf{h}$:

$$\gamma(\mathbf{h}) = \frac{1}{2}E\left[\{Z(\mathbf{x}) - Z(\mathbf{x}+\mathbf{h})\}^2\right] = \frac{1}{2}var\left[Z(\mathbf{x}) - Z(\mathbf{x}+\mathbf{h})\right]. \tag{21}$$

The quantity $\gamma(\mathbf{h})$ is known as the semivariance at lag $\mathbf{h}$. The "semi"refers to the fact that it is half of a variance. The variogram between random variables at identical locations is zero, i.e. $\gamma(\mathbf{0}) = 0$. Using Equations 19, 20 and 21, we can relate the variogram with the covariance function as:

$$\gamma(\mathbf{h}) = C(\mathbf{0}) - C(\mathbf{h}) = \sigma^2 - C(\mathbf{h}). \tag{22}$$

In practice, the pattern of spatial continuity chosen for the random function is usually taken from the spatial continuity evident in the sample data set. Geostatisticians usually define the spatial continuity of the sample data set through the variogram and solve the ordinary kriging system using covariance (Isaaks & Srivastava, 1989). The maximum value reached by the variogram is called the sill. The distance at which the sill is reached is called the range. The vertical jump from zero at the origin to the value of semivariance at extremely small separation distances is called the nugget effect. The estimator of the variogram usually used, known as Matheron's method-of-moments estimator (MME) is (Matheron, 1965; Webster & Oliver, 2007)

$$\gamma(\mathbf{h}) = \frac{1}{2N(\mathbf{h})} \sum_{i=1}^{N(\mathbf{h})} [Z(\mathbf{x}_i) - Z(\mathbf{x}_i+\mathbf{h})]^2, \tag{23}$$

where $z(\mathbf{x}_i)$ is the value of the variable of interest at location $\mathbf{x}_i$ and $N(\mathbf{h})$ is the number of pairs of points separated by the particular lag vector $\mathbf{h}$. Cressie and Hawkins (Cressie & Hawkins, 1980) developed an estimator of the variogram that should be robust to the presence of outliers and enhance the variogram spatial continuity, having also the advantage of not spreading the effect of outliers in computing the maps. This estimator (CRE) is defined as follows (Cressie & Hawkins, 1980):

$$\gamma(\mathbf{h}) = \frac{1}{2} \times \frac{\left\{\dfrac{1}{N(\mathbf{h})} \sum_{i=1}^{N(\mathbf{h})} |Z(\mathbf{x}_i) - Z(\mathbf{x}_i+\mathbf{h})|^{1/2}\right\}^4}{0.457 + \dfrac{0.494}{N(\mathbf{h})} + \dfrac{0.045}{[N(\mathbf{h})]^2}}. \tag{24}$$

Once the sample variogram has been calculated, a function (called the variogram model) has to be fit to it. First, because the matrices $\mathbf{C}$ and $\mathbf{D}$ may need semivariance values for lags that are not available from the sample data. And second, because the use of the variogram does not guarantee the existence and uniqueness of solution to the ordinary kriging system. The most commonly used variogram models are the spherical model, the exponential model, the Gaussian model and the Matern model (Isaaks & Srivastava, 1989).

2.5 Cross-validation

Cross-validation is a procedure used to compare the performance of several competing models (Webster & Oliver, 2007). It starts by splitting the data set into two sets: a modelling set and a validation set. Then the modelling set is used for variogram modelling and kriging on the locations of the validation set. Finally the measurements of the validation set are compared to their predictions (Bivand et al., 2008). If the average of the cross-validation errors (or Mean Error, ME) is close to 0,

$$ME = \frac{1}{m} \sum_{i=1}^{m} \left[Z(\mathbf{x}_i) - \hat{Z}(\mathbf{x}_i) \right]. \tag{25}$$

we may say that apparently the estimates are unbiased ($Z(\mathbf{x}_i)$ and $\hat{Z}(\mathbf{x}_i)$ are, respectively, the measurement and estimate at point x_i and m is the number of measurements of the validation set). A significant negative (positive) mean error can represent systematic overestimation (underestimation). The magnitude of the Root Mean Squared Error (RMSE) is particularly interesting for comparing different models (Wackernagel, 2003; Webster & Oliver, 2007):

$$RMSE = \sqrt{\frac{1}{m} \sum_{i=1}^{m} \left[Z(\mathbf{x}_i) - \hat{Z}(\mathbf{x}_i) \right]^2}. \tag{26}$$

The RMSE value should be as small as possible indicating that estimates are close to measurements. The kriging standard deviation represents the error predicted by the estimation method. Dividing the cross-validation error by the corresponding kriging standard deviation allows to compare the magnitudes of both actual and predicted error (Wackernagel, 2003; Webster & Oliver, 2007). Therefore, the average of the standardized squared cross-validation errors (or Mean Standardized Squared Error, MSSE)

$$MSSE = \frac{1}{m} \sum_{i=1}^{m} \frac{\left[Z(\mathbf{x}_i) - \hat{Z}(\mathbf{x}_i) \right]^2}{\sigma_{R(\mathbf{x}_i)}^2}, \tag{27}$$

should be about one, indicating that the model is accurate. A scatterplot of true versus predicted values provides additional evidence on how well an estimation method has performed. The coefficient of determination R^2 is a good index for summarizing how close the points on the scatterplot come to falling on the 45-degree line passing through the origin (Isaaks & Srivastava, 1989). R^2 should be close to one.

3. Results

3.1 Study site

A map of the study site is shown in Fig. 2(a). Foz do Arelho outfall is located off the Portuguese west coast near Óbidos lagoon. In operation since June 2005, is presently discharging about 0.11 m^3/s of mainly domestic wastewater from the WWTPs of Óbidos, Carregal, Caldas da Rainha, Gaeiras, Charneca and Foz do Arelho, but it can discharge up to 0.35 m^3/s. The total length of the outfall, including the diffuser, is 2150 m. The outfall pipe, made of HDPE, has a diameter of 710 mm. The diffuser, which consists of 10 ports spaced 8 or 12 meters apart, is 93.5 m long. The ports, nominally 0.175 m in diameter, are discharging upwards at an angle of 90° to the pipe horizontal axis; the port height is about 1 m. The outfall direction is southeast-northwest (315.5° true bearing) and is discharging at a depth of about 31 m. In that area the coastline itself runs at about a 225° angle with respect to true north and the isobaths are oriented parallel to the coastline. A seawater quality monitoring program for the outfall has already started in May 2006. Its main purposes are to evaluate the background seawater quality both in offshore and nearshore locations around the vicinity of the sea outfall and to follow the impacts of wastewater discharge in the area. During the campaign the discharge remained fairly constant with an average flowrate of approximately 0.11 m^3/s. The operation area specification was based on the outputs of a plume prediction model (Hunt et al., 2010) which include mixing zone length, spreading width, maximum rise height and thickness. The model inputs are, besides the diffuser physical characteristics, the water column stratification, the current velocity and direction, and the discharge flowrate. Information on density stratification was obtained from a vertical profile of temperature and salinity acquired in the vicinity of the diffuser two weeks before the campaign (see Fig. 3). The water column was weakly stratified due to both low-temperature and salinity variations. The total difference in density over the water column was about 0.13 σ-unit. The current direction of 110° was estimated based on predictions of wind speed and direction of the day of the campaign. A current velocity of 0.12 m/s was estimated based on historic data. The effluent flowrate consider for the plume behavior simulation was 0.11 m^3/s. According to the predictions of the model, the plume was spreading 1 m from the surface, detached from the bottom and forming a two-layer flow. The end of the mixing zone length was predicted to be 141 m downstream from the diffuser. Fig. 2(b) shows the diffuser and a plan view of the AUV operation area (specified according to the model predictions), mainly in the northeast direction from the diffuser, covering about 20000 m^2.

The vehicle collected CTD data at 1.5 m and 3 m depth, in accordance to the plume minimum dilution height prediction. During the mission transited at a fairly constant velocity of 1 m/s (2 knots) recording data at a rate of 16 Hz. Maximum vertical oscillations of the AUV in performing the horizontal trajectories were less than 0.5 m (up and down).

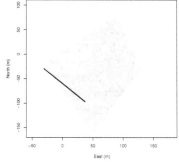

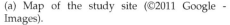

(a) Map of the study site (©2011 Google - Images).   (b) AUV operation area.

Fig. 2. Vicinity of Foz do Arelho sea outfall.

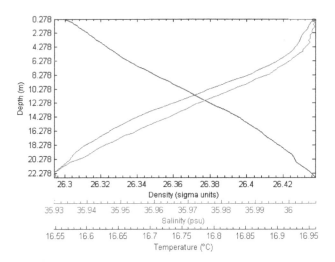

Fig. 3. Vertical STD profile used in the plume behavior simulation.

3.2 Exploratory analysis

In order to obtain elementary knowledge about the temperature and salinity data sets, conventional statistical analysis was conducted (see the results in Table 1 and Table 2). At the depth of 1.5 m the temperature ranged from 15.359°C to 15.562°C and at the depth of 3 m the temperature ranged from 15.393°C to 15.536°C. The mean value of the data sets was 15.463°C and 15.469°C, respectively at the depths of 1.5 m and 3 m, which was very close to the median value that was respectively 15.466°C and 15.472°C. The coefficient of skewness is relatively low (-0.309) for the 1.5 m data set and not very high (-0.696) for the 3 m data set, indicating that in the first case the histogram is approximately symmetric and in the second case that distribution is only slightly asymmetric. The very low values of the coefficient of variation (0.002 and 0.001) reflect the fact that the histograms do not have a tail of high values. At the depth of 1.5 m the salinity ranged from 35.957 psu to 36.003 psu and at the depth of 3 m the salinity ranged from 35.973 psu to 36.008 psu. The mean value of the data sets was 35.991 psu and 35.996 psu, respectively at the depths of 1.5 m and 3 m, which was very close to the median value that was respectively 35.990 and 35.998 psu. The coefficient of skewness is not to much high in both data sets (-0.63 and -1.1) indicating that distributions are only slightly asymmetric. The very low values of the coefficient of variation (0.0002 and 0.0001) reflect the fact that the histograms do not have a tail of high values. The ordinary kriging method works better if the distribution of the data values is close to a normal distribution. Therefore, it is interesting to see how close the distribution of the data values comes to being normal. Fig. 4 shows the plots of the normal distribution adjusted to the histograms of the temperature measured at depths of 1.5 m and 3 m, and Fig. 5 shows the plots of the normal distribution adjusted to the histograms of the salinity measured at depths of 1.5 m and 3 m. The density value in the histogram is the ratio between the number of samples in a bin and the total number of samples divided by the width of the bin (constant). Apart from some erratic high values it can be seen that the histograms are reasonably close to the normal distribution.

	Temperature@1.5 m	Temperature@3.0 m
Samples	20,026	10,506
Mean	15.463°C	15.469°C
Median	15.466°C	15.472°C
Minimum	15.359°C	15.393°C
Maximum	15.562°C	15.536°C
Coefficient of skewness	-0.31	-0.70
Coefficient of variation	0.002	0.001

Table 1. Summary statistics of temperature measurements.

	Salinity@1.5 m	Salinity@3.0 m
Samples	20,026	10,506
Mean	35.991 psu	35.996 psu
Median	35.990 psu	35.998 psu
Minimum	35.957 psu	35.973 psu
Maximum	36.003 psu	36.008 psu
Coefficient of skewness	-0.63	-1.1
Coefficient of variation	0.0002	0.0001

Table 2. Summary statistics of salinity measurements.

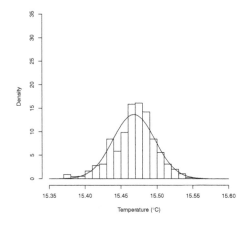
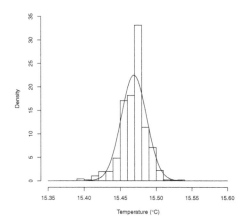

Fig. 4. Histograms of temperature measurements at depths of 1.5 m (left) and 3 m (right).

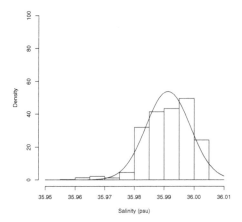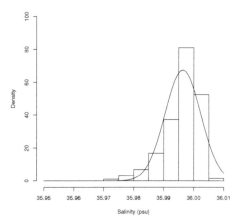

Fig. 5. Histograms of salinity measurements at depths of 1.5 m (left) and 3 m (right).

3.3 Variogram modeling

For the purpose of this analysis, the temperature and the salinity measurements were divided into a modeling set (comprising 90% of the samples) and a validation set (comprising 10% of the samples). Modeling and validation sets were then compared, using Student's-t test, to check that they provided unbiased sub-sets of the original data. Furthermore, sample variograms for the modeling sets were constructed using the MME estimator and the CRE estimator. This robust estimator was chosen to deal with outliers and enhance the variogram's spatial continuity. An estimation of semivariance was carried out using a lag distance of 2 m. Table 3 and Table 4 show the parameters of the fitted models to the omnidirectional sample variograms constructed using MME and CRE estimators. All the variograms were fitted to Matern models (for several shape parameters v) with the exception to the salinity data measured at the depth of 3 m. The range value (in meters) is an indicator of extension where autocorrelation exists. The variograms of salinity show significant differences in range. The autocorrelation distances are always larger for the CRE estimator which may demonstrate the enhancement of the variogram's spatial continuity. All variograms have low nugget values which indicates that local variations could be captured due to the high sampling rate and to the fact that the variables under study have strong spatial dependence. Anisotropy was investigated by calculating directional variograms. However, no anisotropy effect could be shown.

3.4 Cross-validation

The block kriging method was preferred since it produced smaller prediction errors and smoother maps than the point kriging. Using the 90% modeling sets of the two depths, a two-dimensional ordinary block kriging, with blocks of 10×10 m^2, was applied to estimate temperature at the locations of the 10% validation sets. The validation results for both parameters measured at depths of 1.5 m and 3 m depths are shown in Table 5 and Table 6. At both depths temperature was best estimated by the variogram constructed using CRE. Salinity at the depth of 1.5 m was best estimated by the variogram constructed using CRE and at the depth of 3 m was best estimated using the Gaussian model with the MME. The

Depth	Variogram Estimator	Model	Nugget	Sill	Range
1.5	MME	Matern ($v = 0.4$)	0.000	0.001	75.0
	CRE	Matern ($v = 0.5$)	0.000	0.002	80.1
3.0	MME	Matern ($v = 0.3$)	0.000	0.0002	101.3
	CRE	Matern ($v = 0.7$)	0.000	0.002	107.5

Table 3. Parameters of the fitted variogram models for temperature measured at depths of 1.5 and 3.0 m.

Depth	Variogram Estimator	Model	Nugget	Sill	Range
1.5	MME	Matern ($v = 0.6$)	0.436	11.945	134.6
	CRE	Matern ($v = 0.6$)	0.153	10786.109	51677.1
3.0	MME	Matern ($v = 0.8$)	0.338	11.724	181.6
	CRE	Gaussian	0.096	120.578	390.1

Table 4. Parameters of the fitted variogram models for salinity measured at depths of 1.5 and 3 m.

Depth	Method	R^2	ME	MSE	RMSE
1.5	MBK	0.9184	2.0174e-4	8.0530e-5	8.9739e-3
	CBK[a]	0.9211	1.6758e-4	7.7880e-5	8.8248e-3
3.0	MBK	0.8748	1.0338e-4	3.6295e-5	6.0244e-3
	CBK[a]	0.8827	0.6538e-4	3.4008e-5	5.8316e-3

[a] The preferred model.

Table 5. Cross-validation results for the temperature maps at depths of 1.5 and 3 m.

difference in performance between the two estimators: block kriging using the MME estimator (MBK) or block kriging using the CRE estimator (CBK) is not substantial. Fig. 6 shows the omnidirectional sample variograms for temperature at the depth of 1.5 m and 3 m fitted by the preferred models. Fig. 7 shows the omnidirectional sample variograms for salinity at the depth of 1.5 m and 3 m fitted by the preferred models.

Fig. 8 and Fig. 9 show the scatterplots of true versus estimated values for the most satisfactory models. The dark line is the $45°$ line passing through the origin and the discontinuous line is the OLS (Ordinary Least Squares) regression line. These plots show that observed and predicted values are highly positively correlated. The R^2 value for the temperature at the depth of 1.5 m was 0.9211 and the RMSE was 0.0088248°C, and at the depth of 3 m was 0.8827 and the RMSE was 0.0058316°C (Table 5). The R^2 value for the salinity at the depth of 1.5 m was 0.9513 and the RMSE was 0.0016435 psu, and at the depth of 3 m was 0.8982 and the RMSE was 0.0019793 psu (Table 6).

Depth	Method	R^2	ME	MSE	RMSE
1.5	MBK	0.9471	3.1113e-5	2.8721e-6	1.6947e-3
	CBK[a]	0.9513	-3.1579e-5	2.7010e-6	1.6435e-3
3.0	MBK[a]	0.8982	-7.1735e-5	3.9175e-6	1.9793e-3
	CBK	0.7853	-8.1264e-5	8.2589e-6	2.8738e-3

[a] The preferred model.

Table 6. Cross-validation results for the salinity maps at depths of 1.5 and 3 m.

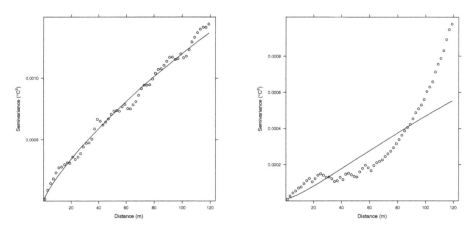

Fig. 6. Variograms for temperature at depths of 1.5 m (left) and 3 m (right).

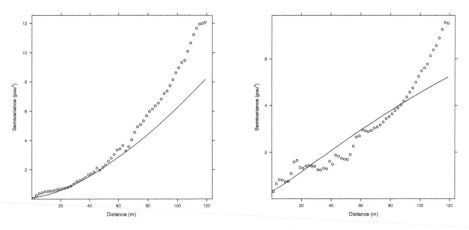

Fig. 7. Variograms for salinity at depths of 1.5 m (left) and 3 m (right).

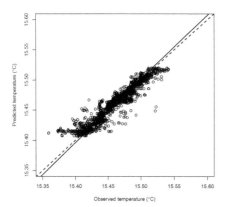
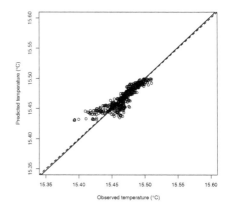

Fig. 8. Predicted versus observed temperature at the depths of 1.5 m (left) and 3 m (right) using the preferred models.

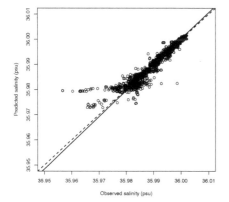
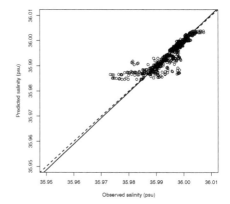

Fig. 9. Predicted versus observed salinity at the depths of 1.5 m (left) and 3 m (right) using the preferred models.

3.5 Mapping

Fig. 10 shows the block kriged maps of temperature on a 2×2 m^2 grid using the preferred models. Fig. 13 shows the block kriged maps of salinity on a 2×2 m^2 grid using the preferred models. In the 1.5 m kriged map the temperature ranges between 15.407°C and 15.523°C and the average value is 15.469°C (the measured range is 15.359°C–15.562°C and the average value is 15.463°C). In the 3 m kriged map the temperature ranges between 15.429°C and 15.502°C and the average value is 15.467°C (the measured range is 15.393°C–15.536°C and the average value is 15.469°C). We may say that estimated values are in accordance with the measurements since their distributions are similar (identical average values, medians, and quartiles). The difference in the ranges width is due to only 5.0% of the samples in the 1.5 m depth map (2.5% on each side of the distribution) and only 5.3% of the samples in the 3.0 m depth map

(3.1% on the left side and 2.2% on the rigth side of the distribution). These samples should then be identified as outliers not representing the behaviour of the plume in the established area. In the 1.5 m kriged map the salinity ranges between 35.960 psu and 36.004 psu and the average value is 35.992 psu, which is in accordance with the measurements (the measured range is 35.957psu – 36.003psu and the average value is 35.991 psu). In the 3 m kriged map the salinity ranges between 35.977 psu and 36.004 psu and the average value is 35.995 psu, which is in accordance with the measurements (the measured range is 35.973psu – 36.008psu and the average value is 35.996 psu). As predicted by the plume prediction model, the effluent was found dispersing close to the surface. From the temperature and salinity kriged maps it is possible to distinguish the effluent plume from the background waters. It appears as a region of lower temperature and lower salinity when compared to the surrounding ocean waters at the same depth. At the depth of 1.5 m the major difference in temperature compared to the surrounding waters is about -0.116°C while at the depth of 3 m this difference is about -0.073°C. At the depth of 1.5 m the major difference in salinity compared to the surrounding waters is about -0.044 psu while at the depth of 3 m this difference is about -0.027 psu. It is important to note that these very small differences in temperature and salinity were detected due to the high resolution of the CTD sensor. (Washburn et al., 1992) observed temperature and salinity anomalies in the plume in the order, respectively of -0.3°C and -0.1 psu, when compared with the surrounding waters within the same depth range. The small plume-related anomalies observed in the maps are evidence of the rapid mixing process. Due to the large differences in density between the rising effluent plume and ambient ocean waters, entrainment and mixing processes are vigorous and the properties within the plume change rapidly (Petrenko et al., 1998; Washburn et al., 1992). The effluent plume was found northeast from the diffuser beginning, spreading downstream in the direction of current. Using the navigation data, we could later estimate current velocity and direction and the values found were, respectively, 0.4 m/s and 70°C, which is in accordance with the location of the plume.

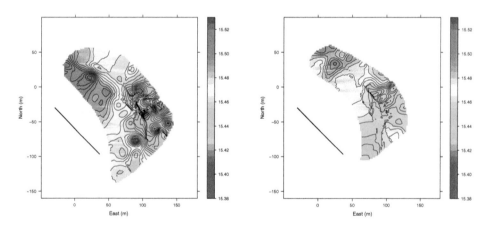

Fig. 10. Prediction map of temperature distribution (°C) at depths of 1.5 m (left) and 3 m (right).

Fig. 12 shows the variance of the estimation error (kriging variance) for the maps of temperature distribution at depths of 1.5 m and 3 m. The standard deviation of the estimation error is less than 0.0195°C at the depth of 1.5 m and less than 0.0111°C at the depth of 3

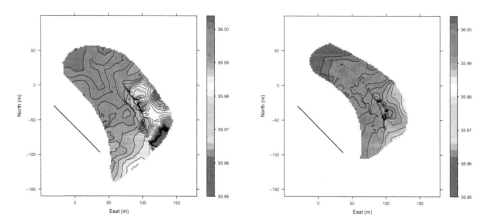

Fig. 11. Prediction map of salinity distribution (psu) at depths of 1.5 m (left) and 3 m (right).

m. Results of the same order were obtained for salinity. It's interesting to observe that, as expected, the variance of the estimation error is less the closer is the prediction from the trajectory of the vehicle. The dark blue regions correspond to the trajectory of MARES AUV.

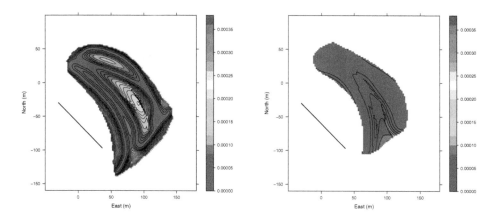

Fig. 12. Variance of the estimation error for the maps of temperature distribution at depths of 1.5 m (left) and 3 m (right)

3.6 Dilution estimation

Environmental effects are all related to concentration C of a particular contaminant X. Defining C_a as the background concentration of substance X in ambient water and C_0 as the concentration of X in the effluent discharge, the local dilution comes as follows (Fischer et al., 1979):

$$S = \frac{C_0 - C_a}{C - C_a}, \tag{28}$$

which can be rearranged to give $C = C_a \left(\frac{S-1}{S}\right) + \left(\frac{1}{S}\right) C_0$. In the case of variability of the background concentration of substance X in ambient water the local dilution is given by

$$S = \frac{C_0 - C_{a0}}{C - C_a}, \tag{29}$$

where C_{a0} is the background concentration of substance X in ambient water at the discharge depth. This expression in 29 can be arranged to give $C = C_a + \left(\frac{1}{S}\right)(C_0 - C_{a0})$, which in simple terms means that the increment of concentration above background is reduced by the dilution factor S from the point of discharge to the point of measurement of C. Using salinity distribution at depths of 1.5 m and 3 m we estimated dilution using Equation 29 (see the contour maps in Fig. 13). We assumed $C_0 = 2.3$ psu, $C_{a0} = 35.93$ psu, $C_a = 36.008$ psu at 1.5 m depth and $C_a = 36.006$ psu at 3 m depth. The minimum dilution estimated at the depth of 1.5 m was 705 and at the depth of 3.0 m was 1164 which is in accordance with Portuguese legislation that suggests that outfalls should be designed to assure a minimum dilution of 50 when the plume reaches surface (INAG, 1998). (Since dilution increases with the plume rising we should expect that the minimum values would be greater if the plume reached surface (Hunt et al., 2010)).

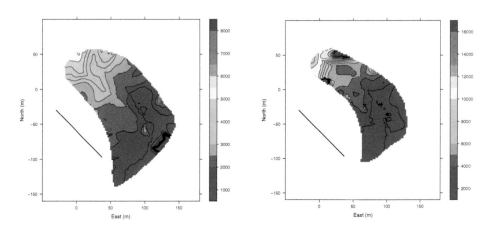

Fig. 13. Dilution maps at depths of 1.5 m (left) and 3 m (right).

4. Conclusion

Through geostatistical analysis of temperature and salinity obtained by an AUV at depths of 1.5 m and 3 m in an ocean outfall monitoring campaign it was possible to produce kriged maps of the sewage dispersion in the field. The spatial variability of the sampled data has been analyzed and the results indicated an approximated normal distribution of the temperature and salinity measurements, which is desirable. The Matheron's classical estimator and Cressie and Hawkins' robust estimator were then used to compute the omnidirectional variograms that were fitted to Matern models (for several shape parameters) and to a Gaussian model. The performance of each competing model was compared using a split-sample approach. In the case of temperature, the validation results, using a two-dimensional ordinary block kriging, suggested the Matern model ($v = 0.5 - 1.5$ m and $v = 0.7 - 3.0$ m) with semivariance estimated by CRE. In the case of salinity, the validation results, using a two-dimensional ordinary block kriging, suggested the Matern model ($v = 0.6 - 1.5$ m and $v = 0.8 - 3.0$ m) with semivariance estimated by CRE, for the depth of 1.5 m, and with semivariance estimated by MME, for the depth of 3 m. The difference in performance between the two estimators was not substantial. Block kriged maps of temperature and salinity at depths of 1.5 m and 3 m show the spatial variation of these parameters in the area studied and from them it is possible to identify the effluent plume that appears as a region of lower temperature and lower salinity when compared to the surrounding waters, northeast from the diffuser beginning, spreading downstream in the direction of current. Using salinity distribution at depths of 1.5 m and 3 m we estimated dilution at those depths. The values found are in accordance with Portuguese legislation. The results presented demonstrate that geostatistical methodology can provide good estimates of the dispersion of effluent that are very valuable in assessing the environmental impact and managing sea outfalls.

5. Acknowledgment

This work was partially funded by the Foundation for Science and Technology (FCT) under the Program for Research Projects in all scientific areas (Programa de Projectos de Investigação em todos os domínos científicos) in the context of WWECO project - Environmental Assessment and Modeling of Wastewater Discharges using Autonomous Underwater Vehicles Bio-optical Observations (Ref. PTDC/MAR/74059/2006).

6. References

Abreu, N., Matos, A., Ramos, P. & Cruz, N. (2010). Automatic interface for AUV mission planning and supervision, *MTS/IEEE International Conference Oceans 2010*, Seattle, USA.

Abreu, N. & Ramos, P. (2010). An integrated application for geostatistical analysis of sea outfall discharges based on R software, *MTS/IEEE International Conference Oceans 2010*, Seattle, USA.

Bellingham, J. (1997). New Oceanographic Uses of Autonomous Underwater Vehicles, *Marine Technology Society Journal* 31(3): 34–47.

Bellingham, J., Goudey, C., Consi, T. R. & Chryssostomidis, C. (1992). A Small Long Range Vehicle for Deep Ocean Exploration, *Proceedings of the International Offshore and Polar Engineering Conference, San Francisco*, pp. 151–159.

Bivand, R. S., Pebesma, E. J. & Gómez-Rubio, V. (2008). *Applied spatial data analysis with R.* ISBN: 97 -0-387-78170-9.

Cressie, N. (1993). *Statistics for spatial data*, New York.

Cressie, N. & Hawkins, D. M. (1980). Robust estimation of the variogram, I, *Jour. Int. Assoc. Math. Geol.* 12(2): 115–125.

Fernandes, P. G., Brierley, A. S., Simmonds, E. J., Millard, N. W., McPhail, S. D., Armstrong, F., Stevenson, P. & Squires, M. (2000). Fish do not Avoid Survey Vessels, *Nature* 404: 35–36.

Fischer, H. B., List, J. E., Koh, R. C. Y., Imberger, J. & Brooks, N. H. (1979). *Mixing in Inland and Coastal Waters*, Academic Press.

Goovaerts, P. (1997). *Geostatistics for natural resources evaluation, Applied Geostatistics Series.* ISBN13: 9780195115383, ISBN10: 0195115384.

Hunt, C. D., Mansfield, A. D., Mickelson, M. J., Albro, C. S., Geyer, W. R. & Roberts, P. J. W. (2010). Plume tracking and dilution of effluent from the Boston sewage outfall, *Marine Environmental Research* 70: 150–161.

INAG (1998). *Linhas de Orientação Metodológia para a Elaboração de Estudos Técnicos Necessários para Cumprir o Artï¿½ 7ï¿½ do D.L. 152/97. Descargas em Zonas Menos Sensíveis*, Instituto da Água. Ministério do Ambiente, Lisboa.

Isaaks, E. H. & Srivastava, R. M. (1989). *Applied geostatistics*, New York Oxford. ISBN 0-19-505012-6-ISBN 0-19-505013-4 (pbk.).

Kitanidis, P. (1997). *Introduction to geostatistics: applications in hydrogeology*, New York (USA).

Matheron, G. (1965). *Les variables régionalisées et leur estimation: une application de la théorie des fonctions aléatoires aux sciences de la nature*, Paris.

Nadis, S. (1997). Real Time Oceanography Adapts to Sea Changes, *Science* 275: 1881–1882.

Petrenko, A. A., Jones, B. H. & Dickey, T. D. (1998). Shape and Initial Dilution of Sand Island, Hawaii Sewage Plume, *Journal of Hydraulic Engineering, ASCE* 124(6): 565–571.

Ramos, P. & Abreu, N. (2010). Spatial analysis of sea outfall discharges using block kriging, *6th International Conference on Marine Waste Water Discharges and Coastal Environment, MWWD2010*, Langkawi, Malaysia.

Ramos, P. & Abreu, N. (2011a). Environmental impact assessment of Foz do Arelho sewage plume using MARES AUV, *IEEE/OES International Conference Oceans 2011*, Santander, Spain.

Ramos, P. & Abreu, N. (2011b). Spatial analysis and dilution estimation of Foz do Arelho Outfall using observations gathered by an AUV, *International Symposium on Outfall Systems*, Mar del Plata, Argentina.

Ramos, P. & Abreu, N. (2011c). Using an AUV for Assessing Wastewater Discharges Impact: An Approach Based on Geostatistics, *Marine Technology Society Journal* 45(2).

Robinson, A., Bellingham, J., Chryssostomidis, C., Dickey, T., Levine, E., Petrikalakis, N., Porter, D., Rothschild, B., Schmidt, H., Sherman, K., Holliday, D. & Atwood, D. (1999). Real-time forecasting of the multidisciplinary coastal ocean with the littoral ocean observing and predicting system (LOOPS), *Proceedings of the Third Conference on Coastal Atmospheric and Oceanic Prediction Processes*, American Meteorological Society, New Orleans, LA.

Stein, M. L. (1999). *Interpolation of spatial data: some theory for kriging*, New York.

Wackernagel, H. (2003). *Multivariate geostatistics: an introduction with applications*, Berlin.

Washburn, L., Jones, B. H., Bratkovich, A., Dickey, T. D. & Chen, M.-S. (1992). Mixing, Dispersion, and Resuspension in Vicinity of Ocean Wastewater Plume, *Journal of Hydraulic Engineering, ASCE* 118(1): 38–58.

Webster, R. & Oliver, M. (2007). *Geostatistics for environmental scientists, 2nd Edition*. ISBN-13: 978-0-470-02858-2(HB).

Yu, X., Dickey, T., Bellingham, J., Manov, D. & Streitlien, K. (2002). The Application of Autonomous Underwater Vehicles for Interdisciplinary Measurements in Massachusetts and Cape Cod Bayes, *Continental Shelf Research* 22(15): 2225–2245.

Permissions

The contributors of this book come from diverse backgrounds, making this book a truly international effort. This book will bring forth new frontiers with its revolutionizing research information and detailed analysis of the nascent developments around the world.

We would like to thank Nuno A. Cruz, for lending his expertise to make the book truly unique. He has played a crucial role in the development of this book. Without his invaluable contribution this book wouldn't have been possible. He has made vital efforts to compile up to date information on the varied aspects of this subject to make this book a valuable addition to the collection of many professionals and students.

This book was conceptualized with the vision of imparting up-to-date information and advanced data in this field. To ensure the same, a matchless editorial board was set up. Every individual on the board went through rigorous rounds of assessment to prove their worth. After which they invested a large part of their time researching and compiling the most relevant data for our readers. Conferences and sessions were held from time to time between the editorial board and the contributing authors to present the data in the most comprehensible form. The editorial team has worked tirelessly to provide valuable and valid information to help people across the globe.

Every chapter published in this book has been scrutinized by our experts. Their significance has been extensively debated. The topics covered herein carry significant findings which will fuel the growth of the discipline. They may even be implemented as practical applications or may be referred to as a beginning point for another development. Chapters in this book were first published by InTech; hereby published with permission under the Creative Commons Attribution License or equivalent.

The editorial board has been involved in producing this book since its inception. They have spent rigorous hours researching and exploring the diverse topics which have resulted in the successful publishing of this book. They have passed on their knowledge of decades through this book. To expedite this challenging task, the publisher supported the team at every step. A small team of assistant editors was also appointed to further simplify the editing procedure and attain best results for the readers.

Our editorial team has been hand-picked from every corner of the world. Their multi-ethnicity adds dynamic inputs to the discussions which result in innovative outcomes. These outcomes are then further discussed with the researchers and contributors who give their valuable feedback and opinion regarding the same. The feedback is then collaborated with the researches and they are edited in a comprehensive manner to aid the understanding of the subject.

Apart from the editorial board, the designing team has also invested a significant amount of their time in understanding the subject and creating the most relevant covers. They scrutinized every image to scout for the most suitable representation of the subject and create an appropriate cover for the book.

The publishing team has been involved in this book since its early stages. They were actively engaged in every process, be it collecting the data, connecting with the contributors or procuring relevant information. The team has been an ardent support to the editorial, designing and production team. Their endless efforts to recruit the best for this project, has resulted in the accomplishment of this book. They are a veteran in the field of academics and their pool of knowledge is as vast as their experience in printing. Their expertise and guidance has proved useful at every step. Their uncompromising quality standards have made this book an exceptional effort. Their encouragement from time to time has been an inspiration for everyone.

The publisher and the editorial board hope that this book will prove to be a valuable piece of knowledge for researchers, students, practitioners and scholars across the globe.

List of Contributors

Ji-Hong Li, Sung-Kook Park, Seung-Sub Oh and Jin-Ho Suh
Pohang Institute of Intelligent Robotics, Republic of Korea

Gyeong-Hwan Yoon and Myeong-Sook Baek
Daeyang Electric Inc., Republic of Korea

Shuxiang Guo and Xichuan Lin
Kagawa University, Japan

Wu Jianguo
Shenyang Institute of Automation Chinese Academy of Sciences, China

Zhang Minge and Sun Xiujun
Tianjin University, China

Lei Wan and Fang Wang
Harbin Engineering University, China

Matko Barisic, Zoran Vukic and Nikola Miskovic
University of Zagreb, Faculty of Electrical Engineering and Computing, Croatia

Oleg A. Yakimenko and Sean P. Kragelund
Naval Postgraduate School Monterey, CA, USA

Matthew Kokegei, Fangpo He and Karl Sammut
Flinders University, Australia

Gunilla Burrowes and Jamil Y. Khan
The University of Newcastle, Australia

Franz Uiblein
Institute of Marine Research, Bergen, Norway and South African Institute of Aquatic Biodiversity, Grahamstown, South Africa

Pedro Patrón, Emilio Miguelañez and Yvan R. Petillot
Ocean Systems Laboratory, Heriot-Watt University, United Kingdom

Patrícia Ramos
Institute for Systems and Computer Engineering of Porto (INESC Porto), School of Accounting and Administration of Porto – Polytechnic Institute of Porto, Portugal

Nuno Abreu

Institute for Systems and Computer Engineering of Porto (INESC Porto)Faculty of Engineering of University of Porto, Portugal

Printed in the USA
CPSIA information can be obtained
at www.ICGtesting.com
JSHW011444221024
72173JS00004B/941

9 781632 400741